T0227977

The Coccidian Parasites of Rodents

Authors

Norman D. Levine
and
Virginia Ivens
College of Veterinary Medicine
Agricultural Experiment Station
University of Illinois
Urbana, Illinois

CRC Press
Taylor & Francis Group
Boca Raton London New York

CRC Press is an imprint of the
Taylor & Francis Group, an **informa** business

First published 1990 by CRC Press
Taylor & Francis Group
6000 Broken Sound Parkway NW, Suite 300
Boca Raton, FL 33487-2742

Reissued 2018 by CRC Press

Library of Congress Cataloging-in-Publication Data

Levine, Norman D.
    The coccidian parasites of rodents/authors, Norman D. Levine and
    Virginia Ivens.
        p.   cm.
    Includes bibliographical references.
    ISBN 0-8493-4898-6
    1. Coccidia.   2. Rodents--Parasites.   I. Ivens, Virginia.
    II. Title.
    QL368.C59L47   1990
    593.1'9--dc20   89-22196

A Library of Congress record exists under LC control number: 89022196

ISBN 13: 978-1-315-89163-7 (hbk)
ISBN 13: 978-1-351-07073-7 (ebk)

Visit the Taylor & Francis Web site at http://www.taylorandfrancis.com and the
CRC Press Web site at http://www.crcpress.com

PREFACE

This monograph is essentially a revision of our earlier volume, *The Coccidian Parasites (Protozoa, Sporozoa) of Rodents* (1965). However, the results of cross-transmission studies, tables giving the known structural characters of the various species of coccidia and figures have been omitted in the interest of economy and to avoid duplication.

Quite a few new species of coccidia have been named since 1965. In the monograph that was published at that time, 225 named species were discussed, including 204 of *Eimeria*, 10 of *Isospora*, 3 of *Wenyonella*, 2 each of *Cryptosporidium* and *Klossiella*, and 1 each of *Dorisa, Caryospora, Tyzzeria*, and *Klossia*. In the interim, it has been found that *Sarcocystis, Frenkelia, Toxoplasma*, and *Besnoitia* are coccidia, and additional members of the previously accepted genera have been named. In addition, we have found a few species that were not included in the earlier assemblage. The present monograph, then, discusses a total of 473 named species of rodent coccidia, of which 372 are *Eimeria*; 39 *Isospora*; 28 *Sarcocystis*; 5 each *Besnoitia, Toxoplasma*, and *Wenyonella*; 4 each *Klossiella* and *Caryospora*; 2 each *Dorisa, Frenkelia, Klossia* and *Cryptosporidium*; and 1 each *Mantonella, Pythonella, Skrjabinella*, and *Tyzzeria*. This is an increase of 252 (112%) over the 1965 monograph. In addition, 260 host species, 95 host genera, and 15 host families are represented. Since, however, Nowak & Paradiso (1983) indicated that there are 29 families, 380 genera, and 1,687 extant species of rodents, this number of coccidian species has been named from only 52% of the families, 25% of the genera, and 15% of the species of rodents. Further, all the coccidian species in most host species have yet to be described, most of the descriptions of oocysts are far from complete, and the life cycles of only a few are known. In addition to the above, there have been 99 reports of coccidia (*Sarcocystis*), *Eimeria, Frenkelia, Isospora, Klossia, Klossiella, Adelina* and "*Coccidium*") without names, and 6 names have been given without descriptions and are therefore *nomina nuda*. There is ample room for future research.

It might also be commented that the numbers of merozoites per meront are those given by the authors. However, it is not known whether these are the numbers seen in a single section (at some level or other) or have been calculated for the whole meront.

THE AUTHORS

Norman Dion Levine, Ph.D. is a Professor Emeritus in the Department of Veterinary Pathobiology, in the College of Veterinary Medicine and the Illinois Agricultural Experiment Station, University of Illinois, Urbana. He is also a former director of the Center for Human Ecology (University of Illinois), a former member of the Department of Zoology (University of Illinois), and an affiliate of the Illinois Natural History Survey.

Dr. Levine received his B.S. degree in Zoology and Entomology from Iowa State University, his Ph.D. degree in Zoology (Physico-Chemical Biology) from the University of California (Berkeley), and an honorary Sc.D. degree from the University of Illinois.

He joined the staff of the University of Illinois and Illinois State Department of Agriculture in 1937, and moved up through the ranks to his present position. He has been President of the American Microscopical Society, Society of Protozoologists, American Society of Parasitologists, Illinois State Academy of Science, and American Society of Professional Biologists, and is an honorary member of the American Microscopical Society, Illinois State Academy of Science, Society of Protozoologists, Phi Sigma (honorary biology fraternity), and World Association for the Advancement of Veterinary Parasitology. He has received the Distinguished Achievement Citation of Iowa State University, and has been elected a Distinguished Veterinary Parasitologist by the American Association of Veterinary Parasitologists. He was Editor of the *Journal of Protozoology* from 1965 to 1971. He has been a member of the editorial boards of the *American Journal of Veterinary Research, Transactions of the American Microscopical Society, American Midland Naturalist*, and *Laboratory Animal Science*. He has been chairman of the National Institutes of Health Tropical Medicine and Parasitology Study Section, a member of the Board of Governors of the American Academy of Microbiology, a Committeeman-at-Large for Section F (Zoological Science) and Council Member of the American Association for the Advancement of Science, a member of the Governing Board of the American Board of Microbiology, and a member of the National Academy of Science-National Reasearch Council. He was one of the organizers of the World Federation of Parasitologists.

Dr. Levine is the author or editor of 15 books and about 570 publications in scientific journals. His special fields of research are protozoology, parasitology, and human ecology.

Virginia R. Ivens is an Associate Professor Emerita of Veterinary Parasitology in the College of Veterinary Medicine at the University of Illinois. She is an honorary member of Phi Zeta (Honor Society of Veterinary Medicine) and a member of Phi Sigma (honorary biology society), and an active member of the American Society of Parasitologists, Society of Protozoologists, Entomological Society of America, and American Institute of Biological Sciences. She was a member of the ASP Translations Committee and chaired the Ninth Annual Conference of Coccidiosis (1972). Professor Ivens has authored with Professor Norman D. Levine monographs dealing with the coccidia of rodents, artiodactyla, and carnivores, respectively, and has contributed to more than 50 scientific publications on the parasites of domestic and wild animals. She is the senior author of the book, *Principal Parasites of Domestic Animals in the United States, Biological and Diagnostic Information* (1978, 1981, and in press). She is listed in *American Men of Science* (1966), *Who's Who of American Women* (1977-88), the *Dictionary of International Biography* (1978-80), and *Contemporary Authors* (1982).

TABLE OF CONTENTS

INTRODUCTION*

According to Nowak & Paradiso (1983), the mammalian order Rodentia contains 29 living families, 380 living genera, and 1687 living species. This is the most successful order of mammals in terms of number of species and probably also in terms of number of individuals.

The taxonomy, general life cycles, and structure of the coccidia have been given by Levine & Ivens (1981). Levine (1988) listed the 4600 species of apicomplexan protozoa already named, including 1301 species of coccidia, and also gave their taxonomy.

The known species of rodent coccidia are discussed below. Details regarding geographic distribution, prevalence, merogony, gamogony, sporulation, cross-transmission, cultivation, pathogenicity, and chemotherapy as well as illustrations can be found in the References; they are omitted here for the sake of economy. Blood protozoa are not included.

* Supported in part by National Institutes of Health grant AI-15367 and National Science Foundation grant GB-30800X.

INTRODUCTION

SYSTEMATIC SECTION

The orders and names of the families and genera of rodents are those of Nowak & Paradiso (1983).

If a species' location in the host is known, it is given. In most cases, however, it is not known, oocysts having been found in the feces or intestinal contents.

The terms used in describing prevalence are: (1) common — in 20% or more of the host animals examined; (2) quite common — in 10—19%; (3) moderately common — in 5—9%; (4) uncommon — in 1—4%; and (5) rare — in less than 1%.

HOST FAMILY SCIURIDAE

HOST GENUS *TAMIAS*

EIMERIA OVATA DUNCAN, 1968

This species has been reported from the eastern chipmunk *Tamias striatus* in North America.

The oocysts are ovoid, 23—31 × 15—19 (mean 26 × 17) µm, with a two-layered wall, the outer layer rough, with small pits, green to colorless, 1—1.2 µm thick, the inner layer dark brown, 0.2—0.4 µm thick, thinner at the small end, without a micropyle or residuum, with a polar granule. The sporocysts are ovoid, 11—13 × 7—9 (mean 12 × 8) µm, with a small Stieda body and a residuum. The sporozoites are stout and slightly curved, and lie at the ends of the sporocyst. The sporulation time is 7—8 d at room temperature in 2% potassium bichromate solution.

Reference. Duncan (1968).

EIMERIA VILASI DORNEY, 1962

This species is common in the eastern chipmunk *Tamias striatus* in North America. It occurs primarily in the posterior quarter of the small intestine.

The oocysts are subspherical, occasionally ellipsoidal or ovoid, 11—23 × 7—19 µm, with a two-layered wall, the outer layer smooth, yellow-green, 0.7 µm thick, the inner layer dark tan, 0.3 µm thick, without a micropyle or residuum, with zero to six (mean 1.5) polar granules. The sporocysts are ellipsoidal (illustrated as elongate ovoid), 10 × 6 µm, with a small Stieda body and a residuum. The sporozoites are elongate, have a clear globule at the larger end, and lie side by side in the sporocysts. The sporulation time is 3—7 d at room temperature in 2.5% potassium bichromate solution.

References. Dorney (1962, 1963, 1966); Duncan (1968).

EIMERIA WISCONSINENSIS DORNEY, 1962

This species is fairly common in the eastern chipmunk *Tamias striatus* in North America. It occurs in the wall of the middle part of the small intestine.

The oocysts are ellipsoidal, 23—35 × 18—27 µm, with a two-layered wall, the outer layer rough, yellow-tan, 1.4—1.5 µm thick, the inner layer colorless to pale pink, 0.3—0.4 µm thick, without a micropyle or residuum, with one or rarely two polar granules. The sporocysts are lemon-shaped, 14—16 × 9—10 (mean 15 × 9) µm, with a Stieda body and a residuum. The sporulation time is 10—28 d at room temperature in 2.5% potassium bichromate solution.

Hill & Duszynski (1986) reported this species from the chipmunk *Eutamias wisconsinensis* in Mexico.

References. Dorney (1962, 1963, 1968); Duncan (1968); Hill & Duszynski (1986).

SARCOCYSTIS SP. ENTZEROTH, SCHOLTYSECK, & CHOBOTAR, 1983

Sarcocysts of this form were found in the muscles of the eastern chipmunk *Tamias striatus* (type intermediate host) in North America. The definitive host is unknown.

Reference. Entzeroth, Scholtyseck, & Chobotar (1983).

HOST GENUS *EUTAMIAS*

EIMERIA ASIATICI LEVINE & IVENS, 1965

Synonym. *Eimeria beecheyi* Henry of Tanabe & Okinami (1940).

This species is fairly common in the Asiatic chipmunk *Eutamias asiaticus* in Japan.

The oocysts were described as ovoid but illustrated as ellipsoidal, 15—22 × 10—12 μm, with a colorless, two-layered wall, the outer layer thick, the inner layer thin, without a micropyle or residuum. The sporocysts have a residuum. The sporulation time is 3—4 d.

References. Levine & Ivens (1965); Tanabe & Okinami (1940).

EIMERIA DORSALIS HILL & DUSZYNSKI, 1986

This species was found in the feces of the chipmunks *Eutamias dorsalis* (Type host), *E. canipes*, *E. merriami*, *E. obscurus*, and *E. townsendii* in North America.

The oocysts are slightly ovoid, 17—24 × 14—20 (mean 22 × 17) μm, with a two-layered wall 1 μm thick, the outer layer smooth, light brown, about ³/₄ of the total thickness and seeming to become thinner at the apical end, the inner layer membranous and dark brown, without a micropyle or residuum, with a bilobed polar body (sometimes two). The sporocysts are slightly ovoid, 10—14 × 6—8 (mean 11.5 × 7) μm, with a small Stieda body, with sub-Stiedal and parastiedal bodies, with a spheroidal, granular residuum which may be membrane-bound. The sporozoites have a large refractile posterior refractile body.

Reference. Hill & Duszynski (1986).

EIMERIA COCHISENSIS HILL & DUSZYNSKI, 1986

This species was found in the feces of the chipmunk *Eutamias dorsalis* (type host), *E. canipes*, *E. obscurus*, and *E. townsendii* in North America.

The oocysts are spherical to subspherical, 15—18 × 14—17 (mean 17 × 15) μm with a smooth, one-layered wall less than 1 μm thick, without a residuum or micropyle, with a large polar body. The sporocysts are ovoid, 6—11 × 4—7 (mean 8 × 6) μm with a small Stieda body, without sub-Stieda and parastieda bodies, with a compact, granular residuum. The sporozoites have a large posterior refractile body.

Reference. Hill & Duszynski (1986).

EIMERIA EUTAMIAE LEVINE, IVENS, & KRUIDENIER, 1957

This species was found in the cliff chipmunk *Eutamias dorsalis* in North America.

The oocysts are ovoid, 24—30 × 19—23 (mean 27 × 21) μm, with a smooth, two-layered wall, the outer layer colorless, 1 μm thick, the inner layer very pale tan, 0.4 μm thick, disappearing at the small end of the oocysts, without a micropyle or residuum, with a polar granule. The sporocysts are lemon-shaped, about 11 × 8 μm, with a fairly thick transparent wall, a Stieda body, and a residuum; most sporocysts seemed to lie with their long axes perpendicular to the long axis of the oocyst, so that they are usually seen end-on.

Reference. Levine, Ivens, & Kruidenier (1957).

EIMERIA TAMIASCIURI LEVINE, IVENS, & KRUIDENIER, 1957

See under *Tamiascurus*. This species was reported from *Eutamias* spp. in Arizona, California, and Mexico by Vance & Duszynski (1985).

Reference. Vance & Duszynski (1985).

SARCOCYSTIS EUTAMIAS TANABE & OKINAMI, 1940

Sarcocysts of this species were found in the muscles of the chipmunk *Eutamias asiaticus* (type intermediate host) in Japan.

They are compartmented, 300—530 × 30—40 μm with a one- or two-layered wall without striations. The bradyzoites are crescentic, with one end rounded and the other pointed, 5—7 × 2—3 μm. The definitive host is unknown.

Reference. Tanabe & Okinami (1940).

HOST GENUS *MARMOTA*

EIMERIA ARCTOMYSI GALLI-VALERIO, 1931

This species was found in the intestine of the marmot *Marmota* (syn., *Arctomys*) *marmota* in Switzerland.

The oocysts are cylindroid, 24 × 20 μm, with a clearly visible, slightly protruding micropyle. (The above description is so incomplete that the characters of the species are uncertain.)

Reference. Galli-Valerio (1931).

EIMERIA MARMOTAE GALLI-VALERIO, 1923

This species was found in the intestine of the marmot *Marmota* (syn., *Arctomys*) *marmota* in Switzerland.

The oocysts were described by Galli-Valerio (1923) as ovoid, with somewhat rounded ends, 51 × 42 μm, with a very distinct micropyle. This description is so incomplete that the characters of the species are uncertain.) However, it is larger than other species from this host.

References. Galli-Valerio (1923); Bornand (1937).

EIMERIA MENZBIERI SVANBAEV, 1963

This species was found in the Menzbier marmot *Marmota menzbieri* in the USSR.

The oocysts are ovoid or short ovoid, 18—28 × 13—24 (mean 24 × 21) μm, with a smooth, yellow-brown or yellow-orange, two-layered wall 1.2—2.1 μm thick, with a micropyle, without a residuum or polar granule. The sporocysts are ovoid or short ovoid, 8—11 × 6—8 (mean 9 × 7) μm, with a Stieda body and residuum. The sporozoites are comma-shaped, with clear globules. The sporulation time is 96-120 h at room temperature.

Reference. Svanbaev (1963).

EIMERIA MONACIS FISH, 1930

Synonyms. *Eimeria dura* Crouch & Becker, 1931; *E. os* Crouch & Becker, 1931 of Svanbaev (1963).

This species occurs in the woodchuck *Marmota monax* (Type host), marmot *M. bobak*, Siberian marmot *M. sibirica*, and Menzbier marmot *M. menzbieri* in North America and the USSR.

The oocysts are spherical, subspherical, or ovoid, 16—25 × 15—23 μm, with a smooth, colorless, one- or two-layered wall, without a micropyle, with or without a polar granule, with a residuum. The sporocysts are ovoid, 7—10 × 5—8 μm, with a Stieda body and residuum. The sporozoites are elongate, with one end narrower than the other, with a clear globule at the large end. The sporulation time is 60—64 h at room temperature in 2% potassium bichromate solution.

References. Crouch & Becker (1931); Dorney (1965); Fish (1930); Fleming et al. (1979); Iwanoff-Gobzem (1934); Svanbaev (1963); McQuistion & Wright 1985 (1984).

EIMERIA OS CROUCH & BECKER, 1931

This species occurs in the woodchuck *Marmota monax* (type host); Siberian marmot *M. sibirica*; and marmots *M. bobak*, *M. baibacina*, and *M. caudata* in North America and the USSR.

The oocysts are ovoid, 20—26 × 18—22 μm (Dorney said they are 34—37 × 23—26 μm), colorless to pale yellow, with a one- or two-layered wall, with a micropyle through which the membrane lining the oocyst wall sometimes protrudes to form a bulb-like swelling, without a residuum, usually without a polar granule. The sporocysts are ovoid to ellipsoidal, 9—13 × 5—8 μm (Dorney [1965] said they are about 15 × 9 μm), with a Stieda body and residuum. The sporozoites are elongate with one end broader than the other, have a clear globule at the broad end, and lie lengthwise head to tail in the sporocysts. The sporulation time is 90—105 h at room temperature in 2% potassium bichromate solution.

References. Crouch & Becker (1931); Dorney (1965); Machul'skii (1941, 1949); McQuistion & Wright 1985 (1984); Musaev & Veisov (1965); Svanbaev (1979).

EIMERIA PERFOROIDES CROUCH & BECKER, 1931

This species occurs in the woodchuck *Marmota monax* in North America.

The oocysts are ellipsoidal, ovoid, or subspherical, 17—25 × 15—20 (mean 23 × 18) μm, with a one-or two-layered wall, without a micropyle, with or without a residuum, with zero to three polar granules. The sporocysts are ellipsoidal, 8—11 × 4—6 μm, with a Stieda body and residuum. The sporozoites are elongate with one end broader than the other, with a clear globule at the broad end, and lie lengthwise head to tail in the sporocysts. The sporulation time is 70 h in 2% potassium bichromate solution at room temperature.

References. Crouch & Becker (1931); Dorney (1965); McQuistion & Wright (1985).

EIMERIA TUSCARORENSIS DORNEY, 1965

This species occurs in the woodchuck *Marmota monax* in North America.

The oocysts are ellipsoidal, 30—36 × 25—30 (mean 33 × 28) μm, with a two-layered wall, the outer layer yellow-tan, finely sculptured, the inner layer clear, 0.5—0.7 μm thick, without a micropyle or residuum, with 1—14 (mean 6.4) polar granules. The sporocysts are lemon-shaped (mean 16 × 10 μm) with a Stieda body and residuum. The sporozoites are elongate and lie lengthwise head to tail in the sporocysts. The sporulation time is more than 27 to more than 49 d at room temperature.

References. Dorney (1965); Fleming et al. (1979); McQuistion & Wright (1985).

EIMERIA TYANSHANENSIS LEVINE & IVENS, 1965

Synonyms. *E. monacis* Fish, 1930 of Svanbaev (1963); *E. tyanchanensis* (Svanbaev, 1963) Levine & Ivens, 1965 of Pellérdy (1974) *lapsus calami.*

This species occurs in the Menzbier marmot *Marmota menzbieri* in the USSR.

The oocysts are ovoid, short ovoid, or spherical, 16—28 × 13—24 (mean 23.5 × 19) μm, with a smooth, yellow-green, double-contoured wall 0.8—1.8 μm thick, without a micropyle, with a residuum and polar granule. The sporocysts are spherical or ovoid, 6—11 × 7.0—7.2 [sic] (mean 8.5 × 6.5 [sic]) μm, without a residuum. The sporozoites are comma-shaped, with a clear globule at the broad end.

References. Svanbaev (1963); Levine & Ivens (1965).

EIMERIA SP. SVANBAEV, 1963

This form occurs in the Menzbier marmot *Marmota menzbieri* (type host) and marmots *M. bobak* and *M. baibacina* in the USSR.

The oocysts are ovoid to short ovoid, 79—89 × 63—68 (mean 84 × 66) μm, with a yellow-brown, three-contoured wall 5—6 μm thick, the outer layer rough, the inner layer radially striated, without a micropyle, residuum, or polar granule. The sporocysts are ovoid, short ovoid, or spherical, 24—30 × 21—27 (mean 27 × 24) μm. The oocysts did not sporulate completely.

References. Svanbaev (1963, 1979).

ISOSPORA SP. SVANBAEV, 1963

This species occurs in the Menzbier marmot *Marmota menzbieri* in the USSR.
The oocysts are short ovoid to spherical, 20—22 × 19—20 (mean 21 × 20) μm,
with a smooth, greenish, double-contoured wall 1 μm thick, without a micropyle
or residuum, with a polar granule. The sporocysts are ovoid, 14—15 × 7—8
(mean 14 × 7) μm. The oocysts did not sporulate completely.
Reference. Svanbaev (1963).

SARCOCYSTIS BAIBACINACANIS UMBETALIEV, 1979

This species was found in the muscles of the marmots *Marmota baibacina, M.
caudata,* and *M. bobac* (Intermediate Hosts) in the Kazakh SSR. It was transmitted to the dog *Canis familiaris,* wolf *C. lupus* , and fox *Vulpes vulpes* (definitive
hosts).
The sarcocysts (muscle meronts) are filiform 3,000—17,000 × 100—300 μm,
and contain banana-, sickle-, or V-shaped bradyzoites 10—18 × 3.5—6 μm. The
sporocysts are ovoid.
Reference. Umbetaliev 1979.

BESNOITIA BESNOITI (MAROTEL, 1913) HENRY, 1913

While the type intermediate host of this species is the ox *Bos taurus,* it has been
found that the marmot *Marmota* sp. can be infected experimentally.
References. See under *Mus.*

SARCOCYSTIS SP. JOYEUX, 1927

Sarcocysts of this form were found in the muscles of the marmot *Marmota
marmota* (type intermediate host) in France. They were not described. The
definitive host and oocysts are unknown.
Reference. Joyeux (1927).

KLOSSIA (?) SP. DORNEY, 1965

This form occurs in the woodchuck *Marmota monax* in North America.
The oocysts are subspherical, without a micropyle, with a residuum; they each
contain 12—14 ovoid sporocysts 10 × 9 μm, without a Stieda body, with a
residuum. The number of sporozoites per sporocyst could not be determined.
Dorney (1965) thought that this might be a spurious parasite of the woodchuck,
perhaps of invertebrate origin.
Reference. Dorney (1965).

HOST *GENUS SPERMOPHILUS*

EIMERIA BECKERI YAKIMOFF & SOKOLOFF, 1935

Synonym. *E. ussuriensis* Yakimoff & Sprinholtz-Schmidt, 1939.
This species occurs commonly in the ground squirrels (susliks) *Spermophilus
pygmaeus* (type host), *S. beldingi, S. eversmanni, S. fulvus, S. maximus,* and *S.
relicutus* in the USSR and USA.

The oocysts are usually ovoid, sometimes spherical, 16—28 × 15—24 µm, without a micropyle, residuum, or polar granule. The sporocysts are ovoid, seldom piriform, 8—12 × 4—7 µm, without a Stieda body, with a residuum.

References. Abenov & Svanbaev (1979, 1982); Svanbaev (1956, 1962, 1979); Veluvolu (1981); Veluvolu & Levine (1984); Yakimoff & Sokoloff (1935); Yakimoff & Sprinholtz-Schmidt (1939); Zolotareff (1938).

EIMERIA BEECHEYI HENRY, 1932

This species occurs in the ground squirrel *Spermophilus beecheyi* in North America and *S. relictus* in the USSR.

The oocysts are ovoid, 16—22 × 10—13 (mean 19 × 16) µm, with a smooth, colorless, one-layered wall 1 µm thick, without a micropyle or residuum, with a polar granule. The sporocysts are without a residuum. The sporulation time is 4—5 d in 2% potassium bichromate solution.

References. Henry (1932); Abenov & Svanbaev (1982).

EIMERIA BELDINGII VELUVOLU IN VELUVOLU & LEVINE, 1984

This species occurs in the ground squirrel *Spermophilus beldingi* in North America.

The oocysts are ellipsoidal, 30—34 × 24—30 (mean 32 × 26) µm, with a two-layered wall, the outer layer 1.5 µm thick, rough, striated, colorless to yellow, the inner layer about 0.4 µm thick, yellowish, without a micropyle or residuum, with two or three polar granules. The sporocysts are ellipsoidal or ovoid, 11—15 × 9—12 (mean 13 × 10) µm, with a Stieda body, without a sub-Stiedal body, with a residuum. The sporozoites are elongate, without clear globules, and lie lengthwise head to tail in the sporocysts. Sporulation occurs outside the host.

References. Veluvolu (1981); Veluvolu & Levine (1984).

EIMERIA BERKINBAEVI ABENOV & SVANBAEV, 1979

This species occurs in the ground squirrel *Spermophilus fulvus* in the USSR.

The oocysts are ovoid or ellipsoidal, 29—39 × 23—31 (mean 33 × 27) µm, tricontoured, the outer layer rough, the middle layer striated, the wall 2.1—3.4 µm thick, without a micropyle or polar granule. The sporocysts are elongate ovoid or piriform, 14—16 × 9—11 (mean 15 × 10) µm, with a residuum, apparently without a Stieda body.

Reference. Abenov & Svanbaev (1979).

EIMERIA BILAMELLATA HENRY, 1932

Synonym. *Eimeria eubeckeri* Hall & Knipling, 1935.

This species occurs commonly in the golden-mantled ground squirrel *Spermophilus lateralis chrysodeirus* (type host) and the ground squirrels *S. armatus, S. beecheyi, S. citellus, S. columbianus, S. franklinii, S. richardsoni, S. variegatus,* and also *S. tridecemlineatus* in North America and Europe. It was found in the small intestine and particularly in the jejunum and ileum of *S. citellus* by Pellérdy & Babos (1953), in the cecum and large intestine of *S. franklinii* by Hall & Knipling (1935), and in the cells of the crypts of Lieberkuehn in the mid-small intestine of experimentally infected *S. armatus* by Anderson & Hess (1980).

The oocysts are ellipsoidal or perhaps ovoid, 25—41 × 21—32 μm, with a two-layered wall, the outer layer brown, thick, and rough, the inner layer clear, thin, and smooth, with a micropyle and polar granule, without a residuum. The sporocysts are elongate ellipsoidal, 13—23 × 9—12 μm, with a Stieda body and residuum. The sporozoites have a clear globule at the large end and do not lie in any particular arrangement in the sporocysts. The sporulation time is 4—11 d.

There are apparently two meront generations. First generation meronts can be seen 6—8 d after inoculation in the jejunum and ileum; mature meronts are 6—19 × 4—14 μm and contain 6—25 merozoites. Gamonts can be seen in the epithelial cells of the lower small intestine and sometimes duodenum. They enlarge the host cell. In *S. armatus*, mature macrogametes are 18—27 × 16—26 μm; mature microgamonts are 40—66 × 25—43 (mean 52 × 33) μm and contain many microgametes 3—5 × about 0.4 μm and a highly basophilic residual mass. The prepatent period in *S. armatus* is 10—11 d and the patent period is 5—21 d.

This species causes diarrhea but seldom death, depending on the oocyst dosage.

References. Anderson & Hess (1980, 1982, 1985); Hall & Knipling (1935); Hammond, Speer, & Roberts (1970); Henry (1932); Hilton & Mahrt (1971); Pellérdy & Babos (1953); Ryšavý (1957); Speer, Hammond, & Anderson (1970); Todd, Hammond, & Anderson (1968); Torbett, Marquardt, & Carey (1982).

EIMERIA CALLOSPERMOPHILI HENRY, 1932

Synonym. *E. callosphermophili* [sic] Svanbaev, 1962 in part.

This species occurs commonly in the golden-mantled ground squirrel *Spermophilus lateralis* (type host) and the ground squirrels *S. armatus*, *S. beecheyi S. beldingi*, *S. columbianus*, *S. franklinii*, *S. fulvus*, *S. maximus*, *S. richardsonii*, *S. spilosoma*, *S. tridecemlineatus*, and the white-tailed prairie dog *Cynomys leucurus* small intestine, especially the jejunum and upper ileum; a few gamonts may be in the cecum in heavy infections. It is in the villar epithelial cells. It has been found in North America and the USSR.

The oocysts are spherical to subspherical, 15—27 × 14—25 (mean 20 × 19) μm, with a slightly rough and pitted, colorless to pale yellowish, one-layered wall about 1.1 μm thick, without a micropyle, with a residuum and polar granule. The sporocysts are lemon-shaped, about 9 × 7 μm, with a Stieda body, without or with a residuum. The sporozoites have a clear globule at the large end and usually a small one just in front of the nucleus. The sporozoites are often at the ends of the sporocysts.

There are apparently two asexual generations. They are in the villar epithelial cells. First generation meronts are mature 2—4 d after inoculation, are 5—9 ×

4—7 (mean 7 × 6) μm, and contain 8—12 (mean 11) merozoites 9—12 × about 2.5—3 μm (mean length 11 μm) in stained smears, and generally lie parallel to each other. Mature second generation meronts can be found 3 d after inoculation. They are 7—14 × 6—12 (mean 10 × 8) μm and contain 8—26 (mean 18) merozoites 5—9 × about 2—2.5 μm (mean length 6 μm).

Gamogony occurs as early as 3 d after inoculation. The macrogametes and microgamonts are in the villar epithelial cells. Mature microgamonts are 11—16 × 9—14 (mean 13 × 10) μm and contain numerous microgametes about 2.5 × 0.3 μm in sectioned material. The prepatent period in experimental infections is 5—6 d and the patent period 9 d.

A host not on the list of natural hosts that has been infected experimentally is *S. variegatus*; mammals that could not be infected are *Eutamias minimus, Meriones unguiculatus*, and *Rattus norvegicus*.

References. Hammond & Anderson (1970); Hammond, Speer, & Roberts (1970); Henry (1932); Hilton & Mahrt (1971); Levine & Ivens (1965, 1970); Levine, Ivens, & Kruidenier (1958); Porchet-Henneré (1972); Roberts, Hammond, & Speer (1970); Roberts, Speer, & Hammond (1970); Scholtyseck & Mehlhorn (1970); Scholtyseck, Mehlhorn, & Friedhoff (1970); Speer & Hammond (1970); Speer, Hammond, & Anderson (1970); Svanbaev (1962, 1979); Todd & Hammond (1968); Torbett, Marquardt, & Carey (1982); Veluvolu (1981); Veluvolu & Levine (1984); Abenov & Svanbaev (1979).

EIMERIA CITELLI KARTCHNER & BECKER, 1930

This species occurs quite commonly in the thirteen-lined ground squirrel *Spermophilus tridecemlineatus* (type host) and the ground squirrels *S. beldingi, S. citellus, S. fulvus, S. maximus, S. pygmaeus, S. relictus*, and *S. undulatus* in North America, Europe, and the USSR. It was found by Kartchner & Becker (1930) in the mucosa of the cecum of *S. tridecemlineatus* and by Pellérdy & Babos (1953) in the small intestine, particularly the jejunum and ileum, of *S. citelli*; the latter also found gamonts sporadically in the epithelial cells of the cecum and colon. It may or may not cause enteritis, diarrhea, or death, but also may have no unfavorable effects.

The oocysts are subspherical, ellipsoidal, or ovoid, 14—33 × 13—21 μm, with a smooth, colorless, three-layered wall, the middle layer thick and the outer and inner layers membranes, without a micropyle or polar granule, with a residuum at first that becomes inconspicuous within 3—4 d. The sporocysts are 5—9 × 4—7 μm, with a small Stieda body and a residuum. The sporozoites are 5—8 × 2—3 μm, with a large clear globule at the broader, more rounded end. Sporulation takes 3 d in 4% potassium bichromate solution.

The mature macrogametes in *S. citellus* are ovoid, 16 × 11 μm. The microgamonts are spherical to ellipsoidal, 12 × 9 μm, and produce 50—70 microgametes 2—3 × 0.3 μm. The prepatent period is 4—5 d.

Rattus norvegicus cannot be infected.

References. Kartchner & Becker (1930); Pellérdy & Babos (1953); Ryšavý

(1957); Svanbaev (1962); Veluvolu (1981); Veluvolu & Levine (1984); Yaki-moff & Sokoloff (1935); Zasukhin & Rauschenbach (1932); Zasukhin & Tiflov (1932, 1933); Zolotareff (1938); Abenov & Svanbaev (1979, 1982); Frank (1978).

EIMERIA DESERTICOLA DAVRONOV, 1973
This species was found in the dark yellow suslik *Spermophilus fulvus* in the USSR.

The oocysts are ovoid, rarely spherical, 17—24 × 15—19 (mean 19.5 × 17) μm, with a dark yellow, one-layered wall 1.5—1.7 μm thick, without a micropyle or polar granule, with a residuum. The sporocysts are ellipsoidal, 8—12 × 5—10 μm, with a Stieda body and residuum. The sporozoites lack a clear globule. The sporulation time is 2—3 d at 27—30°C.

Reference. Davronov (1973).

EIMERIA FRANKLINII HALL & KNIPLING, 1935
This species was found in the Franklin ground squirrel *Spermophilus fran-klinii* in North America.

The oocysts are subspherical to ovoid, 19—24 × 13—18 (mean 21 × 15) μm, with a smooth, colorless, transparent, two-layered wall, without a micropyle, with a residuum and two polar granules. The sporocysts are ellipsoidal to subovoid, 11.5 × 6 μm, with a Štieda body and residuum. Sporulation takes 3—4 d at room temperature in potassium bichromate solution.

Reference. Hall & Knipling (1935).

EIMERIA HOFFMEISTERI LEVINE, IVENS, & KRUIDENIER, 1958
This species was found in the spotted ground squirrel *Spermophilus s. spilosoma* in North America.

The oocysts are subspherical, 15—23 × 13—19 (mean 18—20 × 16—17) μm, with a colorless to pale yellowish, smooth, one-layered wall about 1 μm thick, without a micropyle, with or without a residuum, with a polar granule. The sporocysts are elongate ovoid, about 11.5 × 6 μm, with or without a Stieda body, with a residuum. The sporozoites are either at the ends of the sporocysts or lie somewhat longitudinally in them.

Reference. Levine, Ivens, & Kruidenier (1958).

EIMERIA LARIMERENSIS VETTERLING, 1964
See under *Cynomys*.

EIMERIA LATERALIS LEVINE, IVENS, & KRUIDENIER, 1957
This species has been reported fairly commonly in the golden-mantled ground squirrel *Spermophilus lateralis* (type host) and the ground squirrels *Spermophilus columbianus* and *S. richardsonii* in North America.

The oocysts are ellipsoidal to somewhat ovoid, 28—40 × 24—31 (mean 35 ×

27) μm, with a rough, yellowish brown, one-layered wall 2 μm thick, pitted like a thimble, lined by a thin membrane, without a micropyle, with a residuum and one or more polar granules. The sporocysts are about 16 × 10 μm, with a prominent Stieda body and a residuum.

References. Hilton & Mahrt (1971); Levine, Ivens, & Kruidenier (1957).

EIMERIA MORAINENSIS TORBETT, MARQUARDT, & CAREY, 1982

This species occurs in the golden-mantled ground squirrel *Spermophilus lateralis* in North America.

The oocysts are subspherical, 19—26 × 17.5—21 (mean 20 × 20) μm, with a two-layered, gray to blue-gray, smooth wall, the outer layer about twice as thick as the inner, without a micropyle, without or with an inconspicuous residuum, with a polar granule. The sporocysts are ellipsoidal, 9—14 × 6—9 (mean 12 × 7) μm, with a Stieda body and residuum. The sporozoites are elongate, with a clear globule at each end, and lie lengthwise head to tail in the sporocysts. Sporulation occurs outside the host in 6—7 d. The prepatent period is 8—9 d and the patent period 9 d.

Reference. Torbett, Marquardt, & Carey (1982).

EIMERIA SPERMOPHILI HILTON & MAHRT, 1971

This species occurs commonly in the Richardson ground squirrel *Spermophilus richardsonii* (type host) and Franklin ground squirrel *S. franklinii* in North America.

The oocysts are ovoid to ellipsoidal, 21—30 × 16—24 (mean 25 × 21) μm, with a smooth, colorless to very pale yellow, one-layered wall 1.5 μm thick except at the anterior end where it thins to 0.9 μm, without a micropyle, with a residuum and polar granule. The sporocysts are elongate ovoid, 8—12 × 7—8 (mean 10 × 8) μm, with a Stieda body and residuum. The sporozoites are reniform, narrowing anteriorly, and lie at the ends of the sporocysts.

Reference. Hilton & Mahrt (1971).

EIMERIA SUSLIKI LEVINE & IVENS, 1965

Synonym. *Eimeria ussuriensis* Yakimoff & Sprinholtz-Schmidt, 1939 of Svanbaev (1962).

This species occurs in the susliks *Spermophilus maximus* (type host) and S. *fulvus* in the USSR.

The oocysts are ellipsoidal or elongate ellipsoidal, 20—35 × 23—26 (mean 32 × 25) μm, with a smooth, greenish, lilac, or yellow-brown wall 1—1.2 μm thick, without a micropyle, residuum, or polar granule. The sporocysts are ellipsoidal or short ellipsoidal, 9—13 × 7—10 (mean 10 × 9) μm, without a residuum. The sporozoites are comma-shaped, 6—9 × 3—5 (mean 8 × 4) μm.

References. Levine & Ivens (1965); Svanbaev (1962).

EIMERIA VOLGENSIS ZASUKHIN [SASSUCHIN] & RAUSCHENBACH, 1932

This species has been found in the little suslik or steppe ground squirrel *Spermophilus pygmaeus* (type host), relict suslik *S. relictus*, and Arctic ground squirrel *S. undulatus* in the USSR.

The oocysts are ovoid or piriform, sometimes with a sharply pointed end, 18—32 × 15—27 μm, with a smooth, clear, colorless or greenish, two-layered wall 1.2 μm thick, becoming thinner at the micropylar end, with a micropyle, without a residuum. The sporocysts are piriform, 9—15 × 4—10 μm, with a residuum. The sporulation time is about 3 d.

References. Musaev & Veisov (1965); Zasukhin [Sassuchin] & Rauschenbach (1932); Zolotarev (1938); Abenov & Svanbaev (1932).

EIMERIA YUKONENSIS SAMPSON, 1969

This species was found in the Arctic ground squirrel *Spermophilus undulatus* in North America.

The oocysts are elongate ellipsoidal, rarely ovoid, 21—27 × 12—15 (mean 24.5 × 13) μm, with a smooth, two-layered wall, the outer layer brownish yellow, about 0.5 μm thick, the inner layer colorless, about 0.1 μm thick, often lined by a membrane that is wrinkled at the micropylar end, with a micropyle and polar granule, without a residuum. The sporocysts are 9—12 × 5—6 (mean 11 × 6) μm, with a Stieda body and a residuum. The sporozoites are elongate, with one end narrower than the other, have a clear globule at the broad end, and lie lengthwise head to tail in the sporocysts.

Reference. Sampson (1969).

EIMERIA SP. LEVINE, IVENS, & KRUIDENIER, 1957

This species was found in the golden mantled ground squirrel *Spermophilus lateralis* in North America.

The oocysts are broadly ovoid to ellipsoidal, 19 × 17 μm, with a smooth, pale tan, one-layered wall 0.9 μm thick, without a micropyle, with a residuum and polar granule. The sporocysts are ellipsoidal, without a Stieda body, with a residuum.

Reference. Levine, Ivens, & Kruidenier (1957).

EIMERIA (?) SP. LEVINE, 1952

This species was found in Parry's ground squirrel *Spermophilus parryii* in North America.

Unsporulated oocysts were 21—25 × 19—21 (mean 23—20) μm, with a smooth, two-layered wall, the outer layer colorless, the inner layer dark yellowish, without a micropyle. The oocysts failed to sporulate.

Reference. Levine (1952).

ISOSPORA ASSENSIS SVANBAEV, 1979

Synonym. *I. laguri* Iwanoff-Gobzem, 1935 of Svanbaev (1962) in *Spermophillus fulvus.*

This species was found in the yellow suslik *Spermophilus fulvus* in the USSR.

The oocysts are ovoid, short ovoid, or spherical, 18—29 × 18—24 (mean 23.5 × 21) μm, with a smooth, "double-contoured" wall 1.5—2 μm thick without a micropyle or polar granule, with a residuum early in development only. The sporocysts are apparently ellipsoidal, 8—13 × 7—11 μm, or spherical, without a Stieda body or residuum. The sporozoites are 5—7 × 3—4 μm.

References. Svanbaev (1962, 1979).

ISOSPORA CITELLI LEVINE, IVENS, & KRUIDENIER, 1957

This species occurs in the rock squirrel *Spermophilus variegatus* (Type host) and ground squirrel *S. fulvus* in North America and the USSR.

The oocysts are subspherical, 22—23 × 21—22 (mean 22 × 21.5) μm, with a smooth, two-layered wall, the outer pale tan, 1 μm thick, the inner layer also pale tan, 0.4 μm thick, without a micropyle or residuum, with a polar granule that acts like an oil droplet. The sporocysts are broadly lemon-shaped, about 15 × 10 μm, with a small Stieda body and a residuum. The sporozoites are elongate and lie in no particular order in the sporocysts.

References. Levine, Ivens, & Kruidenier (1957); Abenov & Svanbaev (1979).

ISOSPORA SAMSENSIS SVANBAEV, 1979

Synonym. *Isospora uralicae* Svanbaev, 1956 of Svanbaev (1960) in part.

This species was found in the yellow suslik *Spermophilus fulvus* in the USSR.

The oocysts are ovoid, 25—32 × 22—26 (mean 28 × 24) μm, with a smooth, "double-contoured," greenish wall 1.2—1.5 μm thick, without a micropyle, residuum, or polar granule. The sporocysts are ellipsoidal, 11—13 × 8—9 (mean 12 × 8.5) μm, without a Stieda body or residuum. The sporozoites are 5.5 × 3 μm.

References. Svanbaev (1956, 1960, 1979).

ISOSPORA SPERMOPHILI LEVINE, 1984

Synonym. *Isospora laguri* Iwanoff-Gobzem, 1934 of Svanbaev (1962) in part.

This species was found in the suslik *Spermophilus maximus* in the USSR.

The oocysts are ellipsoidal, subspherical, or spherical, 18—29 × 18—24 (mean 23.5 × 21) μm, with a smooth, yellow-green or brown wall 1.5—2 μm thick, without a micropyle or polar granule, with a residuum. The sporocysts are ellipsoidal or spherical, 8—13 × 7—11 (mean 10 × 6.4 [sic]) μm. The sporozoites are comma-shaped, 5—7 × 3—4 (mean 6 × 3) μm.

References. Iwanoff-Gobzem (1934); Svanbaev (1960); Levine (1984).

SARCOCYSTIS BOZEMANENSIS DUBEY, 1983

This species occurs in the muscles of Richardson's ground squirrel *Spermophilus richardsonii* (intermediate host) in Montana.

The sarcocysts are compartmented, 80—300 × 30—112 (mean 135 × 58) μm, and have a smooth, one-layered wall 0.3—0.6 (mean 0.4) μm thick. The wall is PAS-negative, and the sarcocysts contain many PAS-positive bradyzoites 6—8 × 2 (mean 7 × 2) μm which have few rhoptries. Dubey (1983) saw no metrocytes.

Reference. Dubey (1983).

SARCOCYSTIS CAMPESTRIS CAWTHORN, WOBESER, & GAJADHAR, 1983

This species occurs in the Richardson ground squirrel *Spermophilus richardsoni* (type intermediate host — experimental) and American badger *Taxidea taxus* (type definitive host) in North America (Saskatchewan and Montana). Sporocysts are in the intestine of the badger. Meronts and merozoites are in the endothelial cells, primarily of the capillaries, of the (in order of abundance) lung, tongue, brown fat, myocardium, cerebral cortex, and skeletal muscles of the ground squirrel. Sarcocysts are in skeletal muscles of the ground squirrel. The early stages are pathogenic for the ground squirrel; a dose of 1500 sporocysts killed them in 11—13 d. This species causes hepatitis and phlebitis of the hepatic veins.

The oocysts are sporulated when passed in the feces. They are 14.5—16.5 × 8.5—11 (mean 15 × 10) μm, without a micropyle, polar granule, or residuum. The sporocysts are ellipsoidal, 9—12 × 6—9.5 (mean 10 × 8) μm, without a Stieda body, with a residuum. The prepatent period is 9 d and the patent period at least 13 d.

There are apparently three generations, the first associated with the liver veins about day 4 and the second in the vascular endothelial cells throughout the body 9—11 d after inoculation. The latter are 13—38 × 6—17 (mean 22 × 11) μm and contain 15—58 tachyzoites. Sarcocysts are present in the skeletal muscles beginning on day 30. They are 40—196 × 14—88 (mean 80 × 55) μm at 76 d, still not macroscopic at 113 d, but apparently macroscopic and about 2 × 0.5 mm at 258 d. They are compartmented, and have a two-layered wall, the outer layer 2—3 (mean 2.3) μm thick, with a toothbrush-like fringe composed of many projections, and the inner layer 0.3—1.3 (mean 0.7) μm thick. Metrocytes 4.5—12 × 3—8 (mean 7 × 5) μm are present at all times. The bradyzoites are about 12 × 3.5 μm.

References. Wobeser, Cawthorn, & Gajadhar (1982); Cawthorn, Wobeser, & Gajadhar (1983); Dubey (1983).

SARCOCYSTIS CITELLIVULPES PAK, PERMINOVA, & ESHTOKINA, 1979

This species occurs in the yellow suslik *Spermophilus fulvus* (type intermediate host) and fox *Vulpes vulpes* and corsac *V. corsac* (both experimental definitive hosts) in the USSR. Sporocysts are in the intestine of the fox or corsac. Sarcocysts are in the muscles of the suslik.

The oocysts have not been described. Sporulation occurs in the definitive hosts. The sporocysts are ovoid, smooth, 10—13×7—10 µm. The prepatent period is 7—8 d and the patent period 7—14 d.

Sarcocysts in the suslik are about 30—9000×20—600 µm, and their bradyzoites are 9—15×3—6 µm.

Reference. Pak, Perminova, & Eshtokina, 1979.

KLOSSIA SP. ABENOV & SVANBAEV, 1979

This form was found in the suslik *Spermophilus fulvus* in the USSR.

The oocysts are ellipsoidal, 36—50×31—36 (mean 42×33) µm, with a tricontoured wall 1.8 µm thick, and contain 12 spherical sporocysts 11—14 µm in diameter, each with 4 bean-shaped sporozoites.

Reference. Abenov & Svanbaev (1979).

BESNOITIA BESNOITI (MAROTEL, 1913) HENRY (1913)

While the type intermediate host of this species is the ox *Bos taurus*, it has been found that the suslik *Spermophilus fulvus* can be infected experimentally.

References. See under *Mus*.

BESNOITIA JELLISONI FRENKEL, 1955

The ground squirrel *Spermophilus* sp. is an experimental intermediate host of this species.

References. See under *Peromyscus*.

HOST GENUS *CYNOMYS*

EIMERIA CYNOMYSIS ANDREWS, 1928

Synonym. *Eimeria cynomysi* Andrews, 1928 *lapsus calami.*

This species was found in the prairie dog *Cynomys ludovicianus* in North America.

The oocysts are subspherical, broadly ellipsoidal, or ovoid, 30—37×25—32 µm, with a smooth, yellow-green, two-layered wall 1.4—2.5 µm thick, the outer layer transparent, appearing fibrous, with a very irregular outer surface, the inner layer faint orange-yellow, with a micropyle, without a residuum or polar granule. The sporocysts are seed- or lemon-shaped, 13—18×8—12 µm, with an inconspicuous Stieda body and a residuum. The sporozoites are attenuated reniform, blunt at both ends but with one end slightly larger than the other, without clear globules. The sporulation time is 3—4 d in moist feces at room temperature.

The cat cannot be infected.

References. Andrews (1927); Vetterling (1964).

EIMERIA LARIMERENSIS VETTERLING, 1964

This species occurs in the epithelial cells of the jejunum and ileum of the black-tailed prairie dog *Cynomys l. ludovicianus* (type host); white-tailed prairie dog *C. leucurus* and ground squirrels *Spermophilus armatus, S. beecheyi, S. lateralis, S. tridecemlineatus,* and *S. variegatus* in North America. It is apparently not pathogenic for these hosts.

The oocysts are spherical to ellipsoidal, 27—40 × 23—37 µm, with a mammillated or rough, two-layered wall, the outer layer brown, 1.7 µm thick, the inner layer colorless, thin, without a micropyle or residuum, with a polar granule. The sporocysts are lemon-shaped, 14—18 × 9—13 µm, with a Stieda body, sub-Stiedal body, and residuum. The sporozoites are elongate, 18—23 × about 2—2.5 (mean 20 × 2.3) µm, with two clear globules, and lie lengthwise head to tail in the sporocysts. The sporulation time is 5—9 (mean 8) d at 22—24°C in 2.5% potassium bichromate solution. The prepatent period is 5 d and the patent period 3—7 (mean 6.5) d.

There are three (possibly four) meront generations. First generation meronts are at the tips of the villi from the upper jejunum to the lower ileum, and second generation meronts are in the same location, but extend further down the sides of the villi. All are generally below the host cell nucleus. Mature first generation meronts are present 2.5—5 d after inoculation. They are ellipsoidal, 13—16 × 6—12 (mean 14 × 10) µm, and contain 16—32 (mean 23) merozoites 5—8 × 1.5 (mean 6 × 1.5) µm with a nucleus just behind the middle of the body. Mature second generation meronts can be seen 3.5 d after inoculation. They are 18—31 × 13—20 (mean 23 × 15) µm and contain 22—46 (mean 32) merozoites 7—11 × 1.5—2.5 (mean 8 × 2) µm; they have no residuum.

Macrogametes and microgamonts are in the epithelial cells of the villi of the jejunum and ileum (and occasionally the cecum). They can first be seen 3.5 d after inoculation. The macrogametes are 13—19 × 10—16 (mean 16 × 13) µm. Mature microgamonts are 16—36 × 12—25 (mean 26.5 × 19) µm and contain numerous microgametes 3—5 × about 1 µm (mean length 4 µm in stained preparations).

E. larimerensis has been transmitted from the prairie dog to *S. armatus*, and from several species of *Spermophilus* to several others, but it could not be transmitted to *Eutamias minimus, Rattus norvegicus,* or *Meriones unguiculatus*; although patent infections did not follow attempts at transmitting it to *S. richardsoni*, diarrhea occurred in this host 3—4 d after inoculation (Todd & Hammond, 1968).

References. Roberts & Hammond (1973); Roberts, Speer, & Hammond (1970); Speer, Davis, & Hammond (1972); Speer & Hammond (1970); Speer, Hammond, & Elsner (1973); Todd & Hammond (1968); Vetterling (1964).

EIMERIA LUDOVICIANI VETTERLING, 1964

This species occurs in the black-tailed prairie dog *Cynomys l. ludovicianus* (Type host) and white-tailed prairie dog *C. leucurus* in North America.

The oocysts are subspherical to ellipsoidal, 16—26 × 13—21 (mean 21 × 18) μm, with a smooth, colorless, two-layered wall 0.9 μm thick, without a micropyle, with a residuum and polar granule. The sporocysts are lemon-shaped, 9 × 7 μm, with a Stieda body and residuum. The sporozoites are elongate, without clear globules.

References. Todd & Hammond (1968); Vetterling (1964).

EIMERIA CALLOSPERMOPHILI HENRY, 1932

See under *Spermophilus*.

HOST GENUS *XERUS*

EIMERIA GARNHAMI McMILLAN, 1958

This species occurs in the African ground squirrel (palm rat) *Xerus* (*Euxerus*) *erythropus* in Africa. It occurs in the epithelial cells of the villi of at least the proximal part of the small intestine. It is apparently not pathogenic.

The oocysts are spherical to subspherical, 11—19 × 10—13 μm, with a smooth, pink, two-layered wall 0.7—0.8 μm thick, the outer layer thin, the inner layer thick, without a micropyle or residuum, usually with polar granules. The sporocysts are spherical or subspherical, 6 × 5 μm or 6.5 μm in diameter, without a Stieda body or residuum. The sporozoites are curved, slightly tapered at one end, 6 × 2 μm, with a central nucleus and a clear globule. The sporulation time is 14—16 d at 28—40°C. No sporulation occurs at 21—22°C.

Mature meronts contain six to eight elongate, pointed merozoites 7.5 × 1 μm.

The macrogametes are spherical, 10 μm in diameter. Early microgamonts are up to 8 × 6.5 μm, with multiple, rounded nuclei.

References. McMillan (1958); Vassiliades (1967).

EIMERIA XERI VASSILIADES, 1967

This species occurs in the African ground squirrel (palm rat) *Xerus* (*Euxerus*) *erythropus* in Africa.

The oocysts are ellipsoidal, yellowish to pale brown, 19—23 × 15—18 (mean 21 × 17) μm, with a smooth, transparent, two-layered wall 1.2—1.6 (mean 1.4) μm thick, the outer layer twice as thick as the inner one, without a micropyle or residuum, with a polar granule in 20%. The sporocysts are ovoid, 9—13 × 4.6 μm, without a Stieda body, with a residuum. The sporozoites are elongate, apparently without a clear globule, and lie lengthwise in the sporocysts.

Reference. Vassiliades (1967).

HOST GENUS *SCIURUS*

EIMERIA ANDREWSI YAKIMOFF & GOUSSEFF, 1935

This species was found in the small intestine of the "Eichhörnchen" (presumably *Sciurus* sp.) (type host) and also occurs fairly commonly in *S. vulgaris* in Europe and the USSR.

The oocysts are ovoid, somewhat pointed at both ends, or subspherical to ellipsoidal, 18—27 × 12—17 μm, with a smooth, colorless one- or two-layered wall, without a micropyle or residuum, with a polar granule in some. The sporocysts are ovoid, 7—13 × 4—8 μm, without a Stieda body or with a feebly visible one, with or without a residuum. The sporozoites are 6—8 × 2—3 μm. The sporulation time is 3 d at room temperature or 36—48 h at 25°C in 4% potassium bichromate solution. The prepatent period is 6 d.

References. Golemansky & Duhlinska (1973); Nukerbaeva & Svanbaev (1977); Pellérdy (1954); Ryšavý & Černá (1979); Yakimoff & Gousseff (1935).

EIMERIA ASCOTENSIS **LEVINE & IVENS, 1965**

Synonym. *Eimeria neosciuri* Prasad of Webster (1960).

This species has been found in the gray squirrel *Sciurus* (*Neosciurus*) *carolinensis* (Type host) and fox squirrel *S. niger rufiventer* in Europe and North America. It is in the small intestine, where it causes some inflammation but no diarrhea or abnormal feces.

The oocysts are ellipsoidal, 14—31 × 10—20 (mean 24 × 15) μm, with an apparently smooth, three-layered wall, the outer layer pale yellow and 0.75 μm thick, the middle layer brown and 1 μm thick, and the inner layer pinkish orange and 0.6 μm thick, described as without a micropyle but with an operculum 6—9 μm in diameter, with a rather indistinct, wavy line 5—7 μm long in the two inner layers of the wall at the end opposite the operculum, without a residuum, with a polar granule. The sporocysts are ovoid, 9—10 × 6—7 μm, with a prominent Stieda body and a residuum. The sporozoites are piriform, 9 × 3 μm, with a clear globule at the large end. The sporulation time is 1—3 d at 20—30°C, about 6 d at 15°C and about 8 d at 10°C. The oocysts are killed at 40°C.

The meronts are 8—17 μm in diameter, with 6—11 crescentic or banana-shaped merozoites 4—6 × 1.5 μm.

The macrogametes are spherical, 6—8 μm in diameter (probably immature), and the microgametes are 2.5—3 × 0.6—0.9 μm.

References. Joseph (1973); Lee & Dorney (1971); Levine & Ivens (1965); Webster (1960).

EIMERIA BOTELHOI **CARINI, 1932**

This species was found in the small intestine of the squirrel *Sciurus* (*Guerlinguetus*) *ingrami* in South America.

The oocysts were said to be oval but illustrated as piriform, 36 × 28 μm, with a three-layered wall 3 μm thick, the outer layer rough and yellowish, with a large micropyle at the tapering end, without a residuum or polar granule. The sporocysts are elongate ellipsoidal, 19 × 9 μm, without a Stieda body, with a small residuum. The sporozoites are elongate, comma-shaped. Sporulation takes several days in 1% chromic acid.

Merogony occurs in the mucosal cells of the villi. Meronts are spherical, up to 30 μm in diameter, with a variable number (usually 12—20) of fusiform merozoites 10—12 × 4 μm with a nucleus in the center.

Gamonts and gametes are also in the villar mucosal cells. The microgamonts produce hundreds of intensely staining microgametes 4—5 μm long.

References. Carini (1932).

EIMERIA CONFUSA JOSEPH, 1969

This species has been reported from the epithelial cells of the villi of the jejunum and ileum of the gray squirrel *Sciurus carolinensis* (type host) in North America and England. It has also been transmitted experimentally to the fox squirrel *S. niger rufiventer*. It is minimally if at all pathogenic.

The oocysts are predominantly subspherical, 23—46 × 19—36 μm, with a rough, two-layered wall, the outer layer 2—2.5 μm thick, dark yellow to brownish, the inner layer 0.7 μm thick, clear, without or with a micropyle, without a residuum, with zero to five polar granules. The sporocysts are broadly ellipsoidal or ovoid, 16—20×9—13 (mean 18×11) μm, with a prominent Stieda body and a large residuum. The sporozoites are elongate, with one end blunt and the other pointed, or lanceolate, 15—21 × 4—6 (mean 19 × 5) μm, have a clear globule at each end, and lie lengthwise head to tail in the sporocysts. The sporulation time is 8—17 d in 2% potassium bichromate solution at room temperature.

Mature meronts are ellipsoidal to ovoid, 17—23×14—23 (mean 21×19) μm, with 18—30 banana-shaped merozoites 11—17 × 2—3 (mean 15 × 3) μm.

The mature macrogametes are 27—35 × 20—26 (mean 31 × 26) μm. The mature microgamonts are 29—43 × 17—31 (mean 34 × 25) μm, with hundreds of microgametes 5—6 × about 1 μm. The prepatent period is 6—8 d and the patent period 6—15 d.

This species cannot be transmitted to the red squirrel *Tamiasciurus hudsonicus*.

References. Davidson (1976); Joseph (1969, 1972, 1973, 1975, 1977); Lee & Dorney (1971); Ball & Snow (1984).

EIMERIA FRANCHINII BRUNELLI, 1935

This species was found in the squirrel *Sciurus vulgaris* in Italy.

The oocysts are piriform, 24×15 μm, with a rough, yellow, two-layered, thick wall, with a prominent micropyle at the small end. The oocysts failed to sporulate in 2.5% potassium bichromate solution after 24 d, although four sporoblasts formed in some.

Reference. Brunelli (1935).

EIMERIA GUERLINGUETI ARCAY-DE-PERAZA, 1970

A *nomen nudum*.

EIMERIA KNIPLINGI LEVINE & IVENS, 1965

Synonym. *Eimeria sciurorum* Galli-Valerio of Knipling & Becker (1935).

This species occurs commonly in the cecum, colon, and slightly in the small intestine of the fox squirrel *Sciurus niger rufiventer* in North America.

The oocysts are cylindrical with rounded ends, 15—34 × 10—19 (mean 24 × 14) μm, with a smooth, pinkish to orange, one-layered wall, without an apparent micropyle but with the wall slightly thinner at one end, with a residuum, without a polar granule. The sporocysts are ellipsoidal, 13 × 7 μm, with a Stieda body and large residuum. Sporulation takes 57—70 h in 2.5% potassium bichromate solution.

References. Joseph (1973); Knipling & Becker (1935); Levine & Ivens (1965).

EIMERIA LANCASTERENSIS JOSEPH, 1969

This species occurs commonly in the epithelial cells of the whole small intestine of the gray squirrel *Sciurus carolinensis* (type host) and fox squirrel *S. niger rufiventer* in North America. It is apparently nonpathogenic.

The oocysts are ellipsoidal to ovoid, 12—32 × 12—20 (mean 25 × 15) μm, with a smooth, two-layered wall, the outer layer colorless, about 0.5 μm thick, the inner layer pale yellow, about 1.3 μm thick, without a micropyle or residuum, with a polar granule. The sporocysts are elongate ovoid, 11—16 × 8—10 (mean 14 × 8) μm, with a Stieda body and residuum. The sporozoites are slender and elongate, with one end rounded and the other acuminate, 9—17 × 3 (mean 14 × 3) μm, with a clear globule. The sporulation time is 28—36 h in 2.5% potassium bichromate solution at room temperature.

Mature meronts are ovoid, 9—11 × 7—10 (mean 10 × 8) μm; they contain 5—15 merozoites and some have a residuum. The merozoites are sickle-shaped, with both ends pointed, 8—14 × 1—2 (mean 11 × 2) μm in stained smears.

The macrogametes are ovoid to ellipsoidal, 11—19 × 11—13 (mean 16 × 11.5) μm in stained sections. The microgamonts are 11—17 × 10—14 (mean 14 × 12) μm, with many microgametes 3—4 × about 1 μm and a large residuum.

References. Joseph (1969, 1971, 1972).

EIMERIA LUISIERI (GALLI-VALERIO, 1935) REICHENOW, 1953

Synonyms. *Jarrina luisieri* Galli-Valerio, 1935; *Eimeria (Jarrina) luisieri* (Galli-Valerio, 1935) Reichenow, 1953.

This species was found in the intestine of the alpine squirrel *Sciurus vulgaris* var. *alpina* in Switzerland.

The oocysts are ovoid, with one end round and the other shaped like a bottleneck, 33 × 24 μm, with a thick, rough, yellowish wall, with a very distinct micropyle and residuum. The sporocysts are subspherical, 9 × 7.5 μm. The sporozoites are piriform. The sporulation time is 20 d on moist filter paper in a moist chamber at room temperature.

References. Galli-Valerio (1935): Reichenow (1953).

EIMERIA MIRA PELLÉRDY, 1954

Synonym. *Eimeria piriformis* Lubimov, 1934.

This species occurs in the small intestine of the squirrel *Sciurus vulgaris* in Europe and the USSR.

The oocysts are piriform, with a short bottleneck, 30—45 × 16—30 μm, with a rough, brown, two- or three-layered wall 3 μm thick, with a large micropyle at the end of the bottleneck, without a residuum or polar granule. The sporocysts are cigar-shaped, pointed at both ends, or ellipsoidal, 18—20 × 6—11 μm, with a poorly visible Stieda body and a large residuum. The sporozoites are comma-shaped. Sporulation takes 5—9 d.

The ground squirrel *Spermophilus citellus* and dormouse *Glis qlis* cannot be infected.

References. Golemansky & Duhlinska (1973); Lubimov (1934); Nukerbaeva & Svanbaev (1977); Pellérdy (1954); Ryšavý & Černá (1979).

EIMERIA MOELLERI LEVINE & IVENS, 1965

Synonym. *Eimeria sciurorum* Galli-Valerio of Möller (1923).

This species was found in the tips of the villi of the small intestine, especially the posterior part of the jejunum, of the gray squirrel *Sciurus (Neosciurus) carolinensis* (type host) in the Berlin (Germany) zoo. It has been transmitted experimentally to the European domestic squirrel *S. vulgaris*. It is nonpathogenic.

The oocysts are ellipsoidal, sometimes cylindrical, 22—28 × 14—18 μm, with a smooth, two-layered wall, with a micropyle, without a residuum or polar granule. The sporocysts are 10—14 × 6—8 μm, with a small Stieda body (?) and a residuum. The sporozoites are dumbell-shaped, with one end somewhat pointed, 8—14 × 4—5 μm. The sporulation time is 3 d.

The meronts are more or less rounded, about 18 μm in diameter, with 12 or more merozoites 6—7 × 1—2 μm.

The macrogametes are spherical, 16—20 μm in diameter, with a star- or thornapple-shaped nucleus. The microgamonts are spherical, 16—20 μm in diameter, with a large number of slender, comma-shaped microgametes. The prepatent period is 13 d.

References. Levine & Ivens (1965); Möller (1923).

EIMERIA NEOSCIURI PRASAD, 1960

This species was found in the epithelial cells of the villi of the upper part of the ileum of the gray squirrel *Sciurus (Neosciurus) carolinensis* in England. It is apparently nonpathogenic.

The oocysts are ellipsoidal, 22—28 × 14—18 μm, with a smooth, two-layered wall, the outer layer colorless or pale yellowish, the inner layer dark brown, without a micropyle or residuum, with a polar granule. The sporocysts are ovoid, 11—13 × 5—7 μm, with a Stieda body and residuum. The sporozoites are comma-shaped, 6 × 2—3 μm, with a clear globule at the large end. The sporulation time is 36—48 h at 22°C.

The meronts are ovoid, 10—11 × 4—4.5 μm, with 10—13 nuclei. The merozoites are sausage-shaped, 5—5.5 × 1 μm, with a central nucleus.

The macrogametes are ovoid or cylindroid, 5—17 × 5—12 μm. The microgamonts are asymmetrical, ellipsoidal, or cylindroid, 19—20 × 12 μm with many microgametes arranged haphazardly around a central residuum.

Reference. Prasad (1960).

EIMERIA ONTARIOENSIS LEE & DORNEY, 1971

This species occurs in the gray squirrel *Sciurus carolinensis* (type host), fox squirrel *S. niger rufiventer*, and *S. aberti* in North America and England.

The oocysts are piriform, with a short, bottle-shaped neck, 25—51 × 17—31 μm, with a two-layered wall, the outer layer rough, brown, 2 μm thick, forming a cap over the micropyle, the inner layer clear, smooth, homogeneous, 1 μm thick, with a large micropyle, without a residuum, usually without a polar granule. The sporocysts are elongate ellipsoidal, tapering slightly toward the Stieda body, 17—26 × 7—10 μm, with a Stieda body and residuum. The sporozoites are 26×3 μm, with a clear globule at the large end, and lie lengthwise head to tail in the sporocysts. The sporulation time is 6—8 d at room temperature. The prepatent period is 6—7 d and the patent period about 1 month or less.

An attempt to transmit this species to the red squirrel *Tamiasciurus hudsonicus* failed.

References. Davidson (1976); Duncan (1973); Hill & Duszynski (1986); Joseph (1972, 1973); Lee & Dorney (1971); Ball & Snow (1984).

EIMERIA SCIURORUM GALLI-VALERIO, 1922

Synonym. *Eimeria sciuri* Yakimoff & Terwinsky, 1931.

This species occurs commonly in the small intestine of the European alpine squirrel *Sciurus vulgaris* var. *alpinus* (type host), squirrels *S. vulgaris*, *S. aureogaster hypopyrrhus*, and "Eichhornchen" in Europe, the USSR, and Central America. It causes diarrhea, hyperemia, intestinal inflammation, epithelial desquamation, and even death.

The oocysts of the European and USSR forms are cylindroid, usually with parallel sides, 24—37 × 13—25 μm, with a smooth, thin, colorless, one- or two-layered wall, with a micropyle that is seldom visible if present, without a residuum, with one or two polar granules. The sporocysts are ovoid or piriform, 9—14×6—8 μm, with a Stieda body and residuum. The sporozoites are elongate piriform, 7 μm long. The oocysts in *S. a. hypopyrrhus* in Belize are cylindroid to ellipsoidal, 19—32 × 15—19 (mean 25 × 17) μm, with a smooth, colorless, one-layered wall about 1 μm thick, without a visible micropyle, with a residuum represented by a few minute granules or tenuous threads, apparently without a polar granule; the sporocysts are ovoid, 11—17×8—10 (mean 13×8) μm, with an indefinite Stieda body and a prominent residuum; the sporozoites are elongate and lie lengthwise in the sporocysts. Sporulation takes 2—4 d at room temperature in potassium bichromate solution.

The endogenous stages have apparently not been described in the European or USSR forms. In *S. a. hypopyrrhus* there are two types of meront above the host cell nucleus in the villar epithelial cells. One type has rather coarse cytoplasm and gives rise to 6—20 large merozoites 10—13 μm long. The second type has much smaller and denser nuclei and produces tiny merozoites averaging 3.2 μm long; there are more of these merozoites than in the first form. Both types have residua. The mature microgamonts are about 18 μm in diameter and produce many microgametes 5—6.5 μm long with two long flagella and a residuum. Fertilization takes place within the epithelium, and the oocyst wall is well developed before the oocysts are discharged into the lumen. The prepatent period is 7 d.

The European and USSR forms could not be transmitted to the laboratory rat and the rabbit.

References. Galli-Valerio (1922); Golemanski & Duhlinska (1973); Lainson (1968); Nukerbaeva & Svanbaev (1977); Pellérdy (1954); Yakimoff & Terwinsky (1931).

EIMERIA SERBICA POP-CENITCH & BORDJOCHKI, 1957

Synonyms. *Eimeria serbca* Pop-Cenitch & Bordjochki, 1957 *lapsus calami*; *E. sciurorum* Galli-Valerio, 1922 of Jirovec (1942) and Ryšavý (1954).

This species occurs in the Serbian squirrel *Sciurus vulgaris* (?) (type host) and squirrel *S. vulgaris altaicus* in Europe and the USSR. Its location in the host is unknown; oocysts were found in the feces.

The oocysts were described as oval and illustrated as ellipsoidal, 21—37 × 12—25 μm, yellowish rose, with a wall 0.6—0.9 μm thick, without a micropyle or residuum, with or without a polar granule. The sporocysts are ellipsoidal, 12—13 × 7—8 μm, without a Stieda body or with a poorly visible one, with a residuum. The sporulation time has been reported as 4 d at 16—18°C, at least 12 d at 25°C, and 45 d at 5°C.

References. Jirovec (1942); Nukerbaeva & Svanbaev (1977); Pop-Cenitch & Bordjochki (1957); Ryšavý (1954); Ryšavý & Černé (1979).

EIMERIA SILVANA PELLÉRDY, 1954

This species occurs in the small intestine of the squirrel *Sciurus vulgaris* in Europe and the USSR.

The oocysts are ellipsoidal or subspherical, 14—22 × 12—15 μm, with a smooth, pale wall, without a micropyle or residuum. The sporocysts are ellipsoidal, 9—10 × 5—6 μm, without a Stieda body, with a residuum. Sporulation takes 1—2 d at room temperature.

The prepatent period is 7 d.

References. Golemansky & Duhlinska (1973); Nukerbaeva & Svanbaev (1977); Pellérdy (1954).

EIMERIA SP. BOND & BOVEE, 1958

Synonym. *Eimeria sciurorum* (?) Galli-Valerio of Bond & Bovee (1958).

This species occurs in the gray squirrel *Sciurus carolinensis* in North America. The oocysts are ellipsoidal, 27 × 17 μm, with a small micropyle, with a residuum, apparently without a polar granule. The sporocysts were illustrated as elongate kidney-shaped, without a Stieda body or residuum.

References. Bond & Bovee (1958); Parker, Rigg, & Holliman (1972).

EIMERIA SP. BRUNELLI, 1935

This species was found in the squirrel *Sciurus vulgaris* in Italy.

The oocysts are ovoid, 17.5×10 µm, with a smooth, pearl-gray, two-layered wall. There is a residuum at one pole of the oocyst or sporocyst (the description is not clear as to which). The sporulation time is 4—6 d in 2.5% potassium bichromate solution.

Reference. Brunelli (1935).

EIMERIA SP. D. P. HENRY, 1932

This species was found in the western gray squirrel *Sciurus g. griseus* in California.

The oocysts are ovoid, 22—32 × 16—19 (mean 29 × 19) µm, with a smooth wall, with a micropyle. Sporulated oocysts were not described. The sporulation time is about 2 d.

Reference. Henry (1932).

EIMERIA SP. LEE & DORNEY, 1971

This species was found commonly in the gray squirrel *Sciurus carolinensis* in Ontario.

The oocysts are ellipsoidal, occasionally ovoid, 15—34 × 9—20 (mean 24 × 14) µm, with a two-layered wall 1.4 µm thick, both layers thin, smooth, colorless, without a micropyle but with an operculum represented by two transverse lines seen in 8% of 854 oocysts, and a distinct line at the opposite end in 2%, with a residuum at the ends of the oocysts, usually with a polar granule. The sporocysts are ovoid, 9—18 × 4—9 (mean 14 × 6) µm, with a Stieda body and residuum. The sporozoites are comma-shaped, 16 × 2.5 µm, without a clear globule. The sporulation time is 3—4 d.

Reference. Lee & Dorney (1971).

WENYONELLA HOAREI RAY & DAS GUPTA, 1937

This species was found in the epithelial cells of the small intestine of the Indian squirrel *Sciurus* sp. in India.

Its oocysts are spherical, 14 × 19 µm in diameter, with a two-layered wall, the outer layer thinner than the inner, with or without a micropyle, without a residuum or polar granule. The sporocysts are ovoid, 9—12 × 7—9 (mean 10.5 × 8) µm, with a Stieda body and residuum. The sporozoites are elongate, with one end bluntly pointed, 7 × 1.5 µm, and lie lengthwise head to tail in pairs in the sporocysts. The sporulation time is 7 d.

The meronts have 6—8 merozoites. There are two types of merozoite: "macromerozoites" 8 × 2 µm with granule-containing cytoplasm, and "micromerozoites" 6 × 2 μm with hyaline cytoplasm. The microgamonts contain many biflagellate microgametes.

References. Mandal (1976); Ray & Das Gupta (1937).

SARCOCYSTIS SP. DAVIDSON, 1976

Sarcocysts of this form were reported in the muscles of 5% of 270 gray squirrels *Sciurus carolinensis* in the USA. They were not described.

Reference. Davidson (1976).

BESNOITIA DARLINGI (BRUMPT, 1913) MANDOUR, 1965

The squirrels *Sciurus granatensis* and *S. variegatoides* are experimental intermediate hosts of this species.

References. See under *Mus*.

HOST GENUS *TAMIASCIURUS*

EIMERIA TAMIASCIURI LEVINE, IVENS, & KRUIDENIER, 1957

This species occurs commonly in the small intestine (especially the posterior ileum) of the red or spruce squirrel *Tamiasciurus hudsonicus* (type host) and *T. h. loquax* in North America. Vance & Duszynski (1985) also reported it from the vole *Microtus montanus arizonensis* in Arizona and from chipmunks *Eutamias* spp. in Arizona, California, and Mexico. Hill & Duszynski (1986) reported it from *Eutamias dorsalis*, *E. obscurus*, *Sciurus aberti*, *S. griseus*, *T. hudsonicus mogollensis*, and *T. mearnsi* in Arizona, Mexico, and California.

The oocysts are elongate ellipsoidal, occasionally somewhat ovoid, rarely slightly concave on one side, 17—40 × 10—22 μm, with a smooth, colorless, one-layered wall 1.5 μm thick, without a micropyle or residuum, usually with a polar granule. The sporocysts are elongate ovoid, 10—17 × 4—8 μm, with a Stieda body and residuum. The sporozoites are elongate and bent or comma-shaped. The sporulation time is 3 d.

References. Bullock (1959); Dorney (1961, 1963, 1966); Hill & Duszynski (1986); Levine, Ivens, & Kruidenier (1957); Mahrt & Chai (1972); Soon & Dorney (1969); Vance & Duszynski (1985).

EIMERIA TODDI DORNEY, 1962

This species is moderately common in the red or spruce squirrel *Tamiasciurus hudsonicus* in North America.

The oocysts were described as oval but illustrated as ellipsoidal, 36—45 × 27—36 (mean 40 × 32) μm, with a two-layered wall, the outer layer rough, deep yellow, about 2.4 μm thick, the inner layer colorless, 0.6 μm thick, without a micropyle or residuum, with 0—5 (mean 1.5) polar granules. The sporocysts are ovoid, 16—20 × 7—13 (mean 19 × 11) μm, with a prominent Stieda body and a residuum. The sporozoites have one terminal and one central clear globule. The prepatent period is not more than 12 d.

Reference. Dorney (1962); Hill & Duszynski (1986).

SARCOCYSTIS SP. ENTZEROTH, CHOBOTAR & SCHOLTYSECK, 1983

Sarcocysts of this species were found in the tongue muscles of the red squirrel *Tamiasciurus hudsonicus* in Michigan. Its definitive host is unknown.

The sarcocysts are 80 μm in diameter and several times longer. They are thin-walled, with a primary wall highly folded along short, nonbranching villi 90 × 40 nm, septate, with a few periphral metrocytes 5 × 3 μm with scattered micronemes, mitochondria, and often a pair of daughter cells in endogeny, and with many bradyzoites 6 × 2 μm with a conoid, polar rings, micronemes, rhoptries, subpellicular microtubules, polysaccharide granules, and one or more mitochondria with tubular structure.

Reference. Entzeroth, Chobotar, & Scholtyseck (1983).

HOST GENUS *FUNAMBULUS*

EIMERIA BANDIPURENSIS RAY, BANIK, & MUKHERJEA, 1965

This species occurs quite commonly in the palm squirrel *Funambulus palmarum* (type host) and palm squirrel *F. tristriatus* in India.

The oocysts are spherical, subspherical, or ovoid, 15—25 × 14—22 μm, with a two-layered wall 1.5 μm thick, the outer layer usually colorless and smooth, 1 μm thick, the inner layer clear, 0.5 μm thick, without a micropyle or residuum, with or without a polar granule. The sporocysts are broadly piriform or ovoid, 6—13 × 6—9 μm, with a Stieda body and residuum. The sporozoites are banana-shaped, 15—19 × 2—2.5 μm, with a clear globule at the broad end and often another at the small end. The sporulation time is 2—5 d at 31—34°C.

References. Chowattukunnel (1979); Ray, Banik, & Mukherjea (1965).

EIMERIA MALABARICA CHOWATTUKUNNEL, 1979

This species occurs quite commonly in the south Indian tree squirrel *Funambulus tristriatus* in India.

The oocysts are ellipsoidal to subspherical, 35—45 × 29—37 (mean 40 × 32) μm, with a two-layered wall 2.5—3 μm thick, the outer layer yellowish brown, rough, striated, with a thin superficial region and a thick, brownish region beneath it giving it a two-layered appearance, the inner layer colorless, without a micropyle or residuum, with a polar granule in about $^1/_3$. The sporocysts are ovoid, 14—18 × 11—12 (mean 16 × 11) μm, with a Stieda body and residuum. The sporozoites are elongate, 19—23 × 3—4 (mean 22 × 3) μm, with one end wider than the other, with a clear globule. The sporulation time is 7 d at 32—34°C in 2.5% potassium bichromate solution.

Reference. Chowattukunnel (1979).

DORISA BENGALENSIS (BANDYOPADHYAY & RAY, 1982) LEVINE & IVENS, 1987

Synonym. *Dorisiella bengalensis* Bandyopadhyay & Ray, 1982.

This species occurs in the Indian palm squirrel *Funambulus pennanti* in India.

The oocysts are subspherical or ovoid, 18—19.5 × 16.5—18 µm, with a smooth, two-layered wall 1 µm thick, without a micropyle, residuum, or polar granule. The sporocysts are ellipsoidal, 13.5—14 × 9—10 µm, with a Stieda body and residuum. The sporozoites are rounded, 3 µm in diameter. The sporulation time is 36 h at 30—37°C in 2.5% potassium bichromate solution.

References. Levine (1980); Levine & Ivens (1987); Bandyopadhyay & Ray (1982).

HOST GENUS *FUNISCIURUS*

WENYONELLA UELENSIS VAN DEN BERGHE, 1938

This species was found in the African striped squirrel *Funisciurus anerythrus* in Zaire.

The oocysts are ovoid or ellipsoidal, 26—30 × 19—20 µm, without a micropyle or polar granule, with a residuum. The sporocysts are ellipsoidal, 11 × 8 µm, without a Stieda body, with a residuum. The sporozoites are banana-shaped.

Reference. Van den Berghe (1938).

HOST GENUS *PARAXERUS*

WENYONELLA PARVA VAN DEN BERGHE, 1938

This species was found in the African bush squirrel *Paraxerus (Tamiscus) emini* in Zaire

The oocysts are subspherical, 15 × 13 µm, with a smooth, two-layered wall, without a micropyle, residuum, or polar granule. The sporocysts are ellipsoidal, 8 × 5 µm, without a Stieda body or residuum. The sporozoites are sausage-shaped.

Reference. Van den Berghe (1938).

HOST GENUS *CALLOSCIURUS*

EIMERIA CALLOSCIURI COLLEY, 1971

This species occurs quite commonly above the host cell nuclei in the villar epithelial cells, especially near the tips, of the small intestine, most heavily in its middle, of Prevost's squirrel *Callosciurus prevostii* (type host), the plantain squirrel *C. notatus*, the gray-bellied or golden-backed squirrel *C. caniceps*, and the black-banded squirrel *C. nigrovittatus* in Asia.

The oocysts are ovoid, 24—31 × 20—24 (mean 28 × 22) µm, with a smooth or slightly rough two-layered wall, the outer layer yellowish brown and 1.5 µm thick, the inner layer brown and 0.5 µm thick, with a micropyle, without a residuum or polar granule. The sporocysts are ellipsoidal to slightly ovoid, 14—19 × 6—10 (mean 17 × 8) µm, without a Stieda body, with a residuum. The sporozoites are broadly comma-shaped, have a large clear globule at the broad end, and lie lengthwise head to tail in the sporocysts.

The meronts have 6—12 merozoites about 4—7 × 1 μm, rounded at one end and pointed at the other, with a central nucleus. Mature macrogametes are spherical or ovoid, 10—15 × 6—12 μm. Microgamonts are usually ovoid, 10—16 × 8—12 μm, with many microgametes. The prepatent period is 7 d and the patent period 16 d.

References. Mullin & Colley (1972); Colley (1971).

EIMERIA NOTATI COLLEY & MULLIN, 1971

This species occurs commonly in the plantain squirrel *Callosciurus notatus* in Asia.

The oocysts are ellipsoidal to subspherical, 33—43 × 31—35 (mean 39 × 33) μm, with a rough, irregular, two-layered wall, the outer layer greenish brown and 2 μm thick, the inner layer colorless and 1 μm thick, without a micropyle, with a residuum and polar granule. The sporocysts are ovoid, 15—18 × 9—10 (mean 17 × 10) μm, with a Stieda body and residuum. The sporozoites are elongate, with one end narrower than the other, without clear globules, and lie lenthwise head to tail in the sporocysts.

References. Colley & Mullin (1971); Mullin, Colley, & Welch (1975).

EIMERIA PAHANGI COLLEY & MULLIN, 1971

This species occurs commonly in the plantain squirrel *Callosciurus notatus* in Asia.

The oocysts are ellipsoidal to subspherical, 30—38 × 29—32 (mean 35 × 31) μm, with a rough, two-layered wall, the outer layer yellowish brown and 1.5 μm thick, the inner layer colorless and 1 μm thick, without a micropyle, residuum, or polar granule. The sporocysts are ovoid, 16—17 × 10—12 (mean 17 × 11) μm, with or without a small Stieda body, with a residuum. The sporozoites are elongate, with one end narrower than the other, lack clear globules, and lie lengthwise head to tail in the sporocysts.

References. Colley & Mullin (1971); Mullin, Colley, & Welch (1975).

HOST GENUS *SUNDASCIURUS*

EIMERIA HIPPURI COLLEY & MULLIN, 1971

This species occurs in the horse-tailed squirrel *Sundasciurus hippurus* in Asia.

The oocysts are ellipsoidal, 34—37 × 21—23 (mean 35 × 22) μm, with a smooth, yellow, one-layered wall 1.5 μm thick, without a micropyle or residuum, with several polar granules. The sporocysts are ovoid, 13—15 × 8—10 (mean 14 × 9) μm, with a Stieda body and residuum. The sporozoites are elongate comma-shaped, without clear globules, and lie lengthwise head to tail in the sporocysts.

Reference. Colley & Mullin (1971).

HOST GENUS *PETAURISTA*

EIMERIA MALAYENSIS COLLEY & MULLIN, 1971

This species occurs in the spotted giant flying squirrel *Petaurista elegans* (Type host) and red giant flying squirrel *P. petaurista* in Asia.

The oocysts are ellipsoidal to cylindrical, 34—40 × 19—22 (mean 38 × 20) μm, with a smooth, two-layered wall, the outer layer greenish gray and 1.5 μm thick, the inner layer light brown and 0.5 μm thick, without a micropyle or residuum, with a polar granule. The sporocysts are ellipsoidal to ovoid, 15—18 × 7—10 (mean 16 × 8) μm, without a Stieda body, with a residuum. The sporozoites are elongate, broadly comma-shaped, usually with a clear globule at the broad end, and lie head to tail in the sporocysts.

Reference. Colley & Mullin (1971).

EIMERIA PETAURISTAE RAY & SINGH, 1950

This species occurs in the Himalayan flying squirrel *Petaurista petaurista* (syn., *P. inornatus*) in Asia.

The oocysts are flask-shaped, with a short neck and a dome-shaped "pseu-domicropyle" which forms a transparent cap at the anterior end and which disappears, becoming concave, on sporulation; they are 46—53 × 35—40 μm, with a two-layered wall, the outer layer rugged, deep brown, 4—6 μm thick, easily broken away from the inner layer, which is thin, smooth, and transparent, with a residuum, without a polar granule. The sporocysts are naviculoid, with one end slightly broader than the other, 25—31 × 9—10 μm, without a Stieda body, with a residuum. The sporozoites have a clear globule at each end. The oocysts are so heavy that they cannot be floated up by centrifugation in the usual sugar, sodium chloride, or zinc sulfate solutions. The sporulation time is 10—12 d in potassium bichromate solution.

The local Indian rabbit cannot be infected with this species.

Reference. Ray & Singh (1950).

HOST GENUS *AEROMYS*

EIMERIA AEROMYSIS COLLEY & MULLIN, 1971

This species occurs in the large black flying squirrel *Aeromys tephromelas* in Asia.

The oocysts are ellipsoidal, 26—41 × 18—33 (mean 35 × 22) μm, with a smooth, two-layered wall, the outer layer greenish yellow and 1.5 μm thick, the inner layer dark brown and 0.5 μm thick, without a micropyle or residuum, with several polar granules. The sporocysts are ovoid, 11—15 × 8—10 (mean 13 × 9) μm, without a Stieda body, with a residuum. The sporozoites are elongate with one end narrower than the other, without clear globules, and lie head to tail in the sporocysts.

Reference. Colley & Mullin (1971).

HOST GENUS *GLAUCOMYS*

EIMERIA DORNEYI LEVINE & IVENS, 1965

Synonym. *Eimeria* sp. Dorney, 1962.

This species occurs in the northern flying squirrel *Glaucomys sabrinus macrotis* in North America.

The oocysts are ellipsoidal, rarely ovoid, or truncate at one end, 14—30×10—19 μm, with a smooth, one- or two-layered wall, the outer layer light yellow and the inner layer pale green, without a micropyle or residuum, with a polar granule. The sporocysts are ovoid or piriform, 11—14×5—6 μm, with a Stieda body and residuum. The sporozoites are elongate, with a large clear globule at the large end, and lie lengthwise head to tail in the sporocysts.

References. Dorney (1962); Levine & Ivens (1965); Soon & Dorney (1969).

EIMERIA GLAUCOMYDIS ROUDABUSH, 1937

This species occurs in the intestine of the flying squirrel *Glaucomys volans* in North America.

The oocysts are ellipsoidal, 12—18 × 11—13 (mean 16 × 11.5) μm, with a smooth wall, without a micropyle or residuum, presumably without a polar granule. The sporocysts are ellipsoidal, with a Stieda body, with a residuum. The sporulation time is slightly less than 20 h.

Reference. Roudabush (1937).

EIMERIA PARASCIURORUM BOND & BOVEE, 1958

Synonym. *Eimeria sciurorum* Galli-Valerio of Roudabush (1937).

This species occurs in the intestine of the flying squirrel *Glaucomys volans* in North America.

The oocysts are cylindrical, usually rounded at the ends but sometimes truncate, 18—36×12—20 μm, with a smooth, two-layered, light yellow-brown wall 0.4—0.6 μm thick, without a micropyle, residuum, or polar granule. The sporocysts were described as ovoid and rounded at the end but illustrated as ellipsoidal, 8—13 × 4—7 (mean 11 × 6) μm, without a Stieda body, with a residuum. The sporozoites are piriform, 7—11 × 2—4 (mean 10 × 3) μm. The sporulation time is 22—36 h.

References. Bond & Bovee (1958); Roudabush (1937).

HOST GENUS *HYLOPETES*

EIMERIA HYLOPETIS COLLEY & MULLIN, 1971

This species occurs in the red-cheeked flying squirrel *Hylopetes spadiceus* in Asia.

The oocysts are ellipsoidal or flask-shaped, 30—40×21×28 (mean 35 ×24) μm, with a moderately rough, three-layered wall, the outer layer yellowish brown and 3 μm thick, the middle layer colorless and 1 μm thick, and the inner layer colorless and 0.5 μm thick, with a micropyle, without a residuum, with several polar granules. The sporocysts are ellipsoidal to cylindrical, 14—17 × 6—8 (mean 15×7) μm, without a Stieda body, with a residuum. The sporozoites are elongate, with rounded ends, without a clear globule.

Reference. Colley & Mullin (1971).

HOST FAMILY GEOMYIDAE

HOST GENUS *THOMOMYS*

EIMERIA FITZGERALDI TODD & TRYON, 1970

This species occurs in the northern pocket gopher *Thomomys talpoides* in North America.

The oocysts are ellipsoidal to ovoid, often slightly asymmetrical, flattened at one end, 24—33 × 18—24 (mean 28 × 22) μm, with a two-layered wall about 1.5—2 μm thick, the outer layer slightly rough, brown, about $^3/_4$ of the total thickness, the inner layer colorless, lined by a membrane which is wrinkled at the thinned, flat end of the oocysts, without a distinct micropyle or residuum, with a polar granule. The sporocysts are ovoid, 13—16 × 6—10 (mean 14 × 8) μm, with a Stieda body and a residuum. The sporozoites are elongate, with a large clear globule at the broad end and a small one at the narrow end, and lie lengthwise head to tail in the sporocysts.

References. Todd, Lepp, & Tryon (1971); Todd & Tryon (1970).

EIMERIA THOMOMYSIS LEVINE, IVENS, & KRUIDENIER, 1957

This species occurs in the pocket gopher *Thomomys bottae* (type species) and northern pocket gopher *T. talpoides* in North America.

The oocysts are spherical to subspherical, 13—16 (mean 14) μm in diameter, with a smooth, pale yellowish brown, one-layered wall 0.8 μm thick, lined by a thin membrane, without a micropyle, residuum, or polar granule. The sporocysts are 10 × 6 μm, with a small Stieda body and a few scattered residual granules. The sporozoites are elongate, with a clear globule at the large end, and lie head to tail in the sporocysts.

References. Levine, Ivens, & Kruidenier (1957); Todd, Lepp, & Tryon (1971).

HOST GENUS *GEOMYS*

EIMERIA GEOMYDIS SKIDMORE, 1929

This species occurs in the intestine of the pocket gopher *Geomys bursarius* in North America.

The oocysts are spherical to slightly ovoid, 12—15×12—13 (mean 13×12.5) μm, with a smooth, colorless, two-layered wall 0.5 μm thick, with a micropyle visible in a few oocysts, without a residuum or polar granule. The sporocysts are 5—7 × 4—5 μm, without a Stieda body, with a residuum. The sporulation time is 4 d at room temperature in 2% potassium bichromate solution.

Reference. Skidmore (1929).

HOST GENUS *ORTHOGEOMYS*

EIMERIA ORTHOGEOMYDOS LAINSON, 1968

This species occurs commonly, presumably in the intestine, of the pocket gopher *Orthogeomys grandis scalops* in Central America.

The oocysts are ellipsoidal to subspherical, 11—14 × 9—12 (mean 13 × 11) μm, with a smooth, colorless, two-layered wall about 0.7 μm thick, without a micropyle, with some small, scattered residual granules, apparently without a polar granule. The sporocysts are ovoid, 6 × 4 μm, with a slightly developed Stieda body and a residuum. The sporozoites have a clear globule at the large end and lie lengthwise in the sporocysts. The sporulation time is 7 d at about 26°C in 2% potassium bichromate solution.

Reference. Lainson (1968).

HOST FAMILY HETEROMYIDAE

HOST GENUS *PEROGNATHUS*

EIMERIA PENICILLATI IVENS, KRUIDENIER, & LEVINE, 1959

This species occurs in the pocket mice *Perognathus penicillatus* (type host) and *P. flavus* in North America.

The oocysts are subspherical, ellipsoidal, or slightly ovoid, 16—20 × 14—19 μm, with a smooth, pale brownish yellow or tan, one-layered wall about 0.6 μm thick, without a micropyle, with a residuum and polar granule. The sporocysts are broadly lemon-shaped, 9 × 7 μm, with a small Stieda body and a residuum. The sporozoites lie more or less lengthwise in the sporocysts.

Reference. Ivens, Kruidenier, & Levine (1959).

EIMERIA PEROGNATHI LEVINE, IVENS, & KRUIDENIER, 1957

This species occurs in the rock pocket mouse *Perognathus intermedius* in North America.

The oocysts are ovoid, 19—22 × 15—16 (mean 20 × 15) μm, with a somewhat rough, yellowish brown, one-layered wall 1 μm thick, without a micropyle or polar granule, with a residuum. The sporocysts are 6—7 × 4—5 μm, with a small, flat Stieda body and a small amount of scattered residual material.

Reference. Levine, Ivens, & Kruidenier, 1957.

EIMERIA REEDI ERNST, OAKS, & SAMPSON, 1970

This species occurs uncommonly in the pocket mouse *Perognathus formosus* in North America.

The oocysts are subspherical to ellipsoidal, 18—26 × 17—23 (mean 23 × 21) μm, with a pitted, colorless to light yellow, one-layered wall about 1.2 μm thick, without a micropyle, with a residuum and polar granule. The sporocysts are ovoid, 9—12 × 7—9 (mean 11 × 8) μm, with a prominent Stieda body, small substiedal body, and residuum. The sporozoites are elongate, with a large clear globule at the broad end, and lie more or less at the ends of the sporocysts.

Reference. Ernst, Oaks, & Sampson (1970).

HOST GENUS *DIPODOMYS*

EIMERIA BALPHAE ERNST, CHOBOTAR, & ANDERSON, 1967

This species occurs uncommonly to quite commonly in Ord's kangaroo rat *Dipodomys ordii* (type host) and the kangaroo rats *D. agilis*, *D. merriami*, and *D. spectabilis* in North America.

The oocysts are broadly ellipsoidal to ovoid, 15—18 × 13—15 (mean 17 × 14) μm, with a smooth, two-layered wall, the outer layer pale yellowish brown and about 1 μm thick, the inner layer dark brown and about 0.5 μm thick, without a micropyle, with a residuum and polar granule. The sporocysts are ovoid, 8—9 × 5—7 (mean 9 × 6) μm, with a thin wall, Stieda body, and residuum. The sporozoites have a clear globule. The prepatent period is 6 d and the patent period probably more than 20 d.

References. Ernst, Chobotar, & Anderson (1967); Short, Mayberry, & Bristol (1981); Stout & Duszynski (1983).

EIMERIA CHIHUAHUAENSIS SHORT, MAYBERRY, & BRISTOL, 1981

This species occurs uncommonly in the kangaroo rat *Dipodomys merriami* in North America.

The oocysts are ellipsoidal, 31—35 × 24—28 (mean 33 × 26) μm, with a smooth, two-layered wall 1.5 μm thick, each layer 0.75 μm thick, without a micropyle, with a polar granule and residuum. The sporocysts are ovoid, 11—15 × 8—10 (mean 13 × 9) μm, with a Stieda body, without a sub-Stiedal body, with a residuum. The sporozoites are apparently without clear globules and are variously arranged in the sporocysts.

References. Short, Mayberry, & Bristol (1981); Stout & Duszynski (1983).

EIMERIA CHOBOTARI ERNST, OAKS, & SAMPSON, 1970

This species occurs quite commonly to commonly in the kangaroo rats *Dipodomys merriami* (type host), *D. microps*, *D. ordii*, and *D. agilis* in North America.

The oocysts are ellipsoidal to ovoid, 42—52 × 31—41 (mean 49 × 35) μm, with a smooth, two-layered wall, the outer layer light brown, about 1.2 μm thick, the inner layer light brown, about 1 μm thick, without a micropyle or polar granule, with a residuum. The sporocysts are ovoid, 15—17 × 10—12 (mean 16 × 11) μm, with a small, flattened Stieda body, with a sub-Stiedal body and residuum. The sporozoites are indistinct, with a single large, clear globule.

References. Ernst, Oaks, & Sampson (1970); Short, Mayberry, & Bristol (1981).

EIMERIA DIPODOMYSIS LEVINE, IVENS, & KRUIDENIER, 1958

This species occurs in the kangaroo rats *Dipodomys phillipsi* (type host), *D. merriami*, and *D. ordii* in North America.

The oocysts are ellipsoidal, 47—61 × 38—42 (mean 54 × 40) μm, with a rough, yellowish brown, two-layered wall, the outer layer 3.5 μm thick at the sides and 3 μm thick at the ends, the inner layer 0.7 μm thick, without a micropyle or polar granule, with a large residuum. The sporocysts are ovoid, about 16 × 11 μm, with a Stieda body excavated in the center, with a large amount of residual material.

References. Levine, Ivens, & Kruidenier (1958); Short, Mayberry, & Bristol (1981); Stout & Duszynski (1983).

EIMERIA MERRIAMI STOUT & DUSZYNSKI, 1983

This species occurs in the kangaroo rat *Dipodomys merriami* in North America.

The oocysts are ellipsoidal, 21—27 × 20—26 (mean 24 × 23) μm, with a two-layered wall 1.6 μm thick, the outer layer smooth, light yellow, $^2/_3$ of the wall thickness, the inner layer brown, without a micropyle or polar body, with a residuum. The sporocysts are ovoid, 9—12 × 7—9 (mean 11 × 8) μm, with a Stieda body, sub-Stiedal body, and residuum.

Reference. Stout & Duszynski (1983).

EIMERIA MOHAVENSIS DORAN & JAHN, 1949

This species occurs above the host epithelial cell nuclei in the small intestine and cecum (gamonts and gametes in the cecum only) of the kangaroo rat *Dipodomys panamintinus mohavensis* (Type host) and (all experimentally) the kangaroo rats *D. merriami*, *D. nitratoides*, *D. heermanni*, *D. deserti*, and *D. agilis* in North America.

The oocysts are ellipsoidal, 22—26 × 14—18 (mean 24 × 16) μm, with a smooth, light brown, one-layered wall 0.7—0.9 μm thick, without a micropyle, residuum, or polar granule. The sporocysts are 6—10 × 4—8 μm, without a Stieda body, with a residuum. The sporulation time is 24—60 h at 23-27°C, 64—117 h at 9.5°C, or 44—78 h at 40°C in 2—4% potassium bichromate solution; no sporulation occurs at 2.5 or 45°C.

Small, early meronts 4—10.5 μm in diameter are present 2—5 d after inoculation. They produce 20—35 comma-shaped merozoites 6—9 × 1.5—2 μm with a central nucleus. Larger meronts 8—10 μm in diameter are present 6—9 d after inoculation. They produce 50—75 merozoites 4.5—6.5 × 1—1.5 μm with a central nucleus. Gamonts are produced by both types of merozoite, but principally by the second ones. The macrogametes are spherical, 10—16 μm in diameter; the microgamonts are ellipsoidal, 17—19 × 21—23 μm, and produce a large number of microgametes. The prepatent period is 7 d and the patent period 9—10 d.

The following species and subspecies of *Dipodomys* have been infected experimentally: *D. panamintinus leucogenys, D. p panamintinus, D. p. caudatus, D. merriami, D. nitratoides, D. heermanni, D. deserti,* and *D. agilis.* The following rodents could not be infected: *Perognathus longimembris, P. formosus, Peromyscus boylii, P. maniculatus, P. californicus, P. truei, Onychomys torridus, Neotoma lepida, Spermophilus leucurus, Mus musculus,* and *Rattus norvegicus.*

References. Doran (1951, 1953); Doran & Jahn (1949, 1952).

EIMERIA SCHOLTYSECKI ERNST, FRYDENDALL, & HAMMOND, 1967

This species occurs in Ord's kangaroo rat *Dipodomys ordii* (type host) and also in the kangaroo rats *D. agilis, D. gravipes, D. panamintinus,* and *D. spectabilis* in North America.

The oocysts are broadly ovoid to ellipsoidal, 21—27 × 17—21 (mean 25 × 20) μm, with a smooth, two-layered wall, the outer layer yellowish brown, about 1 μm thick, the inner layer dark brown to black, about 0.5 μm thick, without a micropyle or residuum, with a polar granule. The sporocysts are ovoid, 10—14 × 7—10 (mean 12 × 8) μm, with a small, flattened Stieda body and a large residuum. The sporozoites are elongate, with one end wider than the other, with a clear globule at the broad end and another near the middle of the body; they lie lengthwise head to tail in the sporocysts. The prepatent period is 8—9 (mean 8.2) d.

References. Ernst, Frydendall, & Hammond (1967); Stout & Duszynski (1983).

EIMERIA UTAHENSIS ERNST, HAMMOND, & CHOBOTAR, 1968

This species occurs in the small intestine of the Ord kangaroo rat *Dipodomys ordii* (Type host) and also the kangaroo rats *D. agilis, D. californicus, D. merriami,* and *D. microps* in North America.

The oocysts are subspherical to broadly ellipsoidal, 37—45 × 33—41 (mean 42 × 39) μm, with a two-layered wall, the outer layer rough, pitted, yellowish brown and about 2.3 μm thick, the inner layer bluish, about 0.8 μm thick, without a micropyle or polar granule, with a residuum. The sporocysts are elongate ovoid, 18—23 × 12—14 (mean 21 × 12) μm, with a Stieda body, sub-Stiedal body and residuum. The sporozoites are elongate, with one end broader than the other, with a clear globule at each end, and lie lengthwise head to tail in the sporocysts. Free sporozoites are 22.5 × 5 μm. The sporulation time is 4—7 d at about 22°C in 2.5% potassium bichromate solution.

There are four meront generations. The first generation meronts appear 2.5 d after inoculation in the villar epithelial cells of the anterior $^{1}/_{3}$ of the small intestine. They have a mean diameter of 10 μm and contain 12—16 merozoites 8 × 2 μm. The second generation meronts are in the villar epithelial cells of the middle 1/3 of the small intestine on days 4—5. They are 8 μm in diameter and contain 12—16 merozoites 8 × 1 μm and a residuum. The third generation meronts are in the epithelial cells of the upper half of the villi in the middle $^{1}/_{3}$ of the small intestine, mostly on days 5—6. They are 12 μm in diameter and contain 4—8 merozoites 12 × 2 μm. The fourth generation meronts are in the same location on days 6—7, are 9 μm in diameter, and contain 16—24 merozoites 8 × 1 μm. Only the second generation meronts have a residuum.

Gamonts are present beginning 5 d after inoculation. They are in the villar epithelial cells of the small intestine crypts at first, and then become displaced into the lamina propria. The mature macrogametes are 32.5 × 27 μm. The mature microgamonts are mostly ellipsoidal, 64 × 48 μm, and contain many microgametes. The prepatent period is 8—9 (mean 8.8) d.

References. Ernst & Chobotar (1978); Ernst, Hammond, & Chobotar (1968); Hill & Best (1985); Stout & Duszynski (1983).

EIMERIA (?) SPP. HILL & BEST, 1985

This form was found in *Dipodomys elephantinus, D. heermanni,* and *D. vanustus* feces in California. The oocysts failed to sporulate.

Reference. Hill & Best (1985).

ISOSPORA SP. STOUT & DUSZYNSKI, 1983

This form was found in the kangaroo rat *Dipodomys agilis* in North America.

The oocysts are spherical or subspherical, 21 × 28 × 20—28 (mean 26 × 25) μm, with a two-layered wall 1.6 μm thick, the outer layer smooth, pale yellow, about $^{2}/_{3}$ of the total thickness, without a micropyle or residuum, with a polar granule. The sporocysts are broadly ovoid, 12—19 × 9—13 (mean 15 × 10) μm, with a Stieda body and residuum, without a sub-Stiedal body.

Reference. Stout & Duszynski (1983).

BESNOITIA JELLISONI FRENKEL, 1955

The kangaroo rats *Dipodomys* spp. are intermediate hosts of *B. jellisoni*.
References. See under *Peromyscus*.

HOST GENUS *LIOMYS*

EIMERIA LIOMYSIS LEVINE, IVENS, & KRUIDENIER, 1958

This species occurs in the painted spiny pocket mouse *Liomys pictus* (Type host) and Mexican spiny pocket mouse *L. irroratus* in North America.

The oocysts are subspherical to ellipsoidal, 15—24 × 14—21 μm, with a two-layered wall, the outer layer slightly rough and pitted, pale yellow, 0.9 μm thick, the inner layer practically colorless, 0.3 μm thick, without a micropyle or residuum, with a polar granule. The sporocysts are ellipsoidal to ovoid, about 10 × 7 μm, with a small Stieda body and a residuum. The sporozoites are usually at the ends of the sporocysts.

Reference. Levine, Ivens, & Kruidenier (1958).

EIMERIA PICTI LEVINE, IVENS, & KRUIDENIER, 1958

This species occurs in the painted spiny pocket mouse *Liomys pictus* in North America.

The oocysts are subspherical to ellipsoidal, 22—32 × 19—28 (mean 26 × 22.5) μm, with a two-layered wall, the outer layer rough and pitted, brownish yellow and 1.3 μm thick, the inner layer brownish yellow, 0.4 μm thick, without a micropyle, with a residuum and one or two polar granules. The sporocysts are broadly lemon-shaped, 11—12 × 8—9 μm, with a Stieda body and a residuum. The sporozoites are elongate and lie longitudinally head to tail in the sporocysts.

Reference. Levine, Ivens, & Kruidenier (1958).

HOST FAMILY CASTORIDAE

HOST GENUS *CASTOR*

EIMERIA CAUSEYI ERNST, COOPER, & FRYDENDALL, 1970

This species occurs in the Canadian beaver *Castor canadensis* in North America.

The oocysts are ovoid, 23—30 × 14—18 (mean 26 × 16) μm, with a smooth, rarely pitted, two-layered wall, the outer layer light blue and about 1 μm thick, the inner layer light yellow, lined by a thin membrane, with a micropyle and residuum, without a polar granule. The sporocysts are elongate ovoid, 12—15 × 6—7 (mean 14 × 6) μm, with a Stieda body and residuum. The sporozoites are elongate, with a clear globule at the large end, and lie lengthwise head to tail in the sporocysts.

Reference. Ernst, Cooper, & Frydendall (1970).

EIMERIA SPREHNI YAKIMOFF, 1934

This species occurs commonly in the Canadian beaver *Castor canadensis* in North America.

The oocysts are ellipsoidal, 14—22 × 10—14 μm, with a smooth, pale yellowish, one-layered wall about 1 μm thick, without a micropyle or residuum, with a polar granule. The sporocysts are ellipsoidal, 9—12 × 4—6 μm, with a thin wall, Stieda body, and residuum. The sporozoites are elongate, with a clear globule at the large end, and lie lengthwise head to tail in the sporocysts.

References. Ernst, Cooper, & Frydendall (1970); Yakimoff (1934).

HOST FAMILY MURIDAE

HOST GENUS *ORYZOMYS*

EIMERIA COUESII KRUIDENIER, LEVINE, & IVENS, 1960

This species occurs in the rice rat *Oryzomys c. couesi* in North America.

This oocysts are ellipsoidal, pale yellowish, 20—23 × 17—20 (mean 21 × 18) μm, with a somewhat rough, pitted, heavy, one-layered, weakly striated wall about 1.3 μm thick, without a micropyle or residuum, with a polar granule. The sporocysts are ovoid, 10—14 × 7—8 (mean 12 × 8) μm, with a Stieda body, with or without a residuum. The sporozoites are elongate, with a clear globule at the large end, and lie lengthwise in the sporocysts.

Reference. Kruidenier Levine, & Ivens (1960).

EIMERIA KINSELLAI BARNARD, ERNST, & ROPER, 1971

This species occurs in the marsh rice rat *Oryzomys palustris* in North America.

The oocysts are ellipsoidal, with one side often flattened, 23—31 × 15—19 (mean 27 × 17) μm, with a smooth, two-layered wall, the outer layer light yellowish brown, about 0.8 μm thick, the inner layer colorless to light blue, about 0.8 μm thick, without a micropyle or residuum, with one to four polar granules. The sporocysts are elongate ovoid, 12—15 × 7—9 (mean 14 × 8) μm, with a prominent Stieda body and a residuum. The sporozoites are elongate, with a clear globule at the broad end, and lie lengthwise in the spororocysts.

Reference. Barnard Ernst, & Roper (1971).

EIMERIA OJASTII ARCAY-DE-PERAZA, 1964

This species occurs in the rice rat *Oryzomys albigularis* in South America.

It's oocysts are ellipsoidal, 17 × 13 μm, with a thin, double-contoured wall, without a micropyle or residuum. The sporocysts are ellipsoidal, 7.5 × 6 μm, without a Stieda body, with a small residuum. The sporozoites are reniform. The sporoulation time is 7 d in 2% potassium bichromate solution at room temperature.

Oryzomys c. concolor could not be infected experimentally with this species.

Reference. Arcay-de-Peraza (1964).

EIMERIA ORYZOMYSI CARINI, 1937

This species occurs in the small intestine of the rice rat *Oryzomys* sp. in South America.

The oocysts are ellipsoidal to ovoid, 22—25 × 17—19 μm, with a smooth, light brown, double-contoured wall, without a micropyle, with a residuum, apparently without a polar granule. The sporocysts are ovoid, 11 × 8 μm, with a Stieda body and residuum. The sporozoites lie lengthwise in the sporocysts. The sporulation time is 5—6 d at 20—25°C.

The meronts are in the epithelial cells of the small intestine, especially the duodenum. They form 15—20 banana-shaped merozoites 8×2—2.5 μm (10—11 $\times$ 3 μm when free in the lumen). The microgamonts in the epithelium cells produce a large number of microgametes 2—2.5 μm long. The prepatent period is 5 d.

Mus musculus could not be infected with this species.

Reference. Carini (1937).

EIMERIA PALUSTRIS BARNARD, ERNST, & STEVENS, 1971

This species occurs in the marsh rice rat *Oryzomys palustris* in North America.

The oocysts are ellipsoidal, 19—28 $\times$ 16—21 (mean 24 $\times$ 18) μm, with a rough, pitted, two-layered wall, the outer layer dark brown and about 1 μm thick, the inner layer light brown and about 0.5 μm thick, without a micropyle or residuum, with a polar granule. The sporocysts are ellipsoidal, 12—15 $\times$ 7—9 (mean 14 $\times$ 7.5) μm, with a prominent Stieda body, sub-Stiedal body, and residuum. The sporozoites are elongate, with two clear globules, and lie lengthwise in the sporocysts. The sporulation time is 6 d at room temperature (23—25°C). The prepatent period is 3—4 (mean 3.3) d and the patent period 4—5 (mean 4.7) d.

Reference. Barnard Ernst, & Stevens (1971).

ISOSPORA HAMMONDI BARNARD, ERNST, & STEVENS 1971

This species occurs in the marsh rice rat *Oryzomys palustris* in North America.

The oocysts are ovoid, 24—30 $\times$ 16—21 (mean 27 $\times$ 19) μm, with a smooth, colorless to pinkish, one-layered wall about 1 μm thick, without a micropyle, residuum, or polar granule. The sporocysts are ellipsoidal, 13—18 $\times$ 11—15 (mean 16 $\times$ 12) μm, with a Stieda body and a large residuum. The sporozoites are sausage-shaped, lack clear globules, and lie lengthwise in the sporocysts. The sporulation time is 1.5 d in 2.5% potassium bichromate solution at room temperature (23—25°C). The prepatent period is 5—6 (mean 5.3) d and the patent period 3—4 (mean 3.5) d.

Reference. Barnard, Ernst, & Stevens (1971).

SARCOCYSTIS AZEVEDOI SHAW & LAINSON, 1969

This species was found in the muscles of the rice rat *Oryzomys capito* in South America. The definitive host is unknown.

The sarcocysts in the muscles are long, slender, up to 3 mm $\times$ 79—126 μm (but probably longer), with a very finely striated wall due to minute, transverse, delicate villi 1—2 μm thick, without visible trabeculae. The bradyzoites are rounded at one end and tapered at the other, 4—8 $\times$ 1—2 (mean 6—7 $\times$ 1) μm, with a nucleus near the rounded end.

Reference. Shaw & Lainson (1969).

HOST GENUS *PEROMYSCUS*

EIMERIA ARIZONENSIS LEVINE, IVENS, & KRUIDENIER, 1957

This species occurs commonly in the small intestine of the piñon mouse *Peromyscus truei* (type host), deer mouse *P. maniculatus,* white-footed mouse *P. leucopus,* and cactus mouse *P. eremici* in North America.

The oocysts are ellipsoidal to subspherical, 19—29 × 17—23 µm, with a smooth or slightly pitted, one-layered wall 0.9—1.7 µm thick, lined by a thin membrane, without a micropyle, usually with a residuum, with zero to two polar granules. The sporocysts are lemon-shaped, 11—14 × 7—9 µm, with a Stieda body and residuum. The sporozoites are elongate, with one end broader than the other, and lie lengthwise in the sporocysts.

The meronts and gamonts are in the villar epithelial cells. The macrogametes are up to about 14 × 10 µm. The microgamonts are up to about 15 × 8 µm.

References. Levine & Ivens (1960, 1963); Levine, Ivens, & Kruidenier (1957); Reduker, Hertel, & Duszynski (1985).

EIMERIA CAROLINENSIS VON ZELLEN, 1959

This species occurs in the white-footed mouse *Peromyscus leucopus* in North America.

The oocysts are ellipsoidal, 14—19 × 10—13 (mean 18 × 11) µm, with a smooth, two-layered wall, the outer layer colorless, 1 µm thick, the inner layer dark brown, 0.5 µm thick, without a micropyle or residuum, with a polar granule. The sporocysts are ovoid, about 8.5 × 4.5 µm, with a small Stieda body and a residuum. The sporozoites lie lengthwise in the sporocysts.

Reference. Von Zellen (1959).

EIMERIA DELICATA LEVINE & IVENS, 1960

This species occurs in the deer mouse *Peromyscus maniculatus* in North America.

The oocysts are ellipsoidal, 13—15 × 10—12 (mean 14 × 11) µm, with a smooth, very pale yellowish, one-layered wall about 0.6 µm thick, without a micropyle or residuum, with a polar granule. The sporocysts are elongate ovoid, with a rather pointed end, 8 × 4—5 µm, very thin walled, with a tiny Stieda body, with or without some residual granules. The sporozoites lie lengthwise in the sporocysts.

References. Levine & Ivens (1960); Reduker, Hertel, & Duszynski (1985).

EIMERIA EREMICI LEVINE IVENS, & KRUIDENIER 1957

This species occurs in the cactus mouse *Peromyscus eremicus* in North America.

The oocysts are ellipsoidal, 22—30 × 18—22 (mean 25 × 21) µm, with a smooth, two-layered wall, the outer layer colorless and 1 µm thick, the inner layer pale tan and 0.4 µm thick, without a micropyle, with a residuum and one to three polar granules. The sporocysts are lemon-shaped, about $10 \times 8 \, \mu m$, with a Stieda body and residuum. The sporozoites lie lengthwise in the sporocysts.

Reference. Levine, Ivens, & Kruidenier (1957).

EIMERIA LACHRYMALIS REDUKER, HERTEL, & DUSZYNSKI, 1985

This species occurs in the cactus mouse *Peromyscus e. eremicus* in North America.

The oocysts are ellipsoidal, 27—35 × 17—21 (mean 31 × 19) μm, with a one-layered, smooth, colorless wall 0.8—1.6 (mean 1.5) μm thick, without a micropyle or residuum, with one or two polar granules. The sporocysts are teardrop-shaped, with a small, caplike Stieda body, without a sub-Stiedal body, with a residuum. The sporozoites have "grainy" cytoplasm and two clear globules.

Reference. Reduker, Hertel, & Duszynski (1985).

EIMERIA LANGEBARTELI IVENS, KRUIDENIER, & LEVINE, 1959

This species occurs in the brush mouse *Peromyscus boylii*, white-footed mouse *P. leucopus,* and piñon mouse *P. truei* in North America.

The oocysts are elongate ellipsoidal, 20—23 × 13—14 (mean 21 × 14) μm, with a smooth, pale yellowish, one-layered wall about 0.8 μm thick, without a micropyle or residuum, with a polar granule. The sporocysts are ellipsoidal, 8—10 × 5—6 (mean 9 × 5) μm, with a thin wall, small Stieda body, and presumably residual granules. The sporozoites are elongate, curled inside the sporocysts.

References. Ivens, Kruidenier, & Levine (1959); Reduker, Hertel, & Duszynski (1985).

EIMERIA LEUCOPI VON ZELLEN, 1961

This species occurs in the small intestine of the white-footed mouse *Peromyscus leucopus* in North America.

The oocysts are ellipsoidal, sometimes ovoid, rarely spherical, 14—24 × 14—21 (mean 19 × 17) μm, with a rough, two-layered wall, the outer layer yellowish and 0.5 μm thick, the inner layer dark brown and 1 μm thick, without a micropyle or polar granule, with a residuum. The sporocysts are 11—14 × 6—8 μm, with a Stieda body and residuum. The sporozoites are 4—7 × 2—4 (mean 5 × 2) μm and lie lengthwise in the sporocysts. The sporulation time is 54—128 h on vaseline-ringed slides in N/4 sodium chloride solution, presumably at room temperature. The prepatent period is 5—6 d and the patent period 6 d.

The golden deermouse *Peromyscus nuttalli* could not be infected with this species .

Reference. Von Zellen (1961).

EIMERIA PEROMYSCI LEVINE, IVENS, & KRUIDENIER, 1957

This species occurs in the piñon mouse *Peromyscus truei* and deer mouse *P. maniculatus* in North America.

The oocysts are ellipsoidal, 26—32 × 21—27 (mean 29 × 24) μm, with a two-layered wall, the outer layer rough, yellowish brown, and 1.3 μm thick, the inner layer yellowish brown, 0.4 μm thick, without a micropyle, with a residuum and polar granule. The sporocysts are lemon-shaped, about 159 μm, with a wall about 0.4 μm thick, with a Stieda body and residuum. The sporozoites have one end broader than the other and lie head to tail in the sporocysts.

References. Levine, Ivens, & Kruidenier (1957); Reduker, Hertel, & Duszynski (1985).

EIMERIA ROUDABUSHI LEVINE & IVENS, 1960

This species occurs in the white-footed mouse *Peromyscus leucopus* in North America.

The oocysts are ellipsoidal, 20—26 × 17—20 (mean 22 × 19) μm, with a smooth, almost colorless, one-layered wall about 1.3 μm thick at the sides and about 0.9 μm thick at the ends, without a micropyle, with an atypical, cobwebby residuum and a polar granule. The sporocysts are ovoid, 12—13 × 8 μm, thin-walled, with a small Stieda body and a large residuum. The sporozoites are colorless lie lengthwise head to tail, and are folded back on themselves in the sporocysts.

Reference. Levine & Ivens (1960).

EIMERIA SINIFFI LEVINE & IVENS, 1963

This species occurs in the deermouse *Peromyscus maniculatus* in North America.

The oocysts are cylindrical with rounded ends, 26—31 × 17—19 (mean 28 × 18) μm, with a smooth, colorless to pale yellowish, one-layered wall about 1.2 μm thick, sometimes with a micropyle or operculum, without a residuum, with a polar granule. The sporocysts are ovoid, with one end more or less pointed, 13—15 × 6—7 (mean 14 × 7) μm, without a Stieda body, with a residuum. The sporozoites are elongate and lie lengthwise head to tail in the sporocysts.

Reference. Levine & Ivens (1963).

ISOSPORA CALIFORNICA DAVIS, 1967

This species occurs uncommonly or rarely in the California mouse *Peromyscus californicus* (type host), brush mouse *P. boylii*, deermouse *P. maniculatus,* and piñon mouse *P. truei* in North America.

The oocysts are spherical to ellipsoidal or ovoid, 18—32 × 18—27 (mean 26 × 22) μm, with a smooth, two-layered wall, the outer layer gray-green to light brown, about 1 μm thick, the inner layer generally clear and colorless, about 0.4—0.5 μm thick, with a heavy dark line at its inner margin, without a micropyle, normally without a residuum, with one to several polar granules. The sporocysts are ovoid to lemon-shaped, 13—20 × 8—13 (mean 16 × 10) μm, with a wall about 0.3 μm thick, with a Stieda body, sub-Stiedal body, and residuum. The sporozoites are 12—17 × 2—3 (mean 15 × 3) μm, elongate, with one end rounded and the other tapering, with a clear globule near the broad end, and lie longitudinally in the sporocysts.

Reference. Davis (1967).

ISOSPORA HASTINGSI DAVIS, 1967

This species occurs rarely in the piñon mouse *Peromyscus truei* in North America.

The oocysts are ovoid, 29—33 × 22—24 (mean 31 × 23) μm, with a smooth, tan or yellow-tan, one-layered wall 1.2 μm thick, sometimes with a thin lining membrane, without a micropyle or residuum, with a polar granule. The sporocysts are lemon-shaped, with a wall about 0.4 μm thick, with a Stieda body, sub-Stiedal body, and residuum. The sporozoites are about 14 × 3 μm, with one end rounded and the other tapered, with a central or subcentral nucleus and a recurved broad end; they usually lie longitudinally in the sporocysts.

Reference. Davis (1967).

ISOSPORA PEROMYSCI DAVIS, 1967

This species occurs in the jejunum and ileum, sometimes duodenum, and cecum of the deermouse *Peromyscus maniculatus* (type host), California mouse *P. californicus,* and piñon mouse *P. truei* in North America. About 75% are in the villar epithelial cells and 25% in the submucosa.

The oocysts are elongate or ellipsoidal, 25—43 × 14—28 (mean 35 × 21) μm, with a smooth, pale green, tan, or light brown, one-layered wall about 0.8 μm thick, occasionally with a thin lining membrane, without a micropyle, residuum, or polar granule. The sporocysts are broadly ovoid, 13—21 × 11—15 (mean 16 × 13) μm, with a pale green wall about 0.4 μm thick, without a Stieda body, with a residuum. The sporozoites are sausage-shaped, 11 × 3.5 μm, with a central nucleus.

Meronts are up to 25 × 17 μm and produce up to 32 merozoites. There are probably several generations. One type of meront contains many large nuclei with a prominent dark nucleolus. The other type produces several elongate, gregariniform organisms containing several large nuclei and then 3—16 broad, spindle-shaped merozoites. Macrogametes are up to 38 × 18 μm. Microgamonts are up to 26 × 17 μm and produce about 75—110 comma- or sickle-shaped microgametes 3.6 × 0.3 μm.

Reference. Davis (1967).

SARCOCYSTIS IDAHOENSIS BLEDSOE, 1980

The intermediate host of this species is the deermouse *Peromyscus maniculatus* in North America; first generation meronts are in the hepatocytes, and sarcocysts in the muscles, especially of the tongue. The definitive host is the gopher snake *Pituophis melanoleucus*; gametes, gamonts, oocysts, and sporocysts are in its small intestine. This species is pathogenic for the deermouse, 15,000 sporocysts often causing anorexia, weakness, ataxia, and dyspnea in 5—

6 d, followed by death 6—8 d after inoculation. The mice have acute hepatitis, with enlarged livers and spleens.

The oocysts are dumbbell-shaped to ellipsoidal, smooth, colorless, 22—23 × 13—14 (mean 22 × 14) μm, with a one-layered wall less than 0.5 μm thick, tightly stretched around the sporocysts, without a micropyle, residuum, or polar granule. The sporocysts are ellipsoidal, 12—14 × 10—12 (mean 13 × 11) μm, with a wall about 0.5 μm thick, without a Stieda body, with a residuum. The sporozoites are sausage-shaped with one end slightly pointed, 7 × 2 μm, without a clear globule. Sporulation occurs in the epithelial cells of the snake small and anterior large intestines, begins 9 d after feeding, and lasts up to 14 d. The sporocysts are usually released in the intestinal lumen.

Sporozoites appear in the lumen of the deer mouse intestine 2 h after the mice have been fed sporulated sporocysts. They reach the lungs and liver by 16 h. Meronts in the hepatocytes are mature at 7—8 d, at which time they are 15—26 × 12—22 (mean 18 × 16.5) μm and contain a large number (about 30 forming a rosette in a single section) of tachyzoites. These are crescent-shaped and 6—9 × 1—2 (mean 8 × 1) μm in impression smears or 10 × 1 μm when alive. Sarcocysts can be found in the deermouse tongue 11 d after inoculation. They are ellipsoidal, 12—17 × 8—10 (mean 15 × 9) μm and contain 5—12 metrocytes and no merozoites at 13 d. Merozoites do not appear until about 34 d. At 160 days the sarcocysts are 2—10 × 0.2—1.0 (mean 6 × 0.4) mm, compartmented, and contain both metrocytes about 11 × 7 μm at their periphery and many crescent-shaped bradyzoites 7 × 2 μm with a rounded posterior end and a pointed anterior one.

Gamogony occurs in the mucosal epithelial cells of the small intestine and anterior large intestine of the gopher snake. Macrogametes and microgamonts can be differentiated 5 d after the snake has ingested a deermouse with bradyzoites. The macrogametes are mature at 9 d; they are spherical to ellipsoidal, 13—15 × 10—12 (mean 14 × 11) μm. The mature microgamonts at 11 d are 7—9 × 5—9 (mean 8 × 7) μm and contain many biflagellate microgametes about 6 × 0.5 μm in fresh smears. The prepatent period is perhaps 26 d.

The dog and cat cannot be infected by feeding them infected deermice, nor can *Mus musculus* or *Peromyscus leucopus* be infected by feeding them sporulated sporocysts from gopher snakes. *P. maniculatus* can be infected either with sporocysts from gopher snakes or with tachyzoites from other *P. maniculatus*.

References. Bledsoe (1977, 1979, 1980); Dubey (1983).

SARCOCYSTIS PEROMYSCI DUBEY, 1983

This species occurs in the deermouse *Peromyscus maniculatus* (intermediate host) in Montana.

The oocysts are unknown. The sarcocysts in deermice are septate, 38—1875 × 27—81 μm, with a wall 2—5.5 μm thick, with hairy protrusions. The bradyzoites are 10—14 × 3—3.5 (mean 11 × 3) μm, and contain many prominent granules.

Reference. Dubey (1983).

BESNOITIA JELLISONI FRENKEL, 1955

This species occurs in the connective tissue, subcutaneous tissue, mesenteries, and membranes of many organs, including the bones of the skull of the deermouse *Peromyscus maniculatus* (type intermediate host), cricetid rodent *Microxus torques*, kangaroo rats *Dipodomys ordii*, D. *merriami*, and D. *microps*, and opossum *Didelphis virginiana* (?) in North America and South America It has been transmitted experimentally to the following rodents: house mouse *Mus musculus*, laboratory rat *Rattus norvegicus*, golden hamster *Mesocricetus auratus*, vole *Microtus* sp., ground squirrel *Spermophilus* sp., and guinea pig *Cavia porcellus*. The definitive host is unknown. It does not ordinarily cause clinical signs.

Its oocyst structure is unknown.

Organ meronts (actually pseudocysts with a wall formed by the host cell) are white, spherical, up to 1 mm in diameter, with a thick wall containing giant nuclei. Their merozoites are elongate, $7—9 \times 2—3$ (mean 8×2) μm, with a three-membraned wall, 2 polar rings, a conoid, 20—24 subpellicular microtubules, 80—100 micronemes, 3—5 rhoptries, and 20—30 enigmatic bodies.

This species can be transmitted from one intermediate host to another by injection of merozoites.

References. Akinchina & Zasukhin (1971); Ernst, Chobotar, Oaks, & Hammond (1968); Fayer, Hammond, Chobotar, & Elsner (1969); Frenkel (1953, 1955, 1977); Frenkel & Reddy (1977); Jellison, Glesne, & Peterson (1960); Lindberg & Frenkel (1977); Lunde & Gelderman (1971); Porchet-Henneré, D'Heoghe, Sadak, & Frontier (1985); Scholtyseck, Mehlhorn, & Müller (1974); Sénaud & Mehlhorn (1975, 1977, 1978); Sheffield (1966); Stabler & Welch (1961).

TYZZERIA PEROMYSCI LEVINE & IVENS, 1960

This species occurs quite commonly in the deermouse *Peromyscus maniculatus* (type host) and white-footed mouse *P. leucopus* in North America.

The oocysts are ellipsoidal, $11—17 \times 9—12$ μm, with a smooth, colorless to very pale yellowish, one-layered wall about 0.6 μm thick, without a micropyle, without or with a small residuum, with one or two polar granules. The sporozoites are banana-shaped, crescentic, or lanceolate, about $9—11 \times 3.5$ μm, clustered together in a ball without orientation in any particular direction, sometimes with one or two sporozoites separated from the others.

Reference. Levine & Ivens (1960).

KLOSSIA PERPLEXENS LEVINE, IVENS, & KRUIDENIER, 1955

This species was found in the deermouse *Peromyscus maniculatus* in North America. This may be a pseudoparasite of the deermouse.

The oocysts are ellipsoidal, $42—53 \times 35—44$ (mean 48×40) μm, with a smooth, two-layered wall, the outer layer colorless, about 2 μm thick, the inner layer pale brown, 0.5 μm thick, without a micropyle or polar granule, with a

residuum. The oocysts have many sporocysts, each with four sausage-shaped sporozoites. The sporocysts are spherical, 13—14 μm in diameter, with a colorless wall 0.5 μm thick, without a Stieda body, with a residuum.
Reference. Levine, Ivens, & Kruidenier (1955).

HOST GENUS *BAIOMYS*

EIMERIA BAIOMYSIS LEVINE, IVENS, & KRUIDENIER, 1958

This species occurs in the pygmy mice *Baiomys taylori* (type host) and *B. musculus* in North America.

The oocysts are ellipsoidal, 20—25 × 18—21 (mean 23 × 19) μm, with a quite rough and pitted, yellowish, one-layered wall 1.6 μm thick, without a micropyle, with a residuum and polar granule. The sporocysts are ovoid, about 11—12 × 7—8 μm, with a Stieda body and residuum. The sporozoites lie lengthwise in the sporocysts.

References. Kruidenier, Levine, & Ivens (1960); Levine, Ivens, & Kruidenier (1958).

HOST GENUS *ONYCHOMYS*

EIMERIA ONYCHOMYSIS LEVINE IVENS, & KRUIDENIER, 1957

This species occurs in the northern grasshopper mouse *Onychomys leucogaster* in North America.

The oocysts are subspherical to ellipsoidal, 20—21 × 17—20 (mean 20 × 19) μm, with a slightly rough, pale tan, one-layered wall about 0.5 μm thick, without a micropyle, with a residuum and polar granule. The sporocysts are ovoid, about 11 × 8 μm, with a Stieda body and residuum. The sporozoites lie head to tail in the sporocysts.

Reference. Levine, Ivens, & Kruidenier (1957).

HOST GENUS *AKODON*

EIMERIA AKODONI ARCAY-DE-PERAZA, 1970

A *nomen nudum.*

BESNOITIA JELLISONI FRENKEL, 1955

Akodon (Microxus) torques is an intermediate host of this species.
References. See under *Peromyscus.*

HOST GENUS *ZYGODONTOMYS*

EIMERIA SP. BOYER & SCORZA, 1957

This form occurs in the intestine of the cane mouse *Zygodontomys brevicauda* in South America.

The oocysts are 21 × 13 µm with a very small residuum. The sporocysts have a residuum. The sporozoites are kidney-shaped.

Reference. Boyer & Scorza (1957).

EIMERIA SP. BOYER & SCORZA, 1957

This form occurs in the intestine of the cane mouse *Zygodontomys brevicauda* in South America.

The oocysts are 16 × 12 µm, with a large residuum. The sporozoites are in the form of a closed C.

Reference. Boyer & Scorza (1957).

HOST GENUS *PHYLLOTIS*

EIMERIA PHYLLOTIS GONZALES-MUGABURU, 1942

This species occurs commonly in the epithelial cells of the cecum of the leaf-eared mouse *Phyllotis a. amicus* (?) in South America.

The oocysts are elongate ellipsoidal, rose-colored, 22—30 × 12—16 (mean 26 × 14) µm, with a wall 0.8—1 µm thick, without a micropyle, residuum, or polar granule. The sporocysts are 10—12 × 3—4 µm, with a residuum; the presence of a Stieda body is uncertain. The sporozoites are 9—10 × 2—3 µm. The sporulation time is 2—3 d in 3—5% potassium bichromate solution. The macrogametes are 18—20 µm in diameter. The microgamonts are 20—22 µm in diameter and produce microgametes 3 µm long.

Rattus norvegicus and *Mus musculus* cannot be infected.

Reference. Gonzales-Mugaburu (1942).

EIMERIA WEISSI GONZALES-MUGABURU, 1946

This species occurs quite commonly in the ileum and cecum of the leaf-eared mouse *Phyllotis a. amicus* (?) in South America.

The oocysts are ovoid, reddish, 16—32 × 13—24 (mean 23.5 × 19) µm, with a two-layered wall 1.2 µm thick, without a micropyle, with a residuum, apparently without a polar granule. The sporocysts are ellipsoidal, 10—11 × 6—7 µm, with a residuum. The sporozoites lie longitudinally in the sporocysts. The sporulation time is 8—9 d in 3% potassium bichromate solution.

Mus musculus and *Cavia porcellus* cannot be infected.

Reference. Gonzales-Mugaburu (1946).

HOST GENUS *SIGMODON*

CARYOSPORA SIMPLEX LEGER, 1904

See *Caryospora simplex* under *Mus musculus*.

EIMERIA ROPERI BARNARD, ERNST, & DIXON, 1974

This species occurs commonly in the cecum and colon of the cotton rat *Sigmodon hispidus* in North America.

The oocysts are ellipsoidal, 22—38 × 14—22 (mean 25 × 18) μm, with a two-layered wall, the outer layer smooth, light blue, about 1 μm thick, the inner layer light brown, about 0.7 μm thick, without a micropyle or residuum, with a polar granule. The sporocysts are ovoid, 11—15 × 7—9 (mean 13 × 8) μm, with a Stieda body and residuum. The sporozoites are elongate, with a clear globule at each end, and lie lengthwise head to tail in the sporocysts. The sporulation time is 4 d at 22—25°C in 2.5% potassium bichromate solution. The prepatent period is 6—10 (mean 8.6) d and the patent period 1-6 (mean 2.9) d.

The rice rat *Oryzomys palustris*, *Rattus norvegicus*, and *Mus musculus* cannot be infected with this species.

Reference. Barnard, Ernst, & Dixon (1974).

EIMERIA SIGMODONTIS BARNARD, ERNST, & DIXON, 1974

This species occurs commonly in the mucosal epithelial cells of the cecum and colon of the cotton rat *Sigmodon hispidus* in North America.

The oocysts are ellipsoidal, 12—27 × 9—12 μm, with a two-layered wall, the outer layer smooth, light blue, about 0.7 μm thick, the inner layer light brown, about 0.3 μm thick, without a micropyle or residuum, with a polar granule. The sporocysts are ovoid, 8—11 × 5—8 (mean 10 × 6) μm, with a small Stieda body and a residuum. The sporozoites are elongate, with a clear globule at the large end, and lie lengthwise head to tail in the sporocysts. The sporulation time is 42—58 h at 22-25°C in 2.5% potassium bichromate solution.

There are three asexual generations. First generation meronts are present on days 1—4. They are ellipsoidal to subspherical, 10—14 × 7—10 (mean 12 × 8) μm when alive and contain 10—17 (mean 12) merozoites 12—16 × 2—3 (mean 14 × 3) μm. Second generation meronts are present on days 2—5. They are broadly ellipsoidal to subspherical, 6—9 × 5—8 (mean 8 × 6.5) μm, and contain 5—8 (mean 7) merozoites 6—10 × 2—3 (mean 7.5 × 2.5) μm and a residuum. Third generation meronts are present on days 4—6. They are ovoid to subspherical, 16—20 × 8—11 (mean 17 × 10) μm, and contain 10—21 (mean 12) merozoites 15—19 × 2—3 (mean 17 × 2) μm.

Gamonts are present 4—6 d after inoculation. The macrogametes are 11—16 × 8—11 (mean 13 × 9) μm. The microgamonts are 8—13 × 7—11 (mean 11 × 9) μm and contain numerous microgametes 2—3 × 0.5 (mean 2.6 × 0.5) μm. The prepatent period is 3—7 (mean 4.2) d and the patent period 4—10 (mean 7.9) d.

The rice rat *Oryzomys palustris*, *Rattus norvegicus*, and *Mus musculus* cannot be infected.

References. Barnard, Ernst, & Dixon (1974); Ernst, Todd, & Barnard (1977).

EIMERIA TUSKEGEENSIS BARNARD, ERNST, & DIXON, 1974

This species occurs commonly in the epithelial cells of the small intestine villi of the cotton rat *Sigmodon hispidus* in North America.

The oocysts are ellipsoidal, 24—31 × 17—23 (mean 27 × 20) μm, with a two-layered wall, the outer layer rough, pitted, dark brown, about 1.2 μm thick, the inner layer dark brown, about 0.5 μm thick, without a micropyle, with a residuum and polar granule. The sporocysts are ovoid, 10—15 × 6—9 (mean 12 × 8) μm, with a Stieda body and residuum. The sporozoites are elongate, with a clear globule at each end, and lie lengthwise head to tail in the sporocysts. The sporulation time is 10—11 d at 22-25°C in 2.5% potassium bichromate solution.

There are three asexual generations. Mature first generation meronts are seen 1—2 d after inoculation; they are 13—16 × 12—14 (mean 14 × 13) μm, ellipsoidal to subspherical, and contain 22—30 (mean 24) merozoites 11—12 × 2.5—3 (mean 12 × 2.6) μm, and a small residuum. Mature second generation meronts are seen 60—96 h after inoculation. They are generally ellipsoidal, 20—35 × 18—30 μm, and contain 48—62 (mean 56) merozoites 15—16 × 2—2.5 (mean 16 × 2.1) μm and a small residuum. Mature third generation meronts are seen 84—120 h after inoculation. They are ovoid to subspherical, 10—13 × 9—12 μm, and contain 14—20 (mean 16) merozoites 11—12 × 2.5—3.5 μm.

Macrogametes and microgamonts are seen 84—120 h after inoculation. Mature macrogametes are 20—28 × 12—18 (mean 24 × 17) μm. Mature microgamonts are ellipsoidal, 25—30 × 12—21 (mean 28 × 16) μm and produce numerous randomly arranged flagellated microgametes and a central residuum. Oocysts are present 108—120 h after inoculation. The prepatent period is 4—7 (mean 4.9) d and the patent period 4—7 (mean 5.1) d.

The rice rat *Oryzomys palustris*, *Rattus norvegicus,* and *Mus musculus* cannot be infected with this species.

References. Barnard, Ernst, & Dixon (1974); Current, Ernst, & Benz (1981); Ernst, Current, & Moore (1980).

EIMERIA WEBBAE BARNARD, ERNST & DIXON, 1974

This species occurs commonly in the cotton rat *Sigmodon hispidus* in North America.

The oocysts are subspherical, 10—13 × 9—11 (mean 12 × 10) μm, with a smooth, dark yellow, one-layered wall about 1 μm thick, without a micropyle, residuum, or polar granule. The sporocysts are ellipsoidal, 5—7 × 3—5 (mean 6 × 4) μm, with a Stieda body, substiedal body, and residuum. The sporozoites are elongate, lack clear globules, and lie lengthwise head to tail in the sporocysts. The sporulation time is 3—4 d at 22-25°C in 2.5% potassium bichromate solution. The prepatent period is 4—8 (mean 6.2) d and the patent period 2—6 (mean 4.0) d.

The rice rat *Oryzomys palustris*, *Rattus norvegicus* and *Mus musculus* cannot be infected.

Reference. Barnard, Ernst, & Dixon (1974).

EIMERIA SP. BARNARD, ERNST, & DIXON, 1974

This species occurs uncommonly in the cotton rat *Sigmodon hispidus* in North America.

The oocysts are elongate ellipsoidal, 34—40 × 17—22 (mean 37 × 20) μm, with a smooth, two-layered wall, the outer layer light blue, about 1.3 μm thick, the inner layer light brown, about 0.7 μm thick, lined by a thin membrane, with a micropyle, without a residuum or polar granule. The sporocysts are elongate ellipsoidal, 14—17 × 8—10 (mean 16 × 9) μm, with a Stieda body, substiedal body, and residuum. The sporozoites are elongate, with a clear globule at the broad end, and lie lengthwise head to tail in the sporocysts.

The marsh rice rat *Oryzomys palustris*, *Rattus norvegicus*, and *Mus musculus* cannot be infected. The authors were unable to infect *S. hispidus* experimentally, and therefore felt it inappropriate to give this form a name.

Reference. Barnard, Ernst, & Dixon (1974).

ISOSPORA MASONI UPTON, LINDSAY, CURRENT, & ERNST, 1985

This species occurs in the villar epithelial cells of the jejunum and ileum of the cotton rat *Sigmodon hispidus* in North America.

The oocysts are ellipsoidal, with a thin, membrane-like wall, without a micropyle or residuum, with a polar granule. The sporocysts are ovoid, 7—9 × 5—6 (mean 8 × 5) μm, with a Stieda body, substiedal body, and residuum. The sporozoites are elongate, 6—8 × 1.5—2 μm, with a small anterior and larger posterior clear globule and central nucleus; they lie lengthwise head to tail in the sporocysts. Sporulation occurs in the host. The prepatent period is 4—7 d and the patent period at least 40 d.

Reference. Upton, Lindsay, Current, & Ernst (1985).

ISOSPORA SP. BARNARD, ERNST, & DIXON, 1974

This form was found commonly in the cotton rat *Sigmodon hispidus* in North America.

The oocysts are subspherical to ovoid, 21—28 × 20—25 (mean 25 × 22) μm, with a smooth, two-layered wall, the outer layer reddish brown, about 1.3 μm thick, the inner layer light brown, about 0.6 μm thick, without a micropyle or residuum, with a polar granule. The sporocysts are ovoid, 14—18 × 9—12 (mean 16 × 11) μm, with a Stieda body, residuum and two-part sub-Stiedal body. The sporozoites are elongate, with one end broader than the other, with a clear globule at the broad end, and lie lengthwise in the sporocysts, all pointed in the same direction.

The rice rat *Oryzomys palustris*, *Rattus norvegicus*, and *Mus musculus* cannot be infected. The authors were unable in infect cotton rats, and thought that this might be an avian species and a pseudoparasite of the cotton rat.

Reference. Barnard, Ernst, & Dixon (1974).

ADELINA SP. BARNARD, ERNST & DIXON, 1974

This form was found quite commonly in the cotton rat *Sigmodon hispidus* in North America.

The oocysts are broadly ellipsoidal, subspherical, or spherical, 25—53 × 25—

47 (mean 40 × 34) μm, with a smooth, two-layered wall, the outer layer blue, about 1.3 μm, thick, the inner layer light brown, about 0.7 μm thick, without a micropyle or polar granule, with a residuum. There are 4—19 (mean 11.3) sporocysts per oocyst. They are spherical to subspherical, 10—14×9—14 (mean 12 × 12) μm, without Stieda or sub-Stiedal bodies, with a residuum. The sporozoites are broadly elongate, with both ends about the same size, without clear globules.

The rice rat *Oryzomys palustris*, *Rattus norvegicus*, and *Mus musculus* cannot be infected. The authors were unable to infect *S. hispidus* experimentally and thought that this form was probably a pseudoparasite of the cotton rat.

Reference. Barnard, Ernst, & Dixon (1974).

HOST GENUS *NEOTOMA*

EIMERIA ALBIGULAE LEVINE, IVENS, & KRUIDENIER, 1957

This species occurs in the white-throated woodrat *Neotoma albigula* in North America.

The oocysts are spherical to subspherical, 19—26 × 17—23 (mean 22 × 20) μm, with a two-layered wall, the outer layer rough, colorless, about 1 μm thick, the inner layer pale brownish, about 0.5 μm thick, without a micropyle, with a residuum and one or two polar granules. The sporocysts are ovoid, about 11 × 9 μm, with one end slightly pointed, with a small Stieda body and a residuum. The sporozoites are rounded at the ends of the sporocysts.

References. Levine, Ivens, & Kruidenier (1957); Reduker & Duszynski (1985).

EIMERIA ANTONELLII STRANEVA & GALLATI, 1980

This species occurs commonly in the eastern woodrat *Neotoma floridana* in North America.

The oocysts are spherical, subspherical, or ellipsoidal, 15—24 × 12—21 (mean 18 × 15) μm, with a smooth, yellow, one-layered wall about 1 μm thick, without a micropyle, with a residuum and polar granule. The sporocysts are ovoid to elongate ovoid, 7—10 × 4—7 (mean 8.5 × 6) μm, with a Stieda body and residuum, without a substiedal body. The sporozoites were said to be at the ends of the sporocysts or partly curled around each other, but were drawn lying lengthwise in the sporocysts, with a large posterior clear globule and occasionally a smaller anterior one.

Reference. Straneva & Gallati (1980).

EIMERIA BARLEYI STRANEVA & GALLATI, 1980

This species occurs fairly commonly in the eastern woodrat *Neotoma floridana* in North America.

The oocysts are broadly ellipsoidal, 18—25 × 16—21 (mean 22 × 18) μm, with a rough, pitted, two-layered wall, the outer layer yellow to yellow-brown, about 1.1 μm thick, the inner layer yellow to yellow-brown, about 0.7 μm thick, with a micropyle and polar granule, without a residuum. The sporocysts are ovoid, thin-walled, 10—12 × 7—9 (mean 11 × 8) µm, with a Stieda body, sub-Stiedal body, and residuum. The sporocysts were said to be at the ends of the sporocysts or curled around each other but were drawn as lying lengthwise in the sporocysts, with large posterior and smaller anterior clear globules.

Reference. Straneva & Gallati (1980).

EIMERIA DAVISI IVENS, KRUIDENIER, & LEVINE, 1959

This species occurs in the white-throated woodrat *Neotoma albigula* in North America.

The oocysts are subspherical to ellipsoidal, 22—32 × 21—24 (mean 28 × 23) µm, with a two-layered wall, the outer layer smooth to slightly rough, colorless to pale brownish yellow, about 1.2 µm thick, the inner layer pale brownish to brownish, about 0.4 µm thick, without a micropyle, with a residuum and polar granule. The sporocysts are ovoid, 10—12 × 7—9 (mean 11 × 8) µm, with a medium-sized Stieda body, usually with scattered residual granules. The sporozoites lie lengthwise in the sporocysts.

Reference. Ivens, Kruidenier, & Levine (1959).

EIMERIA DUSII WHEAT & ERNST, 1974

This species occurs commonly in the eastern woodrat *Neotoma floridana* in North America.

The oocysts are subspherical to broadly ellipsoidal, 15—25 × 14—22 (mean 19 × 17) µm, with a two-layered wall, the outer layer light brown, rough, pitted (appearing striated), about 1 µm thick, the inner layer colorless, about 0.5 µm thick, without a micropyle, with a residuum and polar granule. The sporocysts are ovoid, 7—11 × 4—8 (mean 9 × 6) µm, with a prominent Stieda body and a residuum, without a sub-Stiedal body. The sporozoites are elongate, lying toward the ends of the sporocysts, partly curled around each other, with a large posterior clear globule and occasionally a small anterior one. The sporulation time is 3—5 d at 22—25°C in 2.5% potassium bichromate solution.

References. Straneva & Gallati (1980); Wheat & Ernst (1974).

EIMERIA GLAUCEAE WHEAT & ERNST, 1974

This species occurs commonly in the eastern woodrat *Neotoma floridana* in North America.

The oocysts are ellipsoidal, 11—21 × 10—13 (mean 18 × 12) µm, with a two-layered wall, the outer layer light yellow, 0.4 µm thick, the inner layer light yellow, about 0.6 µm thick, without a micropyle or residuum, with a polar granule. The sporocysts are elongate ovoid, 7—11 × 4—6 (mean 10 × 5) µm, with a Stieda body and residuum, without a sub-Stiedal body. The sporozoites are elongate, with a small anterior and large posterior clear globule, and lie lengthwise in the sporocysts, partly curled around each other. The sporulation time is 24—36 h at 22—25°C in 2.5% potassium bichromate solution.

References. Straneva & Gallati (1980); Wheat & Ernst (1974).

EIMERIA LADRONENSIS REDUKER & DUSZYNSKI, 1985

This species occurs commonly in the woodrat *Neotoma albigula* in North America.

This oocysts are ellipsoidal, 19—25 × 13—15 (mean 21 × 14) μm, with a smooth, one-layered wall 0.8 μm thick, without a micropyle or residuum, with one or two polar granules. The sporocysts are tapered at one end, 7—10 × 6—7 (mean 8.5 × 6.5) μm, with a Stieda body, without a sub-Stiedal body, with a residuum. The sporozoites have a clear globule at each end. The prepatent period is 8—9 d and the patent period about 11 d. Reduker & Duszynski (1985) found it in 24% of 21 *N. albigula* in New Mexico.

Reference. Reduker & Dusynski (1985).

EIMERIA NEOTOMAE HENRY, 1932

This species occurs commonly in the woodrat *Neotoma fuscipes* in North America.

The oocysts are ellipsoidal, 16—22 × 13—19 (mean 22 [sic] × 16) μm, with a smooth, transparent, apparently one-layered wall, sometimes with a small micropyle visible, apparently occasionally with a polar granule. The sporocysts are subspherical, about 7.5 × 7 μm, with an extremely thin wall, presumably without a Stieda body. The sporulation time is 36—48 h in 2% potassium bichromate solution in small vials.

Reference. Henry (1932).

EIMERIA OPERCULATA LEVINE, IVENS, & KRUIDENIER, 1957

This species occurs in the Stephens woodrat *Neotoma stephensi* in North America.

The oocysts are ellipsoidal, 31—33 × 19—21 (mean 32 × 20) μm, with a smooth, pale tan, one-layered wall about 1 μm thick, with an operculum about 7 μm in diameter at one end of the oocyst, without a residuum or polar granule. The sporocysts are subspherical, about 7 × 6 μm, without a Stieda body or residuum. The sporozoites are banana-shaped, curled up, about 11 × 2 μm.

Reference. Levine, Ivens, & Kruidenier (1957).

EIMERIA RESIDUA HENRY, 1932

This species occurs commonly in the woodrat *Neotoma fuscipes* in North America.

The oocysts are subspherical, 22—29 × 19—26 (mean 26 × 22) μm with a comparatively thick two-layered wall, the outer layer brown, very rough, the inner layer clear, transparent, without a micropyle, with a large residuum and one or more polar granules. The sporocysts are 10 × 7.5 μm pointed at one end, with a Stieda body. The sporulation time is 8—9 d in 2% potassium bichromate solution in small vials.

Reference. Henry (1932).

EIMERIA STRANGFORDENSIS STRANEVA & GALLATI, 1980

This species occurs commonly in the eastern woodrat *Neotoma floridana* in North America.

The oocysts are broadly ellipsoidal, sometimes subspherical, 25—32 × 20—26 (mean 29 × 24) µm, with a rough, pitted, two-layered wall, the outer layer yellow-brown, about 1.5 µm thick, the inner layer yellow, about 0.6 µm thick, with a dark inner lining membrane, without a micropyle, with a residuum, usually without a polar granule. The sporocysts are subspherical to ovoid, 11—17 × 7—16 (mean 13 × 11) µm, with a slight Stieda body, without a sub-Stiedal body, with a residuum. The sporozoites are at the ends of the sporocysts, with a large posterior clear globule.

Reference. Straneva & Gallati (1980).

DORISA ARIZONENSIS (LEVINE, IVENS, & KRUIDENIER, 1955) LEVINE, 1980

Synonym. *Dorisiella arizonensis* Levine, Ivens, & Kruidenier, 1955.

This species occurs in the desert woodrat *Neotoma lepida* in North America.

The oocysts are spherical or subspherical, 21—28 × 21—22 (mean 22 × 21) µm, with a smooth, two-layered wall, the outer layer colorless, 1 µm thick, the inner layer pale tan, 0.5 µm thick, without a micropyle or residuum, with one to three polar granules. The sporocysts are lemon-shaped, 11—13 × 9—10 (mean 13 × 9) µm, with a thin wall, Stieda body, and residuum. The sporozoites are often oriented more or less lengthwise in the sporocysts, sometimes spiralling somewhat around a large, clear central residuum.

References. Levine (1980); Levine, Ivens, & Kruidenier (1955).

HOST GENUS *CALOMYSCUS*

EIMERIA BAILWARDI GLEBEZDIN, 1971

This species occurs in the hamster mouse *Calomyscus bailwardi* in the USSR.

The oocysts are ellipsoidal, 26—32 × 23—26 (mean 29 × 24) µm, with a smooth, one-layered wall 1.5 µm thick, without a micropyle, residuum, or polar granule. The sporocysts are ellipsoidal to spherical, 11.5—12 × 8.5—9 (mean 11.5 × 9) µm, without a Stieda body, with a residuum.

Reference. Glebezdin (1971).

EIMERIA CALOMYSCUS GLEBEZDIN, 1971

This species occurs in the hamster mouse *Calomyscus bailwardi* in the USSR.

The oocysts are ellipsoidal to spherical, 17—29 × 14.5—26 µm, with a smooth, one-layered wall 1.2—1.5 µm thick, without a micropyle, polar granule, or residuum. The sporocysts are ellipsoidal, 11.5—12 × 8.5—9 (mean 11 × 9) µm, without a Stieda body, with a residuum.

Reference. Glebezdin (1971).

ISOSPORA CALOMYSCUS MUSAEV & VEISOV, 1965

This species occurs in the hamster mouse *Calomyscus bailwardi* in the USSR.

The oocysts are ellipsoidal, 20—23 × 16—20 μm, with a smooth, yellow-brown, one-layered wall 1.5 μm thick, without a micropyle, residuum, or polar granule. The sporocysts are ovoid, 14—17 × 10—15 μm, with a prominent Stieda body and a residuum. The sporozoites are spherical or slightly ellipsoidal, 4—7 × 4—6 μm. The sporulation time is 2 d.

References. Glebezdin (1971); Musaev & Veisov (1965).

PYTHONELLA KARAKALENSIS GLEBEZDIN, 1971

This species occurs in the hamster mouse *Calomyscus bailwardi* in the USSR.

The oocysts are ellipsoidal, 32—46 × 29—41 (mean 42 × 34) μm, with a smooth, one-layered wall 1.5—1.7 μm thick, without a micropyle, residuum, or polar granule. Each oocysts contains 16 spherical sporocysts 12—14.5 (mean 12) μm in diameter, each with four ovoid sporozoites and a residuum, without a Stieda body.

Reference. Glebezdin (1971).

HOST GENUS *CRICETUS*

EIMERIA CRICETI NÖLLER, 1920 EMEND. PELLÉRDY, 1956

Synonym. *Eimeria falciformis* var. *criceti* Nöller 1920

This species occurs in the cecum and colon of the hamster *Cricetus cricetus* in Europe. The meronts and gamonts are in the villar epithelial cells, but not in the crypts.

The oocysts are spherical to subspherical, 11—22 μm in diameter, without a residuum. The sporocysts are quite plump. The sporulation time is 2—4 d at room temperature.

References. Nöller (1920); Pellérdy (1956).

ISOSPORA FREUNDI YAKIMOFF & GOUSSEFF, 1935

This species occurs in the hamster *Cricetus cricetus* in the USSR.

The oocysts are spherical or subspherical, 13—27 × 17—24 μm, with a smooth, double-contoured wall, without a micropyle, residuum, or polar granule. The sporocysts are 14 × 8—9 μm, without a Stieda body or residuum.

Reference. Yakimoff & Gousseff, 1935.

SARCOCYSTIS SP. KRASNOVA, 1971

Sarcocysts of this form occur in the muscles of the hamster *Cricetus cricetus* (Type intermediate host) in the USSR. The definitive host is unknown.

Reference. Krasnova (1971).

HOST GENUS *CRICETULUS*

EIMERIA ARUSICA MUSAEV & VEISOV, 1961

This species occurs uncommonly in the gray hamster *Cricetulus migratorius* in the USSR.

The oocysts are spherical or subspherical, 20—26 μm, in diameter, with a smooth, two-layered wall, the outer layer colorless and 1 μm thick, the inner layer dark brown and 1 μm thick, without a micropyle or polar granule, with a residuum. The sporocysts are ovoid, 10—13 × 6—9 (mean 11 × 7) μm, with a Stieda body and residuum. The sporozoites are comma- or bean-shaped. The sporulation time is 4 d at 25—30°C in 2.5% potassium bichromate solution.

Reference. Musaev & Veisov (1961).

EIMERIA CRICETULI MUSAEV & VEISOV, 1961

This species occurs uncommonly in the gray hamster *Cricetulus migratorius* in the USSR.

The oocysts are ovoid, 30—35 × 24—31 (mean 33 × 266.48 [sic]) μm, with a smooth, colorless, one-layered wall 2 μm thick, without a micropyle or residuum, with a polar granule. The sporocysts are ovoid, 10—15×6—11 (mean 13 × 9) μm, without a Stieda body, with a residuum. The sporozoites are bean-shaped. The sporulation time is 4 d at 25—30°C in 2.5% potassium bichromate solution.

Reference. Musaev & Veisov (1961).

EIMERIA IMMODULATA MUSAEV & VEISOV, 1961

This species occurs fairly commonly in the gray hamster *Cricetulus migratorius* in the USSR.

The oocysts are ovoid or ellipsoidal, 18—24 × 12—19 (mean 21 × 16) μm, with a smooth, yellow-brown, one-layered wall 1—1.5 μm thick, without a micropyle or residuum, sometimes with a polar granule. The sporocysts are ovoid, 6—10 × 4—8 (mean 9 × 6) μm, without a Stieda body, with a residuum. The sporozoites are piriform, rarely bean-shaped, with a clear globule at the broad end. The sporulation time is 3 d at 25—30°C in 2.5% potassium bichromate solution.

Reference. Musaev & Veisov (1961).

EIMERIA JARDIMLINICA MUSAEV & VEISOV, 1961

This species occurs uncommonly in the gray hamster *Cricetulus migratorius* in the USSR.

The oocysts are ovoid, rarely ellipsoidal, 22—27 × 16—21 (mean 24 × 19) μm, with a smooth, colorless, one-layered wall 1.5 μm thick, without a micropyle or polar granule, with a residuum. The sporocysts are ovoid, 10—13 × 6—9 (mean 11 × 7) μm, without a Stieda body, with a residuum. The sporozoites are comma-shaped, with a clear globule at the broad end. The sporulation time is 2 d at 25—30°C in 2.5% potassium bichromate solution.

Reference. Musaev & Veisov (1961).

EIMERIA MATSCHOULSKYI PELLÉRDY, 1974

Synonym. *Eimeria daurica* Machul'skii, 1947 in *Cricetulus barabensis*. This species occurs in the striped hamster *Cricetulus barabensis* in the USSR.

The oocysts are cylindrical or ovoid, 20—30 × 15—18 (mean 24 × 17) μm, with a colorless, one-layered wall 1 μm thick, without a micropyle, residuum, or polar granule. The sporocysts are 7—10 × 7 μm, with a residuum.

References. Machul'skii (1947); Pellérdy (1974).

EIMERIA MIGRATORIA MUSAEV & VEISOV, 1961

This species occurs moderately commonly in the gray hamster *Cricetulus migratorius* in the USSR.

The oocysts are spherical or subspherical, 15—23 (mean 19) μm in diameter, with a smooth, colorless, one-layered wall 1.5 μm thick, without a micropyle or residuum, with two or three polar granules. The sporocysts are ovoid or spherical, 6—11 × 4—9 μm, without a Stieda body, with a residuum. The sporozoites are bean-shaped. The sporulation time is 2 d at 25-30°C in 2.5% potassium bichromate solution.

Reference. Musaev & Veisov (1961).

SARCOCYSTIS CRICETULI PATTON & HINDLE, 1926

Sarcocysts of this species occur in the muscles of the striped hamster *Cricetulus griseus* (type intermediate host) in Asia. The definitive host is unknown.

The sarcocysts are elongate ovoid, whitish, mean 1.5 × 0.2 mm, with elongate, slightly curved bradyzoites with one end rounded and the other pointed, 8—10 × 2 μm.

Reference. Patton & Hindle (1926).

HOST GENUS *MESOCRICETUS*

EIMERIA AMBURDARIANA MUSAEV & VEISOV, 1962

This species occurs in the golden hamster *Mesocricetus auratus* in the USSR.

The oocysts are spherical, 18—24 (mean 21) μm in diameter, with a smooth, colorless, one-layered wall 1.5 μm thick, without a micropyle, with a residuum and polar granule. The sporocysts are ovoid, 8—12 × 6—9 (mean 11 × 8) μm, without a Stieda body, with a residuum.

Reference. Musaev & Veisov (1962).

EIMERIA AURATA MUSAEV & VEISOV, 1962

This species occurs in the golden hamster *Mesocricetus auratus* in the USSR.

The oocysts are spherical, 10—18 (mean 16) μm in diameter, with a smooth, yellow, one-layered wall 1 μm thick, without a micropyle or residuum, with a polar granule. The sporocysts are spherical or ovoid, 4—8 × 4—6 μm, without a Stieda body, with a residuum. The sporozoites lack clear globules.

Reference. Musaev & Veisov (1962).

EIMERIA RAZGOVICA MUSAEV & VEISOV, 1962

This species occurs in the golden hamster *Mesocricetus auratus* in the USSR.

The oocysts are ovoid, 16—22 × 14—18 (mean 21 × 17) μm, with a smooth, two-layered wall, the outer layer smooth, bright yellow, 1 μm thick, the inner layer dark brown, 1 μm thick, without a micropyle or residuum, with a polar granule. The sporocysts are ovoid, 6—8 × 4—5 (mean 7 × 5) μm, without a Stieda body, with a residuum. The sporozoites have clear globules.

Reference. Musaev & Veisov (1962).

"COCCIDIUM" ARCAY, 1982

This form (not otherwise named) was found in the epididymis, seminal vesicle, prostate, penis, uterus, and vagina of golden hamsters (called "*Cricetus cricetus*") in Venezuela. It was thought to be transmitted orally.

Reference. Arcay (1982).

BESNOITIA BESNOITI (MAROTEL, 1913) HENRY, 1913

While the type intermediate host of this species is the ox *Bos taurus*, it has been found that the golden hamster *Mesocricetus auratus* can be infected experimentally.

References. See under *Mus*.

BESNOITIA DARLINGI (BRUMPT, 1913) MANDOUR, 1965

The golden hamster *Mesocricetus auratus* is an experimental intermediate host of this species.

References. See under *Mus*.

BESNOITIA JELLISONI FRENKEL, 1955

The golden hamster *Mesocricetus auratus* is an experimental intermediate host of this species.

References. See under *Peromyscus*.

BESNOITIA WALLACEI (TADROS & LAARMAN, 1976) DUBEY, 1977

The golden hamster *Mesocricetus auratus* is an experimental intermediate host of this species.

References. See under *Rattus*.

BESNOITIA SPP. MATUSCHKA & HÄFNER, 1984

Matuschka & Häfner (1984) found that snakes of the genus *Bitis* are definitive hosts of *Besnoitia* spp. in Africa and that rodents of the genus *Mesocricetus* and other genera are experimental intermediate hosts. The rodents had macroscopic *Besnoitia* cysts up to 2 mm in diameter in their connective tissue. These authors thought that rodents might be reservoir hosts of *Besnoitia* of cattle.

Reference. Matuschka & Häfner (1984).

HOST GENUS *PHODOPUS*

BESNOITIA SPP. MATUSCHKA & HÄFNER, 1984

Matuschka & Häfner (1984) found that snakes of the genus *Bitis* are definitive hosts of *Besnoitia* spp. in Africa and that rodents of the genus *Phodopus* and other genera are experimental intermediate hosts. The rodents had macroscopic *Besnoitia* cysts up to 2 mm in diameter in their connective tissue. These authors thought that rodents might be reservoir hosts of *Besnoitia* of cattle.

Reference. Matuschka & Häfner (1984).

HOST GENUS *SPALAX*

EIMERIA ADIYAMANENSIS SAYIN, 1981

This species occurs quite commonly in the herbivorous mole rat *Spalax ehrenbergi* in Asia.

The oocysts are ovoid to ellipsoidal, 30—34 × 15—19 (mean 33 × 18) μm, with a smooth, yellowish brown wall 1.3 μm thick, with a micropyle and residuum, without a polar granule. The sporocysts are almost ellipsoidal, 11—14 × 6—9 (mean 12 × 8) μm, with a thin wall, Stieda body, and residuum. The sporozoites are greenish-yellow, banana-shaped, with a clear globule at the broad end, and lie lengthwise head to tail in the sporocysts. The sporulation time is 2—3 d.

Reference. Sayin (1981).

EIMERIA CELEBII SAYIN, 1981

This species occurs uncommonly in the herbivorous mole rat Spalax *ehrenbergi* in Asia.

The oocysts are ellipsoidal, 14—18 × 8—10 (mean 16 × 9) μm, with a smooth, pale greenish yellow wall 0.9 μm thick, without a micropyle or residuum, with a polar granule. The sporocysts are ellipsoidal, 7—9 × 4—6 (mean 8 × 5) μm, with a thin wall, Stieda body, and residuum. The sporozoites are colorless, comma-shaped, without clear globules, and lie lengthwise head to tail in the sporocysts. The sporulation time is 2 d.

Reference. Sayin (1981).

EIMERIA ELLIPTICA SAYIN, DINCER, & MERIC, 1977

This species occurs moderately commonly in the mole rat *Spalax leucodon* in Asia.

The oocysts are ellipsoidal, 12—17 × 9—12 (mean 14 × 10) μm, with a smooth, greenish yellow, one-layered wall 0.9 μm thick, without a micropyle or residuum, with two polar granules. The sporocysts are somewhat ellipsoidal, 7—9 × 4—6 (mean 8 × 5) μm, with a thin wall, Stieda body, and residuum. The sporozoites have a clear globule at the broad end. The sporulation time is 3—5 d at 20—22°C in 2.5% potassium bichromate solution.

Reference. Sayin, Dincer, & Meric (1977).

EIMERIA HARANICA SAYIN, 1981

This species occurs commonly in the herbivorous mole rat *Spalax ehrenbergi* in Asia.

The oocysts are elongate ovoid, 35—40 × 18—23 (mean 37 × 20) μm, with a smooth, greenish yellow wall 1.3 μm thick, with a micropyle and residuum, without a polar granule. The sporocysts are elongate ovoid, 15—20 × 8—11 (mean 17 × 9) μm, with a thin wall, Stieda body, and residuum. The sporozoites are greenish yellow, small and wedge-shaped, without clear globules. The sporulation time is 2—3 d.

Reference. Sayin (1981).

EIMERIA LALAHANENSIS SAYIN, DINCER, & MERIC, 1977

This species occurs quite commonly in the mole rat *Spalax leucodon* in Asia.

The oocysts are subspherical to ellipsoidal, 16—24 × 10—14 (mean 20 × 12.5) μm, with a smooth, greenish yellow, one-layered wall 1.2 μm thick, without a micropyle or residuum, with two polar granules. The sporocysts are ellipsoidal, 7—13 × 6—8 (mean 12.5 × 7) μm, with a Stieda body and residuum. The sporozoites are elongate, with a clear globule. The sporulation time is 5—7 d at 20—22°C in 2.5% potassium bichromate solution.

Reference. Sayin, Dincer, & Meric (1977).

EIMERIA LEUCODONICA VEISOV, 1975

This species occurs in the mole rat *Spalax leucodon* in the USSR.

The oocysts are spherical or subspherical, 17—22 × 16—21 (mean 20 × 18.5) μm, with a smooth, colorless, two-layered wall 2.5—3 μm thick (each layer 1.25—1.5 μm thick), without a micropyle, residuum, or polar granule. The sporocysts are ovoid or ellipsoidal, 6—10 × 4—6 (mean 8 × 6) μm, without a Stieda body, with a residuum. The sporozoites are lemon-shaped.

Reference. Veisov (1975).

EIMERIA MARALIKIENSIS VEISOV, 1975

This species occurs in the mole rat *Spalax leucodon* in the USSR.

The oocysts are ovoid or rarely ellipsoidal, 16—24 × 13—18 (mean 22 × 16) µm, with a smooth, colorless, one-layered wall 1.5—2 µm thick, without a micropyle or residuum, with a polar granule. The sporocysts are ovoid or ellipsoidal, 6—10 × 3—5 (mean 8 × 4) µm, without a Stieda body, with a residuum. The sporozoites are elongate, with one end pointed and the other rounded, with a clear globule at the broad end, and lie lengthwise head to tail in the sporocysts.

Reference. Veisov (1975).

EIMERIA MARASENSIS SAYIN, 1981

This species occurs uncommonly in the herbivorous mole rat *Spalax ehrenbergi* in Asia.

The oocysts are ellipsoidal, 32—39 × 17—19 (mean 36 × 18) µm, with a greenish yellow, smooth wall 1 µm thick, with a micropyle and residuum, without a polar granule. The sporocysts are elongate ovoid, 13—15 × 7—9 (mean 14 × 8) µm, with a thin, transparent wall, small Stieda body, and residuum. The sporozoites are greenish yellow, banana-shaped, with a clear globule at the large end, and lie lengthwise head to tail in the sporocysts. The sporulation time is 2—3 d.

Reference. Sayin (1981).

EIMERIA OYTUNI SAYIN 1981

This species occurs uncommonly in the herbivorous mole rat *Spalax ehrenbergi* in Asia.

The oocysts are ovoid ("piriform"), 21—27 × 16—18 (mean 24 × 17) µm, with a yellowish green, smooth wall 1.2 µm thick, with a micropyle, without a polar granule or residuum. The sporocysts are piriform, 10—12 × 6—8 (mean 11 × 7) µm, with a thin wall, small Stieda body, and residuum. The sporozoites are pale green, comma-shaped, with clear globules at both ends, and lie lengthwise head to tail in the sporocysts. The sporulation time is 3 d.

Reference. Sayin (1981).

EIMERIA SPALACIS SAYIN, DINCER, & MERIC, 1977

This species occurs commonly in the mole rat *Spalax leucodon* in Asia.

The oocysts are ellipsoidal to cylindrical, 12—18 × 9—11 (mean 16 × 10) µm, with a smooth, greenish yellow, one-layered wall 0.8 µm thick, without a micropyle or residuum, with a polar granule. The sporocysts are elongate ovoid, 9 × 6 µm, with a thin wall, Stieda body, and residuum. The sporozoites have a clear globule at the broad end. The sporulation time is 3—5 d at 20—22°C in 2.5% potassium bichromate solution.

Reference. Sayin, Dincer, & Meric (1977).

EIMERIA TALIKIENSIS VEISOV, 1975

This species occurs in the mole rat *Spalax leucodon* in the USSR.

The oocysts are ovoid or rarely ellipsoidal, 18—24 × 12—18 (mean 21 × 15) μm, with a smooth, colorless, two-layered wall 2.5—3 μm thick (each layer 1.25—1.5 μm thick), with a thin, bright brown inner membrane, without a micropyle, residuum, or polar granule. The sporocysts are piriform, 7—10×4—6 (mean 9 × 6) μm, with a Stieda body and residuum. The sporozoites lie lengthwise head to tail in the sporocysts.

Reference. Veisov (1975).

EIMERIA TOROSICUM SAYIN, 1981

This species occurs uncommonly in the herbivorous mole rat *Spalax ehrenbergi* in Asia.

The oocysts are spherical to subspherical, 10—12×9—11 (mean 11 × 10) μm, with a pale greenish yellow wall 0.8 μm thick, without a micropyle, polar granule, or residuum. The sporocysts are elongate ovoid, 6—8 × 3—5 (mean 7 ×4) μm, with a thin wall, Stieda body, and residuum. The sporozoites are banana-shaped, with a clear globule at the broad end, and lie lengthwise head to tail in the sporocysts. The sporulation time is 5 d.

Reference. Sayin (1981).

EIMERIA TURKMENICA SAYIN, DINCER, & MERIC, 1977

This species occurs uncommonly in the mole rat *Spalax leucodon* in Asia.

The oocysts are ellipsoidal, 10—14×7—11 (mean 11×9) μm, with a smooth, pale greenish yellow, one-layered wall 0.8 μm thick, without a micropyle or residuum, with one polar granule. The sporocysts are elongate ovoid, 5—9×4—6 (mean 8 ×6) μm, with a thin wall, Stieda body, and residuum. The sporozoites are comma-shaped, with a clear globule at the large end, and lie lengthwise head to tail in the sporocysts. The sporulation time is 3—5 d at 20—22°C in 2.5% potassium bichromate solution.

Reference. Sayin, Dincer, & Meric (1977).

EIMERIA TUZDILI SAYIN, DINCER, & MERIC, 1977

This species occurs uncommonly in the mole rat *Spalax leucodon* in Asia.

The oocysts are spherical to subspherical, 18—23 × 13—18 (mean 20.5 × 16.5) μm, with a radially striated, two-layered wall, the outer layer rough, pitted, yellowish brown, 1.7 μm thick, the inner layer colorless, 0.4 μm thick, without a micropyle or residuum, with two polar granules. The sporocysts are ellipsoidal, 9—12 × 7—9 (mean 11 × 8) μm, with a Stieda body and residuum. The sporozoites have a clear globule. The sporulation time is 6—7 days at 20—22°C in 2.5% potassium bichromate solution.

Reference. Sayin Dincer, & Meric (1977).

EIMERIA URFENSIS SAYIN, 1981

This species occurs quite commonly in the herbivorous mole rat *Spalax ehrenbergi* in Asia.

The oocysts are ellipsoidal, 29—36 × 17—23 (mean 34 × 21) μm, with a smooth, yellowish brown wall 1.5 μm thick, with a micropyle and micropylar cap, without a polar granule or residuum. The sporocysts are elongate ovoid, 13—16 × 6—8 (mean 14 × 7) μm, with a Stieda body and residuum. The sporozoites are banana-shaped, pale yellow, without clear globules, and lie lengthwise head to tail in the sporocysts. The sporulation time is 3—5 d.

Reference. Sayin (1981).

ISOSPORA ANATOLICUM SAYIN, DINCER, & MERIC, 1977

This species occurs uncommonly in the mole rat *Spalax leucodon* in Asia.

The oocysts are spherical, 9—11 × 8—9 (mean 9 × 9) μm, with a smooth, pale greenish yellow, one-layered wall 0.8 μm thick, without a micropyle, residuum, or polar granule. The sporocysts are ovoid, 6—9 × 4—6 (mean 8 × 5) μm, with a thin wall, without a Stieda body, with a residuum. The sporozoites are banana-shaped, without a clear globule. The sporulation time is 7—8 d at 20—22°C in 2.5% potassium bichromate solution.

Reference. Sayin, Dincer, & Meric (1977).

HOST GENUS *OTOMYS*

EIMERIA OTOMYIS DE VOS & DOBSON, 1970

This species occurs moderately commonly in the swamp rat *Otomys irroratus* in Africa.

The oocysts are predominantly ellipsoidal and sometimes subspherical, 15—23 × 13—16 (mean 20 × 15) μm, with a smooth, colorless, apparently one-layered wall about 1 μm thick, without a micropyle or residuum, usually with a polar granule. The sporocysts are ellipsoidal, 9—11 × 6—8 (mean 10 × 7) μm, with a Stieda body, usually without a residuum. The sporozoites are 7 μm long, rounded at one end and tapered at the other, with a clear globule at the broad end, and usually lie lengthwise head to tail in the sporocysts. The sporulation time is 2 d at 28°C in 2% potassium bichromate solution.

The chinchilla *Chinchilla laniger* cannot be infected.

Reference. De Vos & Dobson (1970).

SARCOCYSTIS SP. (MANDOUR & KEYMER, 1972) MARKUS, KILLICK-KENDRICK, & GARNHAM, 1974

This form occurs in the organs of the swamp rat *Otomys kempi* (type intermediate host) in Africa. It was found by Mandour & Keymer (1972) and considered to be a *Sarcocystis* by Markus, Killick-Kendrick, & Garnham (1974). Its definitive host is unknown.

References. Mandour & Keymer (1972); Markus, Killick-Kendrick, & Garnham (1974).

HOST GENUS *TACHYORYCTES*

EIMERIA TACHYORYCTIS VAN DEN BERGHE & CHARDOME, 1956

This species occur uncommonly in the fuku *Tachyoryctes ruandae* in Africa.

The oocysts were described as ovoid but illustrated as ellipsoidal, 23×17 μm, with a colorless, quite thick wall, without a micropyle, with a residuum. The sporocysts are ovoid, 12×8 μm, without a Stieda body, with a residuum. The sporozoites are reniform, 8.5×3.5 μm. Sporulation takes 4—6 d in 1% chromic acid.

The white rat *Rattus norvegicus* and mouse *Mus musculus* cannot be infected.
Reference. Van den Berghe & Chardome (1956).

HOST GENUS *GERBILLUS*

BESNOITIA SPP. MATUSCHKA & HÄFNER, 1984

Matuschka & Häfner (1984) found that snakes of the genus *Bitis* are definitive hosts of *Besnoitia* spp. in Africa and that rodents of the genus *Gerbillus* and other genera are experimental intermediate hosts. The rodents had macroscopic *Besnoitia* cysts up to 2 mm in diameter in their connective tissue. These authors thought that rodents might be reservoir hosts of *Besnoitia* of cattle.
Reference. Matuschka & Häfner (1984).

HOST GENUS *TATERA*

EIMERIA TATERAE MIRZA & AL-RAWAS, 1975

This species occurs commonly in the antelope rat *Tatera indica* in Asia.

The oocysts are ellipsoidal, 25—30 × 18—24 (mean 28×21) μm, with a two-layered wall, the outer layer slightly pitted, orange, about 2 μm thick, the inner layer colorless, about 1 μm thick, with a micropyle, residuum, and one or two polar granules. The sporocysts are ovoid, 12—15 × 8—10 (mean 12.5×9) μm, with a Stieda body and residuum. The sporozoites are bean-shaped, without a clear globule, and lie lengthwise head to tail in the sporocysts. The sporulation time is 4 d at 20—22°C in 2.5% potassium bichromate solution.
Reference. Mirza & Al-Rawas (1975).

EIMERIA SP. FANTHAM, 1926

This form occurs in the ileum of Lobengula's gerbil *Tatera lobengulae* in Africa. The oocysts are thin-walled and frail, 15—17 × 10—13 μm.
Reference. Fantham (1926).

SARCOCYSTIS SP. VILJOEN, 1921

Sarcocysts of this form occur in the muscles of the large naked-soled gerbil *Tatera* sp. in Africa. Its definitive host is unknown.
Reference. Viljoen (1921).

BESNOITIA SP. GUNDERS, 1985

Gunders (1985) found cutaneous nodules due to *Besnoitia* sp. on both ears of a bushveld gerbil *Tatera leucogaster* from Namibia, Africa. The cysts were packed with "cystozoites," some undergoing endodyogeny.

Reference. Gunders (1985).

HOST GENUS *MERIONES*

EIMERIA ACHBURUNICA MUSAEV & ALIEVA, 1961

This species occurs uncommonly in the red-tailed jird *Meriones libycus* (syn., *M. erythrourus*) in the USSR.

The oocysts are ovoid, 14—24 × 12—21 (mean 20 × 17) μm, with a smooth, colorless, one-layered wall 1—1.5 μm thick, with a micropyle and residuum, rarely with a polar granule. The sporocysts are ovoid, rarely spherical, 6—10 × 4—8 μm, without a Stieda body, with a residuum. The sporozoites are bean-shaped, with a clear globule in the broad end.

Reference. Musaev & Alieva (1961).

EIMERIA AKERIANA ISMAILOV & GAIBOVA, 1981

This species occurs in the small intestine (mostly the anterior part) of the jird *Meriones blackleri* in the USSR.

There are three meront generations. First generation meronts are found 36 h after inoculation. They are 8—14 × 6—12 (mean 11 × 9) μm and produce 12—32 (mean 19) merozoites. Second generation meronts are found at 72 h. They are 8—13 × 6—10 (mean 10 × 8) μm and produce 10—26 merozoites. Third generation meronts are found at 132 h. They are 6—11 × 5—10 (mean 8 × 6) μm and produce 4—22 merozoites. Apparently all merozoite generations produce gamonts, gametes, zygotes, and oocysts.

References. Ismailov & Gaibova (1981); Musaev, Ismailov, & Gaibova (1982).

EIMERIA ARABIANA VEISOV, 1961

This species occurs in the Vinogradov jird *Meriones vinogradovi* in the USSR.

The oocysts are ovoid, 16—32 × 12—26 (mean 25 × 20) μm, with a two-layered wall, the outer layer smooth, colorless, 0.7 μm thick, the inner layer brownish, 1.5 μm thick, without a micropyle or polar granule, with a residuum. The sporocysts are ovoid or piriform, 6—14 × 4—10 (mean 10 × 7) μm, without a Stieda body, with a residuum. The sporozoites vary in shape; they lack a clear globule. The sporulation time is 3 d at 25—30°C in 2.5% potassium bichromate solution.

Reference. Veisov (1961).

EIMERIA ARAXENA MUSAEV & VEISOV, 1960

This species occurs in the jird *Meriones tristrami* in the USSR.

The oocysts are ovoid, 20—26 × 16—20 (mean 23 × 19) μm, with a rough, dark brown, one-layered wall 1.5—2 μm thick, without a micropyle, with a residuum and polar granule. The sporocysts are ovoid, rarely ellipsoidal, 9—13 × 5—8 (mean 11 × 6) μm, without a Stieda body, with a residuum. The sporozoites are piriform, with a clear globule in the broad end. The sporulation time is 3.5—4 d at 25-30°C in 2.5% potassium bichromate solution.

Reference. Musaev & Veisov (1960).

EIMERIA ASSAENSIS LEVINE & IVENS, 1965

Synonym. *Eimeria callosphermophili* [sic] Henry of Svanbaev (1962).

This species occurs quite commonly in the tamarisk gerbil *Meriones tamariscinus* in the USSR.

The oocysts are subspherical or spherical, 25—30 × 23—26 (mean 28 × 25) μm, with a smooth, greenish, yellow-green, or yellow-brown wall 1.2—1.4 μm thick, without a micropyle, with a residuum and polar granule. The sporocysts are subspherical or ellipsoidal, 11—13 × 8—10 (mean 11.5 × 9) μm, without a residuum. The sporozoites are comma-shaped.

References. Levine & Ivens (1965); Svanbaev (1962).

EIMERIA ASTRACHANBAZARICA MUSAEV & VEISOV, 1960

This species occurs in the jird *Meriones tristrami* in the USSR.

The oocysts are subspherical, ovoid, rarely spherical, 15—30 × 14—26 (mean 22 × 19) μm, with a smooth, two-layered wall, the outer layer colorless, 1 μm thick, the inner layer yellowish, 1 μm thick, without a micropyle or residuum, with a polar granule. The sporocysts are ovoid, 6—14 × 4—10 (mean 9 × 7) μm, without a Stieda body, with a residuum. The sporozoites are comma-shaped, rarely piriform, with a clear globule at the broad end. The sporulation time is 3—3.5 d at 25—30°C in 2.5% potassium bichromate solution.

Reference. Musaev & Veisov (1960).

EIMERIA BISTRATUM VEISOV, 1961

This species occurs in the Vinogradov gerbil *Meriones vinogradovi* in the USSR.

The oocysts are ovoid or spherical, 14—28 × 12—26 μm, with a two-layered wall, the outer layer smooth, colorless to light yellow, 1 μm thick, the inner layer yellow-brown, 1 μm thick, without a micropyle, residuum, or polar granule. The sporocysts are ovoid or spherical, 4—12 × 4—11 μm, without a Stieda body, with a residuum. The sporozoites vary in shape; they lack a clear globule. The sporulation time is 2—3 d at 25—30°C in 2.5% potassium bichromate solution.

Reference. Veisov (1961); Musaev, Ismailova, & Gaibova (1977).

EIMERIA DISAENSIS MUSAEV & VEISOV, 1960

This species occurs in the Persian jird *Meriones persicus* in the USSR.

The oocysts are ovoid or subspherical, 18—24 × 16—22 (mean 23 × 18) μm, with a smooth, colorless, one-layered wall 1—1.2 μm thick, with a micropyle, without a residuum or polar granule. The sporocysts are ovoid, rarely spherical, 8—12 × 6—10 (mean 11 × 9) μm, without a Stieda body, with a residuum. The sporozoites are ovoid, with a clear globule at the broad end. The sporulation time is 2 d at 25—30°C in 2.5% potassium bichromate solution.

References. Glebezdin (1974); Musaev & Veisov (1960): Dyková & Lom, 1983.

EIMERIA DOGELI MUSAEV & VEISOV, 1965

Synonyms. *Eimeria zemphirica* Veisov, 1965 of Musaev in Pellérdy (1974); *E. dogieli* Veisov, 1964 of Pellérdy (1974).

This species occurs in the Persian jird *Meriones persicus* in the USSR.

The oocysts are ellipsoidal or ovoid, 14—30 × 12—26 (mean 25 × 21) μm, with a two-layered wall 2—3 μm thick, both layers being 1—1.5 μm thick, the outer layer smooth, colorless, rarely yellowish, the inner layer yellow-brown, without a micropyle, with a residuum, sometimes with a polar granule. The sporocysts are ovoid, ellipsoidal, or piriform, 6—14 × 4—9 (mean 10 × 7) μm; the piriform sporocysts have a Stieda body; all sporocysts have a residuum. The sporozoites are comma-shaped or piriform, with a clear globule. The sporulation time is 3—4 d.

Remarks. Pellérdy (1974) said that this name was a homonym of *E. dogieli* Pellérdy, 1963. He said that he wrote Musaev about it, and that Musaev wrote him that the new name should be *E. zemphirica* Veisov, 1965. However, the two names are spelled differently, so *E. dogeli* cannot be a homonym, and we have therefore retained it.

References. Musaev & Veisov (1965); Pellérdy (1974).

EIMERIA DZHAHRIANA MUSAEV & VEISOV, 1960

This species occurs in the epithelial cells of the villi of the small intestine of the jirds *Meriones tristrami* (type host) and *M. blackleri* in the USSR.

The oocysts are spherical or subspherical, 17—23 × 17—20 μm, with a rough, tuberculated, dark brown, one-layered wall 1.5 μm thick, without a micropyle, with a residuum and polar granule. The sporocysts are ovoid, 6—10 × 4—6 (mean 9 × 5) μm, without a Stieda body, with a residuum. The sporozoites vary in shape; they have a clear globule. The sporulation time is 3 d at 25—30°C in 2.5% potassium bichromate solution.

Young and mature generation meronts are seen 46 h after inoculation, second generation meronts at 70 h, and third generation meronts at 94 and 115 h. Macrogametes are present 94 h after inoculation, and both macrogametes and oocysts at 115—166 h.

References. Musaev, Ismailov, & Gaibova (1978); Musaev & Veisov (1960).

EIMERIA EGYPTI PRASAD, 1960

This species occurs in Shaw's jird *Meriones s. shawi* in Africa.

The oocysts are ovoid, 27—29 × 15—17 (mean 28 × 16) μm, with a smooth, light brown, two-layered wall, the outer layer thicker than the inner, with a micropyle and residuum, probably without a polar granule. The sporocysts are piriform, 7—9 × 4—5 μm, with a Stieda body and residuum. The sporozoites have clear globules.

Reference. Prasad (1960).

EIMERIA ERYTHROURICA MUSAEV & ALIEVA, 1961

This species occurs commonly in the cecum and colon of the red-tailed jird *Meriones libycus* (syn., *M. erythrourus*) in the USSR. It is slightly pathogenic, but does not kill infected animals.

The oocysts are usually ovoid, sometimes spherical, 14—32 × 12—26 (mean 21 × 18) μm, with a smooth, bright yellow to bright crimson, one-layered wall 1—2 μm thick, without a micropyle or residuum, with a polar granule. The sporocysts are ovoid or spherical, 6—14 × 6—10 μm, without a Stieda body, with a residuum. The sporozoites are piriform or bean-shaped, with a clear globule at the broad end. The sporulation time is 2—4 d at 25—30°C in 2.5% potassium bichromate solution.

There are three or possibly more meront generations. Young and mature meronts appear at 93-95 h, and merogony lasts up to 7 d. Gamogony begins on the 6th day and lasts until the 11th day. The macrogametes are spherical at first, then ovoid. The microgamonts are ovoid, 10—16 × 6—13 (mean 13 × 10) μm, and the microgametes are comma-shaped, 1—2 (mean 1.5) μm long. Oocysts appear in the intestinal lumen as early as 6.5 d. The prepatent period averages 7 d and the patent period is 13-14 d.

References. Musaev & Alieva (1961, 1963); Musaev & Ismailov (1969); Veisov (1964); Musaev, Gailbova, & Ismailov (1984).

EIMERIA JERSENICA DAVRONOV, 1973

This species occurs in the red-tailed jird *Meriones libycus* (syn., *M. erythrourus*) in the USSR.

The oocysts are ovoid or ellipsoidal, 27—36 × 22—31 (mean 31 × 26) μm, with a smooth, one-layered wall 1.5—1.7 μm thick, without a micropyle, with a residuum and polar granule. The sporocysts are ovoid or ellipsoidal, 12—17 × 7—14 μm, without a Stieda body, with a residuum. The sporozoites have a clear globule and lie at the ends of the sporocysts. The sporulation time is 2—3 d at 28—29°C.

Reference. Davronov (1973).

EIMERIA JURSCHUAENSIS VEISOV, 1961

This species occurs in the Vinogradov gerbil *Meriones vinogradovi* in the USSR.

The oocysts are ovoid, 30—36 × 24—30 (mean 34 × 27) μm, with a rough, dark brown, one-layered wall 2.3 μm thick which appears segmented and contains granules, without a micropyle or polar granule, with a residuum. The sporocysts are ovoid, 12—16×8—12 (mean 15×10) μm, without a Stieda body, with a residuum. The sporozoites are comma-shaped and have a clear globule at the broad end. The sporulation time is 3 d at 25—30°C in 2.5% potassium bichromate solution.

Reference. Veisov (1961).

EIMERIA KARSCHINICA DAVRONOV, 1973

This species occurs in the jird *Meriones meridianus* in the USSR.

The oocysts are ovoid, 27—36 × 24—32 (mean 31 × 28) μm, with a one-layered wall 1.7—2 μm thick, without a micropyle, with a residuum and polar granule. The sporocysts are ovoid or ellipsoidal, 10—17 × 7—12 μm, without a Stieda body, with a residuum. The sporozoites have a clear globule.

Reference. Davronov (1973).

EIMERIA KOSTENCOVI DAVRONOV, 1973

This species occurs in the jird *Meriones meridianus* in the USSR.

The oocysts are ovoid or ellipsoidal, 27—34 × 25—32 (mean 31 × 29) μm, with a rough, one-layered wall 1.7—2 μm thick, without a micropyle, with a residuum and polar granule. The sporocysts are ellipsoidal, 7—17×5—12 μm, with a Stieda body and residuum. The sporozoites lack a clear globule.

Reference. Davronov (1973).

EIMERIA KRILOVI MUSAEV & VEISOV, 1965

This species occurs in the Persian jird *Meriones persicus* in the USSR.

The oocysts are ellipsoidal or ovoid, 28—40 × 22—32 (mean 35 × 29) μm, with a three-layered wall 4 μm thick, the outer and middle layers each 1 μm thick, the inner layer 2 μm thick, the outer layer illustrated as colorless, the middle layer as pale, and the inner layer as dark, without a micropyle or polar granule, with a residuum. The sporocysts are piriform, 10—16×6—12 (mean 13×9) μm, with a Stieda body and residuum. The sporozoites are comma-shaped, with a clear globule, and lie lengthwise head to tail in the sporocysts. The sporulation time is 5—6 d.

References. Glebezdin (1974a); Musaev & Veisov (1965).

EIMERIA LERIKAENSIS MUSAEV & VEISOV, 1960

This species occurs in the Persian jird *Meriones persicus* in the USSR.

The oocysts are ovoid or spherical, 14—22 × 12—20 μm, with a two-layered wall, the outer layer smooth, colorless, 0.5—1 μm thick, the inner layer dark yellow, 1 μm thick, without a micropyle or residuum, with a polar granule. The sporocysts are spherical or ovoid, 4—10×4—8 μm, without a Stieda body, with a residuum. The sporozoites are piriform, illustrated without clear globules. The sporulation time is 3—4 d at 25—30°C in 2.5% potassium bichromate solution.

Reference. Musaev & Veisov (1960).

EIMERIA MARKOVI SVANBAEV, 1956

This species occurs commonly in the tamarisk jird *Meriones tamariscinus* in the USSR.

The oocysts are ovoid, 22—48 × 21—35 (mean 32 × 26) μm, with a smooth, greenish to yellowish green, double-contoured wall 1.4—1.6 μm thick, without a micropyle, residuum, or polar granule. The sporocysts were described as ovoid but illustrated as ellipsoidal, 10—16 × 7—11 (mean 12 × 9) μm, without a Stieda body or residuum. The sporozoites are comma-shaped and lie lengthwise in the sporocysts.

Reference. Svanbaev (1956).

EIMERIA MARTUNICA MUSAEV & ALIEVA, 1961

This species occurs uncommonly in the lower part of the small intestine of the red-tailed jird *Meriones libycus* (syn., *M. erythrourus*) in the USSR.

The oocysts are ovoid, 18—34 × 16—28 (mean 26 × 22) μm, with a two-layered wall 2 μm thick, the outer layer smooth, bright yellow, the inner layer dark yellow, without a micropyle or polar granule, with a residuum. The sporocysts are ovoid, 8—14 × 6—10 (mean 11 × 8) μm, with a Stieda body in some, with a residuum. The sporozoites are piriform or bean-shaped, illustrated without clear globules. The sporulation time is 3—4 d at 25—30°C in 2.5% potassium bichromate solution.

The meronts and gamonts are in the epithelial cells, mainly in the distal part of the villi. There are two asexual generations. The first generation meronts start 10—20 h after oral inoculation of oocysts; these meronts are mature at 51 h, being 11—15 × 9—13 μm and containing 14—52 merozoites. The second generation meronts are mature at 65—75 h. They are 10—12 × 8—10 μm and contain 10—30 merozoites. Macrogametes and microgamonts are most numerous at 92—94 h, and are mature at 117 h. The mature microgamonts are round or ovoid, 10—15 × 8—13 μm. The prepatent period is 96—104 h, and the patent period 7—9 d.

References. Musaev & Alieva (1961, 1963), Musaev & Ismailov (1973): Veisov (1964).

EIMERIA MERIDIANA VEISOV, 1964

This species occurs quite commonly in the meridian jird *Meriones meridianus* in the USSR.

The oocysts are ellipsoidal or ovoid, 20—28 × 14—20 (mean 25 × 20) μm, with a two-layered wall, the outer layer smooth, colorless, 1—1.25 μm thick, the inner layer dark brown, 1—1.25 μm thick, without a micropyle or polar granule, with a residuum. The sporocysts are ovoid or piriform, 8—12 × 4—8 (mean 11 × 7) μm, with a Stieda body and residuum. The sporozoites are comma-shaped, with a clear globule at the broad end.

Reference. Veisov (1964).

EIMERIA MERIONIS MACHUL'SKII, 1949

Synonym. *Eimeria marionis* Machul'skii of Svanbaev (1956, 1962) *lapsus calami.*

This species occurs quite commonly in the Mongolian jird *Meriones unguiculatus* (type host) and tamarisk jird *M. tamariscinus* in the USSR.

The oocysts of the form from *M. unguiculatus* are ovoid, 17—23 × 15—19 (mean 19 × 15) μm, with an orange, double-contoured, one-layered wall 1—1.2 μm thick, with a residuum and polar granule. The sporocysts are 8—10 × 6 μm, with a residuum.

The oocysts of the form from *M. tamariscinus* are ellipsoidal or subspherical, 19—21 × 17—18 (mean 21 × 17.5) μm, with a smooth, yellow-orange or yellow-brown wall 0.9-1.3 μm thick, without a micropyle, with a residuum, usually with a polar granule. The sporocysts are ellipsoidal or subspherical, 9—13 × 6—8 (mean 11 × 7) μm, with a residuum. The sporozoites are comma-shaped, 5—7 × 2—4 (mean 6 × 3) μm.

Remarks. The above description of the form from *M. unguiculatus* is taken from Svanbaev (1956) and Musaev & Veisov (1960). That of the form from *M. tamariscinus* is taken from Svanbaev (1962).

References. Machul'skii (1949); Musaev & Veisov (1960), Svanbaev (1956, 1962).

EIMERIA MUSAJEVI VEISOV, 1961

This species occurs in the Vinogradov jird *Meriones vinogradovi* in the USSR.

The oocysts were described as ovoid but illustrated as ellipsoidal, 21—36 × 19—30 (mean 28 × 25) μm, with a rough, tuberculated, dark brown, one-layered wall 1.5-1.8 μm thick, dotted with small granules, without a micropyle or polar granule, with a residuum. The sporocysts are ovoid, 8—14 × 5—11 (mean 12 × 9) μm, without a Stieda body, with a residuum. The sporozoites are piriform, with a clear globule in the broad end. The sporulation time is 4 d at 25—30°C in 2.5% potassium bichromate solution.

Reference. Veisov (1961).

EIMERIA NEHRAMAENSIS MUSAEV & VEISOV, 1960

This species occurs in the jird *Meriones tristrami* in the USSR.

The oocysts are subspherical or spherical, 14—26 × 12—24 μm, with a smooth, colorless, one-layered wall 1 μm thick, without a micropyle, with a residuum and polar granule. The sporocysts are ovoid, 5—12 × 5—8 (mean 9 × 7) μm, with a Stieda body and residuum. The sporozoites are comma-shaped, with a clear globule at the broad end. The sporulation time is 2 d at 25—30°C in 2.5% potassium bichromate solution.

Reference. Musaev & Veisov (1960).

EIMERIA NORASCHENICA MUSAEV & VEISOV, 1960

This species occurs in the Persian jird *Meriones persicus* in the USSR.

The oocysts are ovoid, rarely subspherical, 16—28 × 14—26 (mean 23 × 21) μm, with a smooth, colorless to yellowish, one-layered wall 1 μm thick, without a micropyle, residuum, or polar granule. The sporocysts are ovoid, 6—12 × 4—10 (mean 10 × 7) μm, without a Stieda body, with a residuum. The sporozoites are lemon-shaped, illustrated without clear globules. The sporulation time is 2 d at 25—30°C in 2.5% potassium bichromate solution.

References. Glebezdin (1974); Musaev & Veisov (1960).

EIMERIA ORDUBADICA MUSAEV & VEISOV, 1965

This species occurs in the Persian jird *Meriones persicus* in the USSR.

The oocysts are ellipsoidal or ovoid, 18—32 × 16—26 (mean 26 × 23) μm, with a smooth, colorless, one-layered wall 1.5—2 μm thick, without a micropyle or polar granule, with a residuum. The sporocysts are ellipsoidal or ovoid, 6—12 × 4—8 (mean 11 × 7) μm, without a Stieda body, with a residuum. The sporozoites are comma-shaped or piriform. The sporulation time is 2—3 d.

References. Glebezdin (1974); Musaev & Veisov (1965).

EIMERIA PESCHANKAE LEVINE & IVENS, 1965

Synonym. *Eimeria kriygsmanni* [sic] Yakimoff & Gousseff of Svanbaev (1962).

This species occurs commonly in the tamarisk jird *Meriones tamariscinus* in the USSR.

The oocysts are ellipsoidal, subspherical or spherical, 19—27 × 19—24 (mean 22 × 21) μm, with a smooth, yellow-green or yellow-brown wall 1.3—2.1 μm thick, without a micropyle or residuum; a polar granule was not always seen. The sporocysts are ellipsoidal or spherical, 7—12 × 6—9 (mean 9 × 8) μm, without a residuum. The sporozoites are comma-shaped.

References. Levine & Ivens (1965); Svanbaev (1962).

EIMERIA POLJANSKII VEISOV, 1961 EMEND. LEVINE & IVENS, 1965

Synonym. *Eimeria poljanski* Veisov, 1961.

This species occurs in the Vinogradov jird *Meriones vinogradovi* in the USSR.

The oocysts are spherical or ovoid, 28—34 × 28—34 μm, with a colorless, two-layered wall, the outer layer rough, containing small granules, 0.6 μm thick, the inner layer corrugated, containing small granules, without a micropyle or polar granule, with a residuum. The sporocysts are ovoid, 10—14 × 6—10 (mean 13 × 9) μm, with a Stieda body and residuum. The sporozoites are comma-shaped, with a clear globule at the broad end. The sporulation time is 4 d at 25—30°C in 2.5% potassium bichromate solution.

References. Levine & Ivens (1965); Veisov (1961).

EIMERIA SADARAKTICA VEISOV, 1961

This species occurs in the Vinogradov jird *Meriones vinogradovi* in the USSR.

The oocysts are ovoid, 18—24 × 16—22 (mean 23 × 19) μm, with a smooth, colorless, one-layered wall 1.4 μm thick, with a micropyle and polar granule, without a residuum. The sporocysts are ovoid, 6—12 × 4—8 (mean 10 × 7) μm, without a Stieda body, with a residuum. The sporozoites are comma-shaped, with a clear globule at the broad end. The sporulation time is 3 d at 25—30°C in 2.5% potassium bichromate solution.

Reference. Veisov (1961).

EIMERIA SALASUZICA MUSAEV & VEISOV, 1960

This species occurs in the Persian jird *Meriones persicus* in the USSR.

The oocysts are ovoid, sometimes subspherical, 22—26 × 20—24 (mean 24 × 21) μm, with a rough, granulated, dark brown, one-layered wall 1.5 μm thick, without a micropyle, with a residuum and polar granule. The sporocysts are ovoid, 10—13 × 7—8 (mean 12 × 7) μm, with a Stieda body and residuum. The sporozoites vary in shape, and were illustrated without clear globules.

References. Glebezdin (1974); Musaev & Veisov (1960).

EIMERIA SCHACHTACHTIANA MUSAEV & VEISOV, 1960

This species occurs in the jird *Meriones tristrami* in the USSR.

The oocysts were described as ovoid but illustrated as ellipsoidal, 16—30 × 14—24 (mean 24 × 20) μm, with a smooth, colorless, one-layered wall 1—1.25 μm thick, without a micropyle or residuum, with a polar granule. The sporocysts were described as ovoid but illustrated as ellipsoidal, 6—16 × 4—10 (mean 11 × 8) μm, without a Stieda body, with a residuum. The sporozoites have a clear globule at the broad end and lie at the ends of the sporocysts. The sporulation time is 2—3 d at 25-30°C in 2.5% potassium bichromate solution.

Reference. Musaev & Veisov (1960).

EIMERIA SCHAMCHORICA MUSAEV & ALIEVA, 1961

This species occurs uncommonly to quite commonly in the duodenum, jejunum, and upper ileum of the red-tailed jird *Meriones libycus* (syn., *M. erythrourus*) in the USSR. It is slightly pathogenic, but does not kill infected animals. The oocysts are ovoid, rarely spherical, 16—32 × 14—28 μm, with a smooth, colorless or sometimes slightly yellow-crimson, one-layered wall 1—2 μm thick, without a micropyle, with a residuum. The sporocysts are spherical, 6—10 μm in diameter or ovoid, 10 × 8 μm, with or without a Stieda body, with a residuum. The sporozoites are piriform or bean-shaped, with a clear globule at the broad end. The sporulation time is 72—120 h.

All meronts are in the villar epithelial cells. There are three generations. First generation meronts are in the duodenum, 11—17 × 8—14 (mean 14 × 11) μm and produce 4—22 (mean 13) merozoites 8—14 × 2—3 (mean 12 × 2) μm. Second generation meronts are in the jejunum and upper ileum, 13—18 × 10—14 (mean 15 × 12) μm and produce 10—33 (mean 19) merozoites 8—13 × 1—3 (mean 10 × 2) μm and a residuum. Third generation meronts are in the jejunum and upper ileum, 14—20 × 10—15 (mean 17 × 13) μm and produce 12—38 (mean 22) merozoites 10—13 × 1—2 (mean 11.5 × 1.3) μm. First generation meronts are seen 5—52 h after inoculation, second generation meronts at 52—68 h and third generation meronts at 88—120 h.

Mature macrogametes and microgamonts are present in the villar epithelial cells 86 to about 180 h after inoculation. The macrogametes are 13—18 × 10—16 (mean 16 × 13) μm. The microgamonts are 10—18 × 7—14 (mean 13 × 11) μm and contain many microgametes 1—3 (mean 2) μm long. Oocysts appear in the intestinal lumen at 110—115 h and in the feces at 118—122 h. The prepatent period is 5 d and the patent period 6—9 d.

References. Musaev & Alieva (1961, 1963); Musaev & Ismailova (1969, 1971); Veisov (1964).

EIMERIA SUMGAITICA MUSAEV & ALIEVA, 1961

This species occurs uncommonly in the red-tailed gerbil *Meriones libycus* (syn., *M. erythrourus*) in the USSR.

The oocysts are ovoid, sometimes subspherical, 16—26 × 14—24 (mean 22 × 19) μm, with a two-layered wall 1.5—2 μm thick, the outer layer smooth, bright yellow, the inner layer dark yellow, without a micropyle or residuum, some with a polar granule. The sporocysts are ovoid, rarely spherical, 6—10 × 4—8 μm, without a Stieda body, with a residuum. The sporozoites are piriform, with a clear globule. The sporulation time is 3—4 d at 25—30°C in 2.5% potassium bichromate solution.

References. Musaev & Alieva (1961, 1963); Veisov (1964).

EIMERIA TAMARISCINI LEVINE & IVENS, 1965

Synonym. *Eimeria musculi* Yakimoff & Gousseff of Svanbaev (1956) in *Meriones tamariscinus.*

This species occurs commonly in the tamarisk jird *Meriones tamariscinus* in the USSR.

The oocysts are subspherical, 19 × 19 μm, with a smooth, greenish, double-contoured wall 1.4 μm thick, without a micropyle, residuum, or polar granule. The sporocysts are ovoid, 7 × 4 μm, without a residuum.

References. Levine & Ivens (1965); Svanbaev (1956).

EIMERIA TASAKENDICA VEISOV, 1961

This species occurs in the Vinogradov jird *Meriones vinogradovi* in the USSR.

The oocysts were described as ovoid and illustrated as ellipsoidal, 19—26 × 17—23 (mean 23 × 21) μm, with a rough, cross-striated, colorless, one-layered wall 1.2—1.4 μm thick, without a micropyle, with a residuum and polar granule. The sporocysts are ovoid, 8—13 × 5—9 (mean 12 × 7) μm, with a Stieda body and residuum. The sporozoites are comma-shaped, with a clear globule in the broad end. The sporulation time is 2 d at 25—35°C in 2.5% potassium bichromate solution.

Reference. Veisov (1961).

EIMERIA TRISTRAMI MUSAEV & VEISOV, 1965

This species occurs in the jird *Meriones blackleri* in the USSR.

The oocysts are ellipsoidal, rarely ovoid, 22—30 × 16—26 (mean 26 × 22) μm, with a two-layered wall 3 μm thick, the outer layer smooth, colorless, 1 μm thick, the inner layer yellow-brown, 2 μm thick, without a micropyle or polar granule, with a residuum. The sporocysts are ovoid or piriform, 8—12 × 6—9 (mean 11 × 8) μm, with a Stieda body and residuum. The sporozoites are comma- or bean-shaped. The sporulation time is 4 d.

Reference. Musaev & Veisov (1965).

EIMERIA UZBEKISTANICA DAVRONOV, 1973

This species occurs in the jird *Meriones meridianus* in the USSR.

The oocysts are spherical, rarely ovoid, 15—22 × 15—20 μm, with a one-layered wall 1.5—1.7 μm thick, without a micropyle, with a residuum and polar granule. The sporocysts are ellipsoidal, 8—10 × 5—9 μm, without a Stieda body, with a residuum. The sporozoites lack a clear globule.

Reference. Davronov (1973).

EIMERIA VAHIDOVI MUSAEV & VEISOV, 1965

This species occurs in the Vinogradov jird *Meriones vinogradovi* in the USSR.

The oocysts are ovoid or ellipsoidal, 18—24 × 14—18 (mean 22 × 17) μm, with a two-layered wall, the outer layer smooth, colorless, 1.5 μm thick, the inner layer yellowish brown, 1.5 μm thick, with a micropyle and polar granule, without a residuum. The sporocysts are ellipsoidal or ovoid, 6—10 × 4—8 (mean 9 × 7) μm, without a Stieda body, with a residuum. The sporozoites are lemon-shaped. The sporulation time is 3—4 d.

Reference. Musaev & Veisov (1965).

EIMERIA VINOGRADOVI VEISOV, 1961

This species occurs in the Vinogradov jird *Meriones vinogradovi* in the USSR.

The oocysts are ovoid or ellipsoidal, 16—34 × 14—30 (mean 22 × 19) μm, with a smooth, colorless, one-layered wall 1—1.5 μm thick, without a micropyle, residuum, or ordinarily polar granule. The sporocysts are ovoid, ellipsoidal, or very rarely spherical, 6—16 × 4—12 μm, without a Stieda body, with a residuum. The sporozoites are comma-shaped, ovoid, or occasionally irregular, with a clear globule in the broad end. The sporulation time is 2 d at 25—30°C in 2.5% potassium bichromate solution.

Reference. Veisov (1961).

EIMERIA ZULFIAENSIS VEISOV, 1961

This species occurs in the Vinogradov jird *Meriones vinogradovi* in the USSR.

The oocysts are ovoid, 17—32 × 16—28 (mean 25 × 20) μm, with a smooth, colorless, one-layered wall 1—1.25 μm thick, without a micropyle, with a residuum, very rarely with a polar granule. The sporocysts are ovoid, 6—13 ×

5—10 (mean 11 × 7) μm, with a Stieda body and residuum. The sporozoites are comma-shaped and have a clear globule at the broad end. The sporulation time is 2 d at 25—30°C in 2.5% potassium bichromate solution.

Reference. Veisov (1961).

ISOSPORA EGYPTI PRASAD, 1960

This species occurs in Shaw's jird *Meriones s. shawi* in Africa.

The oocysts are subspherical, 20—22 × 16—20 (mean 21 × 18) μm, with a light brown, two-layered wall, the outer layer smooth, thicker than the inner one, without a micropyle, residuum, or polar granule. The sporocysts are ovoid, 10— 12 × 6—8 μm, with a Stieda body and residuum. The sporozoites have clear globules.

Reference. Prasad (1960).

ISOSPORA ERYTHROURICA VEISOV, 1964

This species occurs rarely in the red-tailed jird *Meriones libycus* (syn., *M. erythrourus*) in the USSR.

The oocysts are spherical or subspherical, 24—30 μm in diameter, with a two-layered wall, the outer layer smooth, colorless, 1.5 μm thick, the inner layer yellow-brown, 1.5 μm thick, without a micropyle or residuum, with a polar granule. The sporocysts are ellipsoidal or spherical, 12—18 × 10—16 μm, apparently without a Stieda body, with a residuum. The sporozoites are bean- or lemon-shaped.

Reference. Veisov (1964).

ISOSPORA MERIONES MUSAEV & VEISOV, 1965

Synonym. *Isospora merionis* Veisov, 1964 of Pellérdy (1974).

This species occurs in the Vinogradov jird *Meriones vinogradovi* in the USSR.

The oocysts are ellipsoidal, 18—28 × 14—24 (mean 23 × 20) μm, with a two-layered wall, the outer layer smooth, colorless, 1—1.25 μm thick, the inner layer yellow-brown, 1—1.25 μm thick, without a micropyle or residuum, with a polar granule. The sporocysts are piriform, 10—16 × 6—10 (mean 14 × 10 [sic]) μm, with a Stieda body and residuum. The sporozoites are bean-shaped, rarely comma-shaped. The sporulation time is 4 d.

Reference. Musaev & Veisov (1965).

ISOSPORA ORDUBADICA MUSAEV & VEISOV, 1960

This species occurs in the Persian jird *Meriones persicus* in the USSR.

The oocysts are ovoid or subspherical, 18—20 × 14—18 (mean 20 × 17) μm, with a smooth, colorless, one-layered wall 1 μm thick, without a micropyle, residuum, or polar granule. The sporocysts were described as ovoid but illustrated as ellipsoidal, 10—12 × 8—10 (mean 12 × 10) μm, without a Stieda body, with a residuum. The sporozoites are relatively small and ovoid. The sporulation time is 2—3 d at 25—30°C in 2.5% potassium bichromate solution.

References. Glebezdin (1974); Musaev & Veisov (1960).

ISOSPORA TAMARISCINI LEVINE, 1985

Synonym. *Isospora laguri* Iwanoff-Gobzem of Svanbaev (1962) in *Meriones tamariscinus.*

This species occurs commonly in the tamarisk jird *Meriones tamariscinus* in the USSR.

The oocysts are ellipsoidal or subspherical, yellow-green, 21—30 × 20—26 (mean 27 × 19.7 [sic]) μm, with a smooth wall 1.5—1.9 μm thick, without a micropyle or polar granule, with a residuum. The sporocysts are ellipsoidal or ovoid, 12—14 × 7—10 (mean 13 × 9) μm, without a residuum. The sporozoites are comma-shaped or ellipsoidal, 5—6 × 2—3 (mean 5.5 × 3) μm.

References. Svanbaev (1962); Levine (1985).

ISOSPORA VANADICA MUSAEV & VEISOV, 1965

This species occurs in the Persian jird *Meriones persicus* in the USSR.

The oocysts are ellipsoidal or ovoid, 20—28 × 14—24 (mean 24 × 19) μm, with a two-layered wall, the outer layer smooth, yellowish, 1.5 μm thick, the inner layer dark brown, 1.5 μm thick, without a micropyle, with a residuum and polar granule. The sporocysts are ovoid, 10—16 × 6—12 (mean 14 × 9) μm, without a Stieda body, with a residuum. The sporozoites are bean- or lemon-shaped. The sporulation time is 4—5 d.

Reference. Musaev & Veisov (1965).

ISOSPORA VINOGRADOVI MUSAEV & VEISOV, 1965

This species occurs in the Vinogradov jird *Meriones vinogradovi* in the USSR.

The oocysts are ellipsoidal, rarely ovoid, 22—28 × 18—24 (mean 27 × 22) μm, with a smooth, colorless, one-layered wall 2—2.5 μm thick, without a micropyle or polar granule, with a residuum. The sporocysts are piriform, 12—16 × 8—10 (mean 15 × 9) μm, with a Stieda body and residuum. The sporozoites are comma-shaped. The sporulation time is 3 d.

Reference. Musaev & Veisov (1965).

MANTONELLA MERIONES GLEBEZDIN, 1971

This species occurs in the red-tailed jird *Meriones libycus* (syn., *M. erythrourus*) in the USSR.

The oocysts are ellipsoidal, 26—29 × 23—26 (mean 27 × 24) μm, with a smooth, colorless, one-layered wall 1.5 μm thick, without a micropyle or residuum, with a polar granule. The sporocysts are ovoid, 20—23 × 14.5—17 (mean 22.5 × 17) μm, with a Stieda body, without a residuum. The sporozoites were described as ovoid and illustrated as ellipsoidal, 12—17 × 6—9 μm.

References. Glebezdin (1971); Glebezdin & Babich (1974).

BESNOITIA BESNOITI (MAROTEL, 1913) HENRY, 1913

While the type intermediate host of this species is the ox *Bos taurus*, it has been found that the gerbil *Meriones tristrami shawi* can be infected experimentally.

References. See under *Mus.*

BESNOITIA WALLACEI (TADROS & LAARMAN, 1976) DUBEY, 1977

The Mongolian gerbil *Meriones unguiculatus* is an experimental intermediate host of this species.

References. See under *Rattus*.

BESNOITIA SPP. MATUSCHKA & HÄFNER, 1984

Matuschka & Häfner (1984) found that snakes of the genus *Bitis* are definitive hosts of *Besnoitia* spp. in Africa and that rodents of the genus *Meriones* and other genera are experimental intermediate hosts. The rodents had macroscopic *Besnoitia* cysts up to 2 mm in diameter in their connective tissue. These authors thought that rodents might be reservoir hosts of *Besnoitia* of cattle.

Reference. Matuschka & Häfner (1984).

HOST GENUS *RHOMBOMYS*

EIMERIA ABIDZHANOVI DAVRONOV, 1973

This species occurs in the large gerbil *Rhombomys opimus* in the USSR.

The oocysts are ellipsoidal, 25—32 × 20—27 (mean 27 × 23) μm, with a grayish yellow, one-layered wall 1.7—2 μm thick, without a micropyle, with a residuum and polar granule. The sporocysts are ovoid, 8—14 × 5—12 μm, with a Stieda body and residuum. The sporozoites have a clear globule.

Reference. Davronov (1973).

EIMERIA BADCHISICA GLEBEZDIN, 1969

This species occurs in the large gerbil *Rhombomys opimus* in the USSR.

The oocysts are spherical or ellipsoidal, 17—26 × 17—23 μm, with a smooth, colorless or slightly greenish, one-layered wall 1.2—1.5 μm thick, without a micropyle, with a residuum, sometimes with a polar granule. The sporocysts are ellipsoidal, 5—12 × 4—9 (mean 9 × 6) μm, without a Stieda body, with a residuum. The sporozoites are bean-shaped, apparently without clear globules, and lie at the ends of the sporocysts.

Reference. Glebezdin (1969).

EIMERIA CONEVI GLEBEZDIN, 1969

This species occurs in the large gerbil *Rhombomys opimus* in the USSR.

The oocysts are ellipsoidal or spherical (illustrated as subspherical), 26—35 × 23—32 (mean 30 × 29) μm, with a rough, striated, dark brown, one-layered wall 1.5—1.7 μm thick, without a micropyle, with a residuum and polar granule. The sporocysts are 9—15 × 6—12 (mean 12 × 7) μm, with a Stieda body and residuum. The sporozoites are elongate, with a clear globule at the broad end, and lie lengthwise head to tail in the sporocysts.

Reference. Glebezdin (1969).

EIMERIA SULTANOVI DAVRONOV, 1973

This species occurs in the large gerbil *Rhombomys opimus* in the USSR.

The oocysts are spherical, ovoid, or ellipsoidal, 19—31 × 15—29 (mean 26 × 24) μm, with a one-layered wall 1.7—2 μm thick, without a micropyle, with a residuum and polar granule. The sporocysts are ellipsoidal or ovoid, 3—14 × 3—10 μm, without a Stieda body or residuum. The sporozoites lack a clear globule.

Reference. Davronov (1973).

HOST GENUS *CLETHRIONOMYS*

EIMERIA CERNAE LEVINE & IVENS, 1965

Synonym. *Eimeria schueffneri* Yakimoff & Gousseff of Cerná (1962)

This species occurs commonly in epithelial cells of the cecum, colon, and rectum of the red-backed vole *Clethrionomys glareolus* (type host) and vole *C. rutilus* in Europe and the USSR.

The oocysts are ellipsoidal, 12—23 × 10—17 μm, with a very thin wall, without a micropyle or residuum, sometimes with a polar granule. The sporocysts are mostly ellipsoidal, occasionally spherical, 8—15 × 4—7 μm, without or with a very small Stieda body, without or with a residuum. The sporozoites lie lengthwise in the sporocysts. The sporulation time is 2—4 d in 1.5% potassium bichromate solution.

There are three meront generations. First generation meronts are present 2 d after inoculation. They are 6—8 × 5—7 μm and contain two to eight merozoites. Second generation meronts are present at 3 d. They are 6—9 × 5—8 μm and contain 12—20 (mean 16) merozoites. Third generation meronts are present at 4 d. They are 8—12 × 7—10 μm and contain 14—21 (mean 18) merozoites 13 × 2 μm with the nucleus near one end. Mature macrogametes are 11—15 × 6—12 μm, and mature microgamonts are 8—15 × 6—11 μm and produce many microgametes. The prepatent period is 6 d and the patent period 4—6 d.

References. Arnastauskene (1977); Cerná (1962); Golemanski (1979); Levine & Ivens (1965); Lewis & Ball (1982, 1983); Ball & Lewis (1984).

EIMERIA CLETHRIONOMYIS STRANEVA & KELLEY, 1979

This species occurs commonly in the vole *Clethrionomys gapperi* in North America.

The oocysts are ellipsoidal, with one or both ends flattened, 16—21 × 14—16 (mean 19 × 15) μm, with a two-layered wall, the outer layer smooth, clear to yellow, about 0.6 μm thick, the inner layer clear to yellow, about 0.4 μm thick, without a distinct micropyle but the oocyst wall thins at one or both ends, with a "terminal cap" at one and often both ends, without a residuum, with a polar granule. The sporocysts are elongate ovoid, 9—12 × 5.5—7 (mean 11 × 6) μm, with a Stieda body, without a sub-Stiedal body, with a residuum. The sporozoites have a large posterior and a somewhat smaller anterior clear globule and usually lie lengthwise, partly curled around each other, in the sporocysts.

Reference. Straneva & Kelley (1979).

EIMERIA GALLATII STRANEVA & KELLEY, 1979

This species occurs moderately commonly in the vole *Clethrionomys gapperi* in North America.

The oocysts are ellipsoidal, sometimes asymmetrical, 21—32 × 17—24 (mean 28 × 19) μm, with a two-layered wall, the outer layer smooth, clear, about 1.5 μm thick, the inner layer clear to yellow brown, about 0.4 μm thick, with a micropyle, cobwebby residua at both ends of the oocyst, and two or three polar granules. The sporocysts are ovoid, 12—15 × 8—10 (mean 13.5 × 9) μm, with a wall 0.4 μm thick, with a Stieda body, without a sub-Stiedal body, with a residuum. The sporozoites have a large posterior and smaller anterior clear globule and usually lie lengthwise, often curled around each other, in the sporocyts.

Reference. Straneva & Kelley (1979).

EIMERIA MARCONII STRANEVA & KELLEY, 1979

This species occurs moderately commonly in the vole *Clethrionomys gapperi* in North America.

The oocysts are ellipsoidal, sometimes subspherical, 11—15 × 9—12 (mean 13 × 11) μm, with a smooth, yellow, one-layered wall about 0.8 μm thick, without a micropyle or thinning of the oocyst wall, without a residuum, with one or two polar granules. The sporocysts are elongate ovoid, 7—8 × 3—5 (mean 8 × 4) μm, with a thin wall, a small, dark Stieda body, and a residuum, without a sub-Stiedal body. The sporozoites rarely have clear globules; they are curled around each other and occasionally lie lengthwise in the sporocysts.

Reference. Straneva & Kelley (1979).

EIMERIA PILEATA STRANEVA & KELLEY, 1979

This species occurs moderately commonly in the vole *Clethrionomys gapperi* in North America.

The oocysts are subspherical to spherical, 20.5—29.5 × 20—25 (mean 25 × 22.5) μm, with a two-layered wall, the outer layer rough, pitted, and striated, clear to yellow, about 1.1 μm thick, the inner layer yellow to yellow brown, about 0.6 μm thick, without a micropyle or thinning of the wall, with a residuum and one or two polar granules. The sporocysts are ellipsoidal, slightly pointed at both ends, 11—15 × 7—9 (mean 13 × 8) μm, with a Stieda body with Stiedal cap, with a residuum, without a sub-Stiedal body. The sporozoites have a large posterior and smaller anterior clear globule; they lie lengthwise, partly curled around each other, in the sporocysts.

Reference. Straneva & Kelley (1979).

EIMERIA RYSAVYI LEVINE & IVENS, 1965

Synonyms. *Eimeria apodemi* Pellérdy of Ryšavý (1957); *E. hindlei* Yakimoff & Gousseff of Ryšavý (1954) and Cerná (1962) in part.

This species occurs uncommonly to quite commonly in the small intestine (the sexual stages in the duodenum, jejunum, and occasionally ileum) of the red-backed vole *Clethrionomys glareolus* in Europe.

The oocysts are broadly ellipsoidal, 23—39×18—23 (mean 25×20) μm, with a smooth, rather thick, pale yellowish brown wall, without a micropyle or residuum, without or with a polar granule. The sporocysts are ellipsoidal, 11—14.5 × 6—8.5 μm, without or with a small Stieda body, with a residuum. The sporozoites have a clear globule and lie lengthwise in the sporocysts. The sporulation time is 4—5 d in 2% potassium bichromate solution.

Mature macrogametes are 16—23 × 17.5 μm and mature microgamonts 29—44 × 18—33 μm.

References. Cerná (1962); Golemanski (1979); Levine & Ivens (1965); Lewis & Ball (1983); Ryšavý (1954, 1957); Ball & Lewis (1984).

EIMERIA SCHIWICKI ARNASTAUSKENE, 1977

This species occurs in the intestine of the vole *Clethrionomys rutilus* in the USSR.

The oocysts are ellipsoidal, sometimes ovoid, 18—21 × 15—18 μm, with a smooth, colorless, two-layered wall 2—2.5 μm thick, without a micropyle or residuum, with a polar granule. The sporocysts were described as ovoid but illustrated as ellipsoidal, 10 × 8 μm, without a Stieda body or residuum. The sporozoites were described as ovoid but illustrated as elongate ellipsoidal, have a clear globule at one end, and lie lengthwise in the sporocyts.

The white laboratory mouse *Mus musculus* and red-backed vole *Clethrionomys glareolus* cannot be infected with this species.

Reference. Arnastauskene (1977).

EIMERIA (?) SP. (RYŠAVÝ, 1954) COMB. NOV.

Synonym. *Eimeria falciformis* (Eimer) of Rysavy (1954) and Cerná & Daniel (1956) in part.

This form occurs in the red-backed vole *Clethrionomys glareolus* in Europe. The oocyst structure is uncertain. Its wall is more than 1 μm thick.

References. Cerná & Daniel (1956); Rysavy (1954).

EIMERIA (?) SP. CERNA & DANIEL, 1956 EMEND. LEVINE & IVENS, 1965

This form occurs uncommonly in the red-backed vole *Clethrionomys glareolus* in Europe. Its location in the host is unknown; oocysts were found in the intestine.

The oocysts are broadly ovoid to spherical, 11—14 × 10—13 μm, with a very delicate wall, without a micropyle. They failed to sporulate.

References. Černá & Daniel (1956); Levine & Ivens (1965).

ISOSPORA CLETHRIONOMYDIS GOLEMANSKI & YANKOVA, 1973

This species occurs rarely in the small intestine of the red-backed vole *Clethrionomys glareolus* in Europe.

The oocysts are spherical, 23—27 (mean 25) µm in diameter, with a thin, colorless, or light yellowish wall, without a micropyle, residuum, or polar granule. The sporocysts are ovoid, 21—23 × 11—12 µm, with a Stieda body and residuum. The sporozoites are elongate, with one end tapered and the other rounded, and have a clear globule at the broad end.

Reference. Golemanski & Yankova (1973).

ISOSPORA FLATECA DZERZHINSKII & SVANBAEV, 1980

A *nomen nudum.*

SARCOCYSTIS CLETHRIONOMYELAPHIS MATUSCHKA, 1986

This species occurs in the muscles of the bank vole *Clethrionomys glareolus* (Type host) and also the common vole *Microtus arvalis*, root vole *M. oeconomus*, and Günther's vole *M. guentheri* in Europe (Germany). Its definitive host is the Aesculapian snake *Elaphe longissima*.

Its sarcocysts (muscle meronts) are up to 1200—4500 × 70—170 µm and their wall has short (about 3 µm long) protrusions. The bradyzoites are 10—12 × 2 µm. Its sporocysts in the snake are 8—9 × 11—12 (mean 9 × 11.5) µm.

Reference. Matuschka, 1986.

SARCOCYSTIS SP. WIESNER, 1980

This species occurs commonly in Tengmal's owl *Aegolius funereus* (Definitive host) and also in the bank or red-backed vole *Clethrionomys glareolus* (Intermediate host) in Europe. It was found in the owl by Wiesner (1980), and what may have been the same species was found in the vole by Catar et al. (1967) and Sebek (1975); the last called it *"Sarcocystis muris?"* Wiesner (1980) transmitted it from the owl to the vole. Sporocysts are in the owl feces and sarcocysts in the vole muscles.

The field mouse *Apodemus tauricus* and house mouse *Mus musculus* cannot be infected.

References. Čatár, Zachar, Valent, Vráblik, Hynie-Holková, & Pavlina (1967); Šebek (1975); Wiesner (1980).

FRENKELIA GLAREOLI (ERHARDOVA, 1955) TADROS, BIRD & ELLIS, 1972

Synonyms. *Toxoplasma glareoli* Erhardova, 1955; *Isospora buteonis* Henry, 1932 of Scholtyseck (1954) in *Buteo buteo*; *Frenkelia clethrionomyobuteonis* Rommel & Krampitz, 1975; *Frenkelia buteonis* (Henry, 1932) Dubey, 1977; *Endorimospora buteonis* (Henry, 1932) Tadros & Laarman, 1976; M-organism of Šebek (1965).

This species occurs in the buzzard *Buteo buteo* (definitive host) and rarely to commonly in the bank or red-backed vole *Clethrionomys glareolus* (type intermediate host), northern red-backed vole *C. rutilus,* and large-toothed, red-backed vole *C. rufocanus* in Europe and the USSR. Sporocysts were found in the feces of the buzzard, and sarcocysts in the liver and brain of the voles.

First generation meronts occur in the bank vole hepatocytes. They cause focal necrosis of the liver cells, followed by inflammation, splenomegaly, haemosiderosis, erythropoiesis, and lymphoid hyperplasia. Second generation meronts in the brain compress the nervous tissue and cause focal necrosis and inflammation of the brain.

The sporocysts in buzzard feces are 10—17 × 8—13 μm, with a residuum.

The first generation meronts are in the Kupffer cells and hepatocytes of the liver of the voles, and have 20—30 banana-shaped merozoites 6—8 × 2 μm plus a residuum. The second generation meronts are in the gray and white matter of the brain, and are compartmented, up to 400 μm in diameter, with a two-layered wall, with metrocytes 4—40 μm in diameter and bradyzoites 4—9 μm long with 5—8 rhoptries and 50—70 micronemes. The prepatent period is 7—9 d and the patent period 7—57 d.

Microtus arvalis, M. agrestis, Apodemus sylvaticus, A. flavicollis, and *Mus musculus* cannot be infected and become intermediate hosts. *Falco tinnunculus, Accipiter gentilis, Asio otus, Tyto alba, Strix aluco,* and the dog and cat cannot be infected and become definitive hosts. Infection can be transmitted from bank vole to bank vole by intraperitoneal injection of merozoites in a suspension of liver cells, and congenital transmission has been found in one of three *C. glareolus* born in the laboratory to a captured wild female.

References. Dubey (1977), Erhardova (1955); Geisel, Kaiser, Krampitz, & Rommel (1978); Kalyakin, Kovalevsky, & Nikitina (1973); Kepka & Scholtyseck (1970); Krampitz, Rommel, Geisel, & Kaiser (1976); Mehlhorn & Scholtyseck (1974); Rommel & Krampitz (1975); Rommel, Krampitz, Göbel, Geisel, & Kaiser (1976); Scholtyseck (1954, 1973); Scholtyseck, Mehlhorn, & Müller (1973, 1974); Šebek (1975); Skofitschon & Kepka (1982); Tadros, Bird, & Ellis (1972), Tadros & Laarman (1976, 1978).

FRENKELIA SP. FRANK, 1978

Frank (1978) reported this form in *Clethrionomys glareolus* around Lake Neusiedler in northern Burgenland, Austria. He did not describe it.

Reference. Frank (1978).

CRYPTOSPORIDIUM MURIS TYZZER, 1907

This species was reported by Elton et al. (1931) from *Clethrionomys glareolus* in England. See under *Mus* for further discussion.

HOST GENUS *ALTICOLA*

EIMERIA ARGENTATA DZERZHINSKII & SVANBAEV, 1980

A *nomen nudum.*

EIMERIA BASSAGENSIS SVANBAEV, 1979

Synonym. *Eimeria arvicolae* Galli-Valerio, 1905 of Svanbaev (1958) in *Alticola strelzovi*.

This species occurs moderately commonly in the flat-skulled vole *Alticola strelzovi* in the USSR.

The oocysts are subspherical or ovoid, 22—29 × 19—24 (mean 25 × 22) μm, with a smooth, "double-contoured" wall 1.1—1.3 μm thick, without a micropyle or residuum, with a polar granule. The sporocysts are ellipsoidal, 6—10 × 5—6 (mean 7 × 6) μm, without a Stieda body, with a residuum. The sporozoites are 3—5 × 3 μm, apparently without clear globules. The sporulation time is 3 d at 25—28°C in 2% potassium bichromate solution.

References. Svanbaev (1958, 1979).

HOST GENUS *ARVICOLA*

EIMERIA BATABATENSIS LEVINE & IVENS, 1965

Synonym. *Eimeria arvicolae* Galli-Valerio of Musaev & Veisov (1960).

This species occurs in the water vole *Arvicola terrestris* in the USSR.

The oocysts are ovoid, 14—20 × 10—16 (mean 19 × 15) μm, with a smooth, "double-contoured" wall 1 μm thick, without a micropyle, residuum, or polar granule. The sporocysts are ovoid, rarely spherical, 4—8 × 4—6 μm, without a Stieda body, with a residuum. The sporozoites are comma-shaped, with a small clear globule at the large end. The sporulation time is 2 d at 25—30°C in 2.5% potassium bichromate solution.

References. Levine & Ivens (1965); Musaev & Veisov (1960).

EIMERIA BOHEMICA RYŠAVÝ, 1957

This species occurs quite commonly in the water vole *Arvicola terrestris* in Europe.

The oocysts are ellipsoidal, pale yellowish brown, 19—23 × 11—13 (mean 20 × 12) μm, with an apparently smooth wall, with a micropyle and residuum, apparently without a polar granule. The sporocysts are ovoid, 9.5 × 6 μm, with a residuum. The sporulation time is 3 d at 24°C in 2% potassium bichromate solution.

Reference. Ryšavý (1957).

EIMERIA TALISCHAENSIS MUSAEV & VEISOV, 1960

This species occurs in the water vole *Arvicola terrestris* in the USSR.

The oocysts are spherical, colorless, 18—22 (mean 21) μm in diameter, with a smooth, tri-contoured and two-layered wall 1.5—2 μm thick, each layer 1 μm thick, the outer layer colorless, the inner layer light yellowish brown, without a micropyle or residuum, with a polar granule. The sporocysts are spherical or ovoid, 4—8 × 2—8 μm, without a Stieda body, with a residuum. The sporozoites are comma-shaped, rarely piriform, with a clear globule at the broad end. The sporulation time is 2 d at 25—30°C in 2.5% potassium bichromate solution.

Reference. Musaev & Veisov (1960).

EIMERIA TERRESTRIS MUSAEV & VEISOV, 1960

This species occurs in the water vole *Arvicola terrestris* in the USSR.

The oocysts are ovoid, rarely ellipsoidal, 18—22 × 12—16 (mean 21 × 16) μm, with a smooth, tri-contoured and two-layered wall 1.5—2 μm thick, the outer layer colorless to yellowish, the inner layer dark yellow or brown, with a micropyle, without a residuum or polar granule. The sporocysts are usually ovoid, rarely spherical, 6—8×4—6 (mean 8×6) μm, without a Stieda body, with a residuum. The sporozoites are piriform, rarely comma-shaped. The sporulation time is 3—3.5 d at 25—30°C in 2.5% potassium bichromate solution.

Reference. Musaev & Veisov (1960).

ISOSPORA BATABATICA MUSAEV & VEISOV, 1960

This species occurs in the water vole *Arvicola terrestris* in the USSR.

The oocysts are almost spherical or ovoid, colorless, 20—24 × 19—21 (mean 23 × 21) μm, with a smooth, double-contoured, one-layered wall 1 μm thick, without a micropyle, residuum, or polar granule. The sporocysts are ovoid, 9—14×6—9 (mean 13 × 8) μm, with a prominent Stieda body and a residuum. The sporozoites are ovoid. The sporulation time is 2 d at 25—30°C in 2.5% potassium bichromate solution.

Reference. Musaev & Veisov (1960).

SARCOCYSTIS SP. KRASNOVA, 1971

Sarcocysts of this form were found in the muscles of the water vole *Arvicola terrestris* (type intermediate host) in the USSR. The definitive host is unknown.

Reference. Krasnova (1971).

FRENKELIA SP. (DOBY, JEANNES, & RAULT, 1965); LEVINE & IVENS, 1987

Synonym. M-organism of Dobey, Jeannes, & Rault (1965) in *Arvicola sapidus*.

This organism was found in the brain of the water vole *Arvicola sapidus* (type immediate host) in Europe. The definitive host is unknown.

References. Doby, Jeannes, & Rault (1965); Levine & Ivens (1987).

HOST GENUS *MICROTUS*

EIMERIA ABUSCHEVI VEISOV, 1962

This species was found in the vole *Microtus majori* in the USSR.

The oocysts are ovoid, colorless, 22—31 × 16—27 (mean 29 × 25) μm, with a smooth, one-layered wall 2 μm thick, without a micropyle or polar granule, with a residuum. The sporocysts are piriform or ovoid, 8—13 × 4—9 (mean 11

× 7) μm, with a Stieda body and residuum. The sporozoites have no clear globules.

Reference. Veisov (1962).

EIMERIA ARVICOLAE (GALLI-VALERIO, 1905) REICHENOW, 1921

Synonyms. *Coccidium arvicolae* Galli-Valerio, 1905; *Eimeria arvalis* Iwanoff-Gobzem, 1934; *E. musculi* Yakimoff & Gousseff of Svanbaev (1956) from *Microtus arvalis*.

This species occurs commonly in the intestine of the snow vole *Microtus* (syn., *Arvicola*) *nivalis* (Type host) and European vole *M. arvalis* in Europe and the USSR.

The oocysts are spherical, 14—18 μm in diameter, with a clear, double-contoured wall (*Galli-Valerio*, 1905 in *M. nivalis*); ovoid, 13—28 × 11—21 (mean 20 × 16) μm, without oocyst or sporocyst residua (Iwanoff-Gobzem, 1934 in *M. arvalis*); ovoid, greenish, 22.5 × 19 μm, with a smooth, double-contoured wall 1.5 μm thick, without a micropyle, residuum, or polar granule (Svanbaev, 1956 in *M. arvalis*); or spherical, 20 μm in diameter (Svanbaev, 1956 in *M. arvalis*). The sporocysts of Svanbaev's first form are ovoid, 12 × 9 μm, without a residuum. The sporocysts of Svanbaev's second form are 13 × 9 μm.

We suspect that more than one species is actually encompassed by this name, but we leave it to someone else to straighten the matter out.

References. Galli-Vallerio (1905); Iwanoff-Gobzem (1934); Litvenkova (1969); Mikeladze (1971); Svanbaev (1956).

EIMERIA BICRUSTAE VEISOV, 1962

This species occurs in the narrow-skulled vole *Microtus majori* in the USSR.

The oocysts are ovoid or ellipsoidal, 16—28 × 12—22 (mean 21 × 15) μm, with a two-layered wall, the outer layer smooth, yellowish, 1 μm thick, the inner layer dark brownish, 1 μm thick, without a micropyle, residuum, or polar granule. The sporocysts are ovoid or piriform, 6—10 × 4—6 (mean 9 × 5) μm, with a Stieda body and residuum. the sporozoites have clear globules.

Reference. Veisov (1962).

EIMERIA CHETAE ARNASTAUSKENE, 1980

This species occurs in the vole *Microtus middendorfi* in the USSR.

The oocysts are ovoid or almost spherical, 27—30 × 18—28 μm, with a two-layered wall up to 2 μm thick, the outer layer smooth, brown, the inner layer yellow, without a micropyle, sometimes with a polar granule, with a residuum. The sporocysts were described as ellipsoidal but illustrated as spindle-shaped, yellow, without a Stieda body or sub-Stiedal body, with a residuum. The sporozoites were described as bean-shaped but illustrated as elongate, have a clear globule at one end, and lie lengthwise in the sporocysts. Sporulation occurs outside the host.

Reference. Arnastauskene (1980).

EIMERIA CHUDATICA MUSAEV, VEISOV, & ALIEVA, 1963

This species occurs in the social vole *Microtus socialis* in the USSR.

The oocysts are ovoid or ellipsoidal, 10—23 × 8—21 (mean 211× 17) μm, with a smooth, colorless, one-layered wall 1—1.5 μm thick, without a micropyle or residuum, with a polar granule. The sporocysts are ovoid or rarely spherical, 4—11 × 4—9 μm, without a Stieda body, with a residuum. The sporozoites have a clear globule.

Reference. Musaev, Veisov & Alieva (1963).

EIMERIA COAHULLIENSIS VANCE & DUSZYNSKI, 1985

This species occurs in the vole *Microtus mexicanus subsimus* in Mexico.

The oocysts are ellipsoidal, slightly flattened at the end opposite the micropyle, 27—34 × 18—22 (mean 30 × 20) μm, with a two-layered wall 1.5—2.3 (mean 2) μm thick, the outer layer rough, golden-amber, $^3/_4$ of the total thickness, the inner layer membranous and colorless, with a micropyle and polar body, without a residuum. The sporocysts are ovoid, 13—18 × 8—10 (mean 14 × 9) μm, with a Stieda body and residuum. The sporocysts have two clear globules.

Reference. Vance & Duszynski (1985).

EIMERIA CORREPTIONIS VEISOV, 1962

This species occurs in the narrow-skulled vole *Microtus majori* in the USSR.

The oocysts are ellipsoidal or ovoid, 18—26 × 14—20 (mean 21 × 17) μm, with a smooth, yellowish, one-layered wall 1.5 μm thick, with a micropyle, without a residuum or polar granule. The sporocysts are ovoid or piriform, 6—11 × 4—7 (mean 9 × 5) μm, apparently without a Stieda body, with a residuum. The sporozoites are without clear globules.

Reference. Veisov (1962).

EIMERIA CUBINICA MUSAEV, VEISOV, & ALIEVA, 1963

This species occurs in the social vole *Microtus socialis* in the USSR.

The oocysts are ovoid to ellipsoidal, 14—24 × 10—18 (mean 21 × 15) μm, with a smooth, colorless, one-layered wall 1.5—2 μm thick, with a micropyle, without a residuum or polar granule. The sporocysts are ovoid or piriform, 6—11 × 4—9 (mean 9 × 5) μm, with a Stieda body and residuum. The sporozoites have clear globules.

Reference. Musaev, Veisov, & Alieva (1963).

EIMERIA CUSARICA MUSAEV, VEISOV, & ALIEVA, 1963

This species occurs in the social vole *Microtus socialis* is the USSR.

The oocysts are ovoid, 24—31 × 16—28 (mean 29 × 23) μm, with a smooth, colorless, one-layered wall 2—2.5 μm thick, without a micropyle or residuum, with a polar granule. The sporocysts are ovoid, 8—13×6—11 (mean 11 ×9) μm, with a Stieda body and residuum. The sporozoites are without clear globules.

Reference. Musaev, Veisov, & Alieva, 1963.

EIMERIA DERENICA VEISOV, 1963

This species occurs moderately to quite commonly in the common European vole *Microtus arvalis* in the USSR.

The oocysts are ovoid or ellipsoidal, 20—38 × 14—30 (mean 29 × 21) μm, with a smooth, colorless, one-layered wall 2—3 μm thick, without a micropyle or residuum, with a polar granule. The sporocysts are ovoid or ellipsoidal, 8—16 × 4—12 (mean 12.5 × 9) μm, without a Stieda body, with a residuum. The sporozoites have clear globules.

References. Litvenkova (1969); Mikeladze (1971); Veisov (1963).

EIMERIA DZULFAENSIS MUSAEV & VEISOV, 1959

This species occurs moderately commonly in the social vole *Microtus socialis* in the USSR.

The oocysts are spherical, sometimes slightly ovoid, 21—25 × 20—24 μm, with a two-layered wall, the outer layer rough, dark brown, 1.2 μm thick, the inner layer 0.2 μm thick, without a micropyle, with a residuum and polar granule. The sporocysts are ovoid with a pointed end, 11—14 × 6—9 (mean 12.5 × 8) μm, with a Stieda body and residuum. The sporozoites are piriform, with a clear globule at the broad end. The sporulation time is 3 d at 25—30°C in 2.5% potassium bichromate solution.

Reference. Musaev & Veisov (1959).

EIMERIA GOMURCHAICA VEISOV, 1963

This species occurs in the common European vole *Microtus arvalis* in the USSR.

The oocysts are ovoid or ellipsoidal, 20—34 × 18—30 (mean 28.5 × 24) μm with a smooth colorless one-layered wall 2—3 μm thick without a micropyle or polar granule, with a residuum. The sporocysts are ovoid or piriform, 8—16 × 6—12 (mean 12.5 × 9) μm, with a Stieda body and residuum. The sporozoites have clear globules.

Reference. Veisov (1963).

EIMERIA GREGALICA DZERZHINSKII & SVANBAEV, 1980

A *nomen nudum*.

EIMERIA GUENTHERII GOLEMANSKY, 1978

This species occurs commonly in the vole *Microtus guentheri* in Europe.

The oocysts are ovoid, 10—23 × 13—16 μm, with a smooth, colorless wall about 1 μm thick, without a micropyle or residuum, with two to four polar granules. The sporocysts are elongate ovoid, 11 × 6 μm, with a Stieda body and residuum. The sporozoites are elongate with one end rounded and the other pointed, have a clear globule at the broad end, and lie lengthwise head to tail in the sporocysts.

References. Golemansky (1978, 1979).

EIMERIA HADRUTICA MUSAEV, VEISOV & ALIEVA, 1963

This species occurs in the social vole *Microtus socialis* in the USSR.

The oocysts are ovoid or ellipsoidal, 16—32 × 14—24 (mean 21 × 17) µm, with a two-layered wall, the outer layer smooth, yellowish brown, 1—1.5 µm thick, the inner layer colorless, 1—1.5 µm thick, without a micropyle, residuum, or polar granule. The sporocysts are ovoid, rarely spherical, 6—14 × 4—12 µm, without a Stieda body, with a residuum. The sporozoites have clear globules.

Reference. Musaev, Veisov, & Alieva (1963).

EIMERIA IRADIENSIS VEISOV, 1963

This species occurs in the common European vole *Microtus arvalis* in the USSR.

The oocysts are ovoid, 20—32 × 16—24 (mean 25 × 20.5) µm, with a rough, granulated, dark yellow, one-layered wall 2 µm thick, without a micropyle, residuum, or polar granule. The sporocysts are ovoid, 8—12 × 4—8 (mean 11 × 7) µm, with a Stieda body and residuum. The sporozoites are without clear globules.

References. Mikeladze (1971); Veisov (1963).

EIMERIA IWANOFFI VEISOV, 1963

Synonym. *Eimeria ivanovi* Vejsov, 1963 of Mikeladze (1971, 1973) and Musaev & Veisov (1965).

This species occurs uncommonly in the common European vole *Microtus arvalis* in the USSR and Europe.

The oocysts are ovoid or ellipsoidal, 12—30 × 8—26 (mean 21.5 × 17.5) µm, with a two-layered wall, the outer layer smooth, yellow-brown, 1.25 µm thick, the inner layer colorless, 1.25 µm thick, without a micropyle or residuum, with a polar granule. The sporocysts are ovoid or spherical, 4—14 × 4—10 µm, without a Stieda body, with a residuum. The sporozoites are without clear globules.

References. Golemanski (1979); Mikeladze (1971, 1973); Musaev & Veisov (1965); Veisov (1963).

EIMERIA KOLABSKI MUSAEV & VEISOV, 1965

This species occurs in the social vole *Microtus socialis* in the USSR.

The oocysts are ellipsoidal or ovoid, 24—32 × 20—28 (mean 29 × 24) µm, with a rough, radially striated, dark yellow-brown, one-layered wall 2—2.5 µm thick, without a micropyle, residuum, or polar granule. The sporocysts are ellipsoidal or ovoid, 10—13 × 6—10 (mean 12 × 9) µm, with a residuum. The sporozoites are comma-shaped, with a clear globule at the broad end. The sporulation time is 4 d.

Reference. Musaev & Veisov (1965).

EIMERIA KOLANICA VEISOV, 1963

This species occurs in the common European vole *Microtus arvalis* in the USSR.

The oocysts are ovoid, 16—26 × 14—22 (mean 21 × 19) μm, with a two-layered wall, the outer layer dark brown, 1 μm thick, the inner layer colorless, 1 μm thick, with a micropyle, without a residuum or polar granule. The sporocysts are spherical, 6—10 (mean 8) μm in diameter, without a Stieda body, with a residuum. The sporozoites are without clear globules.
Reference. Veisov (1963).

EIMERIA KOTUJI ARNASTAUSKENE, 1980

This species occurs in the vole *Microtus middendorfi* in the USSR.

Oocysts are ellipsoidal, 21—23 × 14—16 μm, with a smooth, colorless to rose, thin, two-layered wall, without a micropyle, with a residuum and polar granules. The sporocysts are spindle-shaped, 10—11 × 6—8 μm, without a Stieda body or sub-Stiedal body, with a residuum. The sporozoites are elongate and lie lengthwise head to tail in the sporocysts. Sporulation occurs outside the host.
Reference. Arnastauskene (1980).

EIMERIA LUTEOLA ARNASTAUSKENE, 1980

This species occurs in the vole *Microtus middendorfi* in the USSR.

The oocysts are ovoid, 20—21 × 14—18 μm, with a smooth, yellow, one-layered wall 1.5 μm thick, without a micropyle, polar granule, or residuum. The sporocysts are ellipsoidal (described as oval), 9 × 6—7 μm, without a Stieda body or sub-Stiedal body, with a residuum. The sporozoites are elongate and lie lengthwise head to tail in the sporocysts. Sporulation occurs outside the host.
Reference. Arnastauskene (1980).

EIMERIA MAJORICI VEISOV, 1962

Synonym. *Eimeria arvalis* (Iwanoff-Gobzem 1935) of Veisov (1963).

This species occurs in the narrow-skulled vole *Microtus majori* (type host) and common European vole *M. arvalis* in the USSR.

The oocysts are ovoid or ellipsoidal, 16—28 × 12—24 (mean 23 × 17) μm, with a smooth, colorless, one-layered wall 1.5—2 μm thick, without a micropyle, residuum, or polar granule. The sporocysts are piriform or ovoid, 6—12 × 4—8 (mean 9 × 7) μm, with a Stieda body and residuum. The sporozoites have clear globules.
References. Veisov (1962, 1963).

EIMERIA MICROPILIANA MUSAEV, VEISOV, & ALIEVA, 1963

This species occurs in the social vole *Microtus socialis* in the USSR.

The oocysts are ovoid or subspherical, 20—25 × 16—11 (mean 23 × 19) μm, with a two-layered wall, the outer layer smooth, dark yellow, 1.5 μm thick, the inner layer colorless, 1 μm thick, with a micropyle, without a residuum or polar granule. The sporocysts are ovoid, 6—11 × 4—7 (mean 10 × 5.5) μm, without a Stieda body, with a residuum. The sporozoites are without clear globules.
Reference. Musaev, Veisov, & Alieva (1963).

EIMERIA MICROTINA MUSAEV & VEISOV, 1959

This species occurs in the social vole *Microtus socialis* in the USSR.

The oocysts are spherical, sometimes slightly ovoid, 12—18 × 11—16 μm, with a smooth, colorless, apparently one-layered wall 0.8 μm thick, without a micropyle or polar granule, with a residuum. The sporocysts are ovoid, 5—8 × 4—5 (mean 7 × 4) μm, with a Stieda body and residuum. The sporulation time is 3 d at 25—30°C in 2.5% potassium bichromate solution.

Reference. Musaev & Veisov (1959) .

EIMERIA MIDDENDORFI ARNASTAUSKENE, 1977

This species occurs in the intestine of the vole *Microtus middendorfi* in the USSR.

The oocysts are broadly ellipsoidal (described as oval to subspherical), 13—16 × 11—14 μm, with a smooth, colorless, one-layered wall 1.5 μm thick, without a micropyle, residuum, or polar granule. The sporocysts are ellipsoidal, 7 × 4—5 μm, without a Stieda body, with a residuum. The sporozoites are elongate, kidney-shaped, apparently without a clear globule, and lie lengthwise in the sporocysts.

This species could not be transmitted to *Microtus arvalis* or *Mus musculus*.

Reference. Arnastauskene (1977).

EIMERIA MONOCRUSTAE VEISOV, 1963

Synonym. *Eimeria monochrustae* Vejsov, 1963 of Mikeladze (1973) *lapsus calami.*

This species occurs rarely in the common European vole *Microtus arvalis* in the USSR.

The oocysts are ovoid or subspherical, 22—32 × 18—28 (mean 27 × 24) μm, with a rough, yellowish brown, one-layered wall 2.8 μm thick, without a micropyle or polar granule, with a residuum. The sporocysts are ovoid, 8—16 × 6—10 (mean 12 × 8) μm, with a Stieda body and residuum. The sporozoites have clear globules.

References. Mikeladze (1973); Veisov (1963).

EIMERIA OCHROGASTERI BALLARD, 1970

This species occurs uncommonly in the prairie vole *Microtus ochrogaster* and also in *M. mexicanus fulviventer* and *M. p. pennsylvanicus* in North America.

The oocysts are spherical, subspherical, or ellipsoidal, 18—29 × 16—24 (mean 24 × 20.5) μm, with a two-layered wall 1—2 μm thick, the outer layer rough, pitted, yellow-brown, twice as thick as the inner layer, the inner layer clear, elastic, without a micropyle, with a residuum and polar granule. The sporocysts are ovoid, 11—14 × 7—9 (mean 12 × 8) μm, with a Stieda body and residuum. The sporozoites are elongate, somewhat comma-shaped, with a large clear globule at the large end and a small one at the anterior end, and lie lengthwise head to tail in the sporocysts.

Reference. Ballard (1970); Vance & Duszynski (1985).

EIMERIA PITYMYDIS GOLEMANSKI & YANKOVA, 1973

This species occurs quite commonly in the pine vole *Microtus* (syn., *Pitymys*) *subterraneus* in Europe.

The oocysts are ovoid, 16—19 × 13—15 (mean 18 × 14) μm, with a colorless, two-layered wall 1—1.5 μm thick, without a micropyle or residuum, sometimes with a polar granule. The sporocysts are ovoid, apparently without a Stieda body, with a residuum. The sporozoites are elongate, with one end tapered and the other rounded, with a clear globule in the broad end, and lie lengthwise head to tail in the sporocysts.

References. Golemanski (1979); Golemanski & Yankova (1973).

EIMERIA PRIMBELICA VEISOV, 1963

This species occurs in the common European vole *Microtus arvalis* in the USSR.

The oocysts are ovoid, 20—40 × 16—34 (mean 24 × 23) μm, with a two-layered wall 2—3 μm thick, the outer layer smooth and yellowish brown, the inner layer colorless, without a micropyle, with a residuum and polar granule. The sporocysts are ovoid, 6—18 × 4—12 (mean 11 × 8) μm, without a Stieda body, with a residuum. The sporozoites have clear globules.

Reference. Veisov (1963).

EIMERIA SAXEI VANCE & DUSZYNSKI, 1985

Synonym. *Eimeria wenrichi* species B of Saxe, Levine, & Ivens, 1960.

This species occurs in the voles *Microtus pennsylvanicus, M. californicus, M. mexicanus,* and *M. oregoni* in North America (Pennsylvania, California, Washington, Mexico).

The oocysts are ellipsoidal to subspherical, 11—14 × 10—12 (mean 13 × 11) μm, with a smooth wall less than 1 μm thick, without a micropyle or residuum, with a polar granule. The sporocysts are ovoid, 6—9 × 3—5 (mean 7.5 × 4) μm, with a Steida body and residuum.

References. Saxe, Levine, & Ivens (1960), Vance & Duszynski (1985).

EIMERIA SCHELKOVNIKOVI MUSAEV, 1967

This species occurs commonly in Schelkoynikov's vole *Microtus* (syn., *Pitymys*) *schelkovnikovi* in the USSR.

The oocysts are ellipsoidal, 16—24 × 12—18 (mean 20 × 17) μm, with a smooth, colorless, one-layered wall 1—1.5 μm thick, without a micropyle, residuum, or polar granule. The sporocysts are ellipsoidal, 5—10 × 3—7 (mean 7 × 5) μm, without a Stieda body, with a residuum. The sporozoites are bean-shaped.

Reference. Musaev (1967).

EIMERIA SUBSIMI VANCE & DUSZYNSKI, 1985

This species occurs in the vole *Microtus mexicanus subsimus* in Mexico.

The oocysts are ovoid to subspherical, 22—28 × 17—21 (mean 25 × 19) μm, with a two-layered wall 1.5—1.8 (mean 1.6) μm thick, the outer layer slightly sculptured, comprising $^3/_5$ of the total thickness, without a micropyle or residuum, with a polar granule. The sporocysts are ellipsoidal, 13—15 × 6—8 (mean 14 × 7) μm, with a Stieda body, substiedal body, and residuum. The sporozoites lie head to tail in the sporocysts and contain clear globules.

Reference. Vance & Duszynski (1985).

EIMERIA TAIMYRICA ARNASTAUSKENE, 1977

This species occurs in the intestine of the vole *Microtus middendorfi* in the USSR.

The oocysts are broadly ellipsoidal (described as oval), sometimes asymmetrical, 21—28 × 19—25 μm, with a smooth, yellowish, one-layered wall 2.5 μm thick, without a micropyle, residuum, or polar granule. The sporocysts are ellipsoidal (described as oval), 10—13 × 5—8 μm, without a Stieda body or residuum. The sporozoites are elongate, apparently without clear globules, and lie lengthwise head to tail in the sporocyts.

This species cannot be transmitted to *Microtus arvalis* or *Mus musculus*.

Reference. Arnastauskene (1977).

EIMERIA TAMIASCIURI LEVINE, IVENS & KRUIDENIER, 1957

See under *Tamiasciurus*. This species was reported from one vole *Microtus montanus arizonensis* in Arizona by Vance & Duszynski (1985).

Reference. Vance & Duszynski (1985).

EIMERIA TOLUCANDENSIS VANCE & DUSZYNSKI, 1985

This species oocurs in the vole *Microtus m. mexicanus* in Mexico.

This oocysts are subspherical, 23—26 × 19—23 (mean 25 × 20) μm, with a two-layered wall 2—3 (mean 2.3) μm thick, the outer layer multilaminar, $^2/_3$ of the total thickness, the inner layer smooth, without a micropyle or residuum, with a polar granule. The sporocysts are ellipsoidal, 10—13 × 7—9 (mean 11 × 8) μm, with an inconspicuous Stieda body, without a sub-Stiedal body, with a residuum. The sporozoites have one or two clear globules.

Reference. Vance & Duszynski (1985).

EIMERIA WENRICHI SAXE, LEVINE, & IVENS, 1960

This species occurs commonly in the meadow mouse *Microtus pennsylvanicus* (type host), beach vole *M. breweri*, and voles *M. mexicanus* and *M. montanus* in North America.

The oocysts are ellipsoidal to ovoid, 16—22 × 12—16 (mean 19 × 14) μm, with a smooth, one-layered wall, without a micropyle or residuum, with a polar granule. The sporocysts of the large oocysts are 9—11 × 5—8 (mean 10 × 6) μm and of the small ones 6—8 × 4 (mean 7 × 4) μm, ovoid, with a Stieda body and residuum. The sporulation time is 2—3 d at 24—27°C in 1% chromic acid solution.

References. Saxe, Levine, & Ivens (1960), Winchell (1977); Vance & Duszynski (1985).

EIMERIA ZUVANDICA VEISOV, 1963

This species occurs in the common European vole *Microtus arvalis* in the USSR.

The oocysts are ovoid, rarely ellipsoidal, 16—26 × 12—22 (mean 22 × 16.5) μm, with a smooth, colorless, one-layered wall 1—1.5 μm thick, with a micropyle and polar granule, without a residuum. The sporocysts are ovoid, 6—10 × 4—8 (mean 9 × 7) μm, without a Stieda body, with a residuum. The sporozoites have clear globules.

Reference. Veisov (1963).

EIMERIA SP. (RYŠAVÝ, 1954) LEVINE & IVENS, 1965

This form occurs in the vole *Microtus arvalis* in Europe.

The structure of the oocysts is uncertain, it was reported as similar to those of *E. falciformis*.

References. Levine & Ivens (1965); Ryšavý (1954).

ISOSPORA ARVALIS MIKELADZE, 1973

This species occurs rarely in the common European vole *Microtus arvalis* in the USSR.

The oocysts are subspherical, 10—12 × 8—12 (mean 12 × 10) μm, with a smooth, colorless, one-layered wall 2 μm thick, without a micropyle, residuum, or polar granule. The sporocysts are ovoid, 6—8 × 4—6 (mean 7 × 5) μm, illustrated without a Stieda body, without a residuum. The sporozoites were illustrated without clear globules.

References. Mikeladze (1971, 1973).

ISOSPORA MCDOWELLI SAXE, LEVINE, & IVENS, 1960

This species occurs in the meadow mouse *Microtus pennsylvanicus* in North America.

The oocysts are spherical to subspherical, 9—11 × 8—10 (mean 10 × 9) μm, with a one-layered wall, without a micropyle, residuum, or polar granule. The sporocysts are more or less ellipsoidal, 6—8 × 4—5 (mean 7 × 5) μm, with a residuum. The sporulation time is 3—5 d at 24—27°C in 1% chromic acid solution.

Reference. Saxe, Levine, & Ivens (1960).

ISOSPORA MEXICANASUBSIMI VANCE & DUSZYNSKI, 1985

This species occurs in the vole *Microtus mexicanus subsimus* in Mexico.

The oocysts are spherical or nearly so, 21—26 × 21—26 (mean 24 × 23) μm, with a two-layered wall 1.5 μm thick, the outer layer lightly pitted, the inner layer smooth, somewhat darker than the outer one, without a micropyle, residuum, or polar granule. The sporocysts are ovoid, 12—16 × 10—12 (mean 15 × 11) μm, with a Stieda body, substiedal body and residuum.

Reference. Vance & Duszynski (1985).

CARYOSPORA MICROTI SAXE, LEVINE, & IVENS, 1960

This species occurs in the meadow mouse *Microtus pennsylvanicus* in North America.

The oocysts are spherical to subspherical, 9—11 × 8—10 (mean 10 × 9) μm, with a one-layered wall, without a micropyle, residuum, or polar granule. The sporocysts are spherical to subspherical, 7—9 × 6—7 (mean 7 × 6.5) μm, without a Stieda body, with a residuum. The sporulation time is 10 d at 24—27°C in 1% chromic acid solution.

Reference. Saxe, Levine, & Ivens (1960).

SARCOCYSTIS CERNAE LEVINE, 1977

Synonym. *Sarcocystis* sp. Černá & Loučková, 1976.

This species occurs in the muscles of the common European vole *Microtus arvalis* (type intermediate host) and kestrel *Falco tinnunculus* (type definitive host) in Europe.

The oocysts are presumably ellipsoidal, 19—20 × 13—14 μm, with an extremely thin wall, without a residuum or polar granule. The sporocysts are ovoid or ellipsoidal, 13—16 × 10—11 μm, without a Stieda body, with a residuum. Sporulation takes place in the kestrel intestine.

The pre-muscle meronts occur in the liver of *M. arvalis*; they grow to about 19—23 × 13—15 μm and produce a ring of merozoites 5—7 × 2 μm by endopolygeny. These merozoites are free or in macrophages 6—7 d after inoculation. They become meronts and produce a new generation of merozoites here. Sarcocysts develop in the muscles but not in the brain. They are 60—100 × 50—80 μm, not compartmented, and have a very thin wall and bradyzoites 8—9 × 2—2.5 μm.

The prepatent period is 7—8 d.

Mus musculus cannot be infected.

References. Černá (1977, 1983); Černá & Ally (1979); Černá, Kolarova, & Sulc (1978); Černá & Loučková, (1976); Levine (1977); Kutzer, Frey, & Kotremba (1980); Tadros (1981); Hoogenboom, Daan, & Laarman (1984).

SARCOCYSTIS CLETHRIONOMYELAPHIS MATUSCHKA, 1986

See under *Clethrionomys*.

SARCOCYSTIS MICROTI DUBEY, 1983

This species occurs in the skeletal muscles of the meadow vole *Microtus pennsylvanicus* (type intermediate host) and probably longtail vole *M. longicaudus* in North America. Its definitive host is unknown.

The sarcocysts are 319 × 100 μm, compartmented, with a wall 2.8 μm thick bearing protrusions with a central core. The bradyzoites in the sarcocysts are elongate, with micronemes restricted to the anterior end.

The cat cannot be infected.

Reference. Dubey (1983).

SARCOCYSTIS MONTANAENSIS DUBEY, 1983

This species occurs in the skeletal muscles of the meadow vole *Microtus pennsylvanicus* (type intermediate host) and probably longtail vole *M. longicaudus* in North America. Its definitive host is unknown.

The sarcocysts in the vole are 459 × 199 μm, compartmented, with a smooth wall 1.2 μm thick, without protrusions. Their bradyzoites are elongate, 11 × 3 μm, with micronemes extending to the posterior end.

The cat cannot be infected.

Reference. Dubey (1983).

SARCOCYSTIS PITYMYSI SPLENDORE, 1918

This species was reported from the muscles of the vole *Microtus* (*Pitymys*) *savii* in Europe (Italy). The definitive host is unknown. Its bradyzoites are 5—9 × 2—5 μm (Kalyakin & Zasukhin, 1975).

References. Kalyakin & Zasukhin (1975); Splendore (1918).

SARCOCYSTIS PUTORII (RAILLIET & LUCET, 1891) TADROS & LAARMAN, 1978

Synonyms. *Coccidium bigeminum* var. *putorii* Railliet & Lucet, 1891; *Isospora putorii* (Railliet & Lucet, 1891) Becker, 1934; *Endorimospora putorii* (Railliet & Lucet, 1891) Tadros & Laarman, 1976.

This species occurs in the liver, other visceral organs, and muscles of common European vole *Microtus arvalis* (type intermediate host) and short-tailed vole *M. agrestis* and the ferret *Mustela putorius* var. *furo* (type definitive host), common European weasel *Mustela nivalis*, stoat *M. erminia*, and probably mink *M. vison* in Europe.

The oocysts are unknown. The sporocysts are 10—13 × 7—10 μm, with a residuum. Sporulation occurs in the *Mustela* intestine.

The sarcocysts in the muscles of *M. arvalis* 3 months after inoculation are clearly visible to the naked eye as long, white threads up to several millimeters long and 150 μm in diameter. They appear smooth-walled, but fresh, micro-isolated sarcocysts have many short, blunt projections giving the wall a rough, bristly appearance. The sarcocysts are compartmented, with a wall about 3.5 μm thick. Their bradyzoites are 13—17 × 2—4 μm.

References. Becker (1934); Černá & Ally (1979); Railliet & Lucet (1891); Tadros & Laarman (1976, 1978).

SARCOCYSTIS MURIS? OF SEBEK, 1975

Sarcocysts of this species occur rarely in the muscles of the common European vole *Microtus arvalis* (type intermediate host) and short-tailed vole *M. agrestis* in Europe. Its definitive host is unknown.

This form cannot be transmitted from *M. agrestis* to clean voles by feeding bradyzoites or by feeding feces of voles or cats fed infected voles, nor can it be transmitted to *Mus musculus* or *Mesocricetus auratus* by parenteral inoculation of bradyzoites.

References. Šebek (1975); Tadros (1974).

SARCOCYSTIS SP. ŠEBEK, 1960

Sarcocysts of this species occur in the muscles of the short-tailed vole *Microtus agrestis* (Type intermediate host) in Europe. Its definitive host is unknown. The sarcocysts in the muscles are 0.5—1.5 × 0.1—0.3 mm and contain bradyzoites 7—15 × 2—6 μm.

References. Sebek (1960); Tadros (1976).

SARCOCYSTIS SP. ŠEBEK, 1960

Sarcocysts of this form occur in the muscles of the common European vole *Microtus arvalis* (type intermediate host) in Europe and the USSR. Its definitive host is unknown. The bradyzoites in its sarcocysts are 7—15 × 2—6 μm.

References. Krasnova (1971); Šebek (1960, 1963).

SARCOCYSTIS (?) SP. (ŠEBEK, 1975)

Sarcocysts of this species occur rarely in the muscles of *Microtus arvalis* and moderately commonly in the muscles of *M. agrestis* in Europe. It was said to be intermediate between *Sarcocystis* and *Frenkelia*.

Reference. Šebek (1975).

BESNOITIA JELLISONI FRENKEL, 1955

The vole *Microtus* sp. is an experimental intermediate host of this species.

References. See under *Peromyscus*.

BESNOITIA WALLACEI (TADROS & LAARMAN, 1976) DUBEY, 1977

The vole *Microtus montebelli* is an experimental intermediate host of this species.

References. See under *Rattus*.

FRENKELIA MICROTI (FINDLAY & MIDDLETON, 1934) BIOCCA, 1968

Synonyms. *Toxoplasma microti* Findlay & Middleton, 1934; M-organism of Sebek (1975).

This species occurs rarely in the liver, other organs, and later in the brain and other organs of the European field vole *Microtus agrestis* (type intermediate host), vole *M. arvalis*, American vole *M. modestus*, and experimentally field mouse *Apodemus sylvaticus*, field mouse *A. flavicollis*, vole *A. agrarius*, golden hamster *Mesocricetus auratus*, Norway rat *Rattus norvegicus*, laboratory mouse *Mus musculus*, multimammate rat *Praomys* (syn., *Mastomys*) *natalensis*, common hamster *Cricetus cricetus*, chinchilla *Chinchilla laniger,* and domestic rabbit *Oryctolagus cuniculus*. Its definitive host is the buzzard *Buteo buteo* (type definitive host), in which its sporocysts are found in the feces. It has been found in Europe and North America.

The sporocysts are 12—15×9—12 (mean 12×10) μm, without a Stieda body, with a residuum. Sporulation occurs in the buzzard intestine.

The brain meronts are up to 1 mm in diameter, lobulated, and thin-walled. They occur in many other locations as well. They contain both metrocytes and bradyzoites. The latter are crescent- or spindle-shaped, 4—10×1—3 μm. Young and mature meronts may be found in the liver parenchymal cells of *M. agrestis* on days 6 and 7, respectively, after inoculation. The prepatent period is 7—8 d and the patent period 5—7 weeks.

The owls *Asio otus, Tyto alba,* and *Strix aluco* and falcon *Falco tinnunculus* can be infected from *M. agrestis M. agrestis, M. arvalis, Apodemus sylvaticus, A. flavicollis, A. agrarius, Mesocricetus auratus, Rattus norvegicus,* at least one but not all strains of *Mus musculus, Praomys natalensis,* one strain of *Cricetus cricetus, Chinchilla laniger,* and *Oryctolagus cuniculus* have been infected experimentally with sporocysts from the buzzard, but *Clethrionomys glareolus,* another strain of *Cricetus cricetus, Cricetulus griseus, Meriones unguiculatus, Cavia porcellus, Erinaceus europaeus,* and *Ovis aries* could not be. Buzzards cannot be infected by oral inoculation of sporocysts from other buzzards.

References. Biocca (1968); Findlay & Middleton (1934); Geisel (1979); Krampitz & Rommel (1977); Rommel & Krampitz (1978); Šebek (1975); Tadros (1981).

HOST GENUS *LAGURUS*

ISOSPORA LAGURI IWANOFF-GOBZEM, 1934

Synonym. *Isospora laguris* Iwanoff-Gobzem, 1934 *lapsus calami.*

This species occurs moderately commonly in the steppe lemming *Lagurus lagurus* in the USSR.

The oocysts are ovoid, 24—32 × 16—22 μm, with a thick wall, without a micropyle or polar granule, with a residuum. The sporocysts are 16—21 × 8—13 μm, without a residuum. The sporulation time is 3 d.

Reference. Iwanoff-Gobzem (1934) .

ISOSPORA TERES IWANOFF-GOBZEM, 1934

Synonym. *Isospora feres* Iwanoff-Gobzen, 1934 of Musaev & Veisov (1965) *lapsus calami.*

This species occurs moderately commonly in the steppe lemming *Lagurus lagurus* in the USSR.

The oocysts are spherical, 24—36 μm in diameter, without a micropyle or residuum, with a polar granule. The sporocysts are 16—21 × 8—13 μm, without a residuum.

References. Iwanoff-Gobzem (1934); Musaev & Veisov (1965).

HOST GENUS *ONDATRA*

EIMERIA ONDATRAZIBETHICAE MARTIN, 1930 EMEND. LEVINE & IVENS, 1965

Synonyms. *Eimeria ondatrae-zibethicae* Martin, 1930; *E. stiedae* (Lindemann) Kisskalt & Hartmann of Law & Kennedy (1932).

This species occurs commonly in the jejunum and liver of the muskrat *Ondatra zibethica* in North America and the USSR.

The oocysts are mostly ellipsoidal, but also spherical, ovoid, or cylindrical, 19—35 × 13—27 μm, with a brownish, greenish, or yellow-green, very thick, smooth, double-contoured wall, occasionally radially striated, sometimes with one end flattened, with or without a micropyle and residuum. The sporocysts are spherical, ellipsoidal, or ovoid, 12—17 × 8—11 μm, with a Stieda body and residuum. The sporozoites are comma-shaped or piriform, 8—11 × 3—5 (mean 9.5 × 4) μm. The sporulation time is 5 d at 21—22°C in 5% potassium bichromate solution.

The gamonts are apparently in the epithelial cells of the jejunum. The macrogametes are about 17—30 × 15 μm and the microgamonts are about 11—23 × 10—22 μm.

This species can be highly pathogenic, and may kill young muskrats or adults subjected to drought conditions. There is intense hemorrhagic enteritis of the jejunum, the intestinal epithelium is desquamated and may be almost denuded, and the intestinal lumen may be filled with blood and debris. The liver may contain many small white foci up to 2 mm in diameter due to focal necrosis and vascular congestion.

References. Brumpt (1942); Frank (1978); Law & Kennedy (1932); Le-Compte (1933); Levine & Ivens (1965); Martin (1930); Shillinger (1938); Svanbaev (1962).

SARCOCYSTIS SP. (RYAN, WYAND & NIELSEN, 1982) LEVINE & IVENS, 1987

Synonym. *Hammondia* sp. Ryan, Wyand & Nielsen, 1982.

This form occurs commonly in the skeletal muscles of the muskrat *Ondatra zibethica* (type intermediate host) in North America. Its experimental definitive host is the mink *Mustela vison*.

The oocysts in the mink feces are subspherical, 11.5—12 × 10—11 (mean 12 × 11) μm, with a two-layered wall and a micropyle, without a residuum or polar granule. The sporocysts are 8—9 × 6—7 (mean 9 × 6.5) μm, without a Stieda body, with a residuum. The sporulation time is 4 d at room temperature. The prepatent period in the mink is 6—8 d and the patent period 4—6 d.

The cat and dog could not be infected by feeding them infected muskrats, or the rat, mouse, or mink by feeding them sporulated oocysts.

The sarcocysts are septate and contain both metrocytes and merozoites.

References. Ryan, Wyand, & Nielsen (1982);Levine & Ivens (1987).

FRENKELIA SP. (KARSTAD, 1963) LEVINE & IVENS, 1987

Synonym. *Toxoplasma microti* in *Ondatra zibethica* of Karstad (1963).

This form occurs in the brain of the muskrat *Ondatra zibethica* (type intermediate host) in North America. Its definitive host is unknown.

The brain meronts are lobulated, apparently compartmented, with crescentic merozoites having a dense basophilic nucleus at one end.

This form could not be transmitted to laboratory mice or a juvenile muskrat by intraperitoneal inoculation of muskrat brain suspensions.

References. Karstad (1963); Levine & Ivens (1987).

HOST GENUS *LEMMUS*

EIMERIA CHATANGAE ARNASTAUSKENE, 1980

This species occurs in the lemming *Lemmus sibiricus* in the USSR.

The oocysts are spherical or slightly ovoid, 13—14 × 11—13 μm, with a smooth, colorless, very delicate, one-layered wall, without a micropyle, residuum, or polar granule. The sporocysts are spindle-shaped, 8—9 × 5—6 μm, without a Stieda body or sub-Stiedal body, with a residuum. The sporozoites were described as oval but illustrated as ellipsoidal, and lie more or less at the ends of the sporocysts. Sporulation occurs outside the host.

Reference. Arnastauskene (1980).

EIMERIA NATIVA ARNASTAUSKENE, 1980

Synonym. *Eimeria* sp. Arnastauskene, 1977.

This species occurs in the lemming *Lemmus sibiricus* in the USSR.

The oocysts are ellipsoidal, verging on the cylindrical, 19—22 × 10—12 μm, with a smooth, one-layered wall, without a micropyle, residuum, or polar granule. The sporocysts are ellipsoidal, 11—12 × 6—7 μm, with a Stieda body and residuum, without a sub-Stiedal body. The sporozoites are elongate, with one end smaller than the other, and lie lengthwise head to tail in the sporocysts. Sporulation occurs outside the host.

References. Arnastauskene (1977, 1980).

SARCOCYSTIS SP. KALYAKIN, 1975 IN KALYAKIN & ZASUKHIN, 1975

Sarcocysts of this form occur in the muscles of the lemming *Lemmus sibiricus* (type intermediate host) in the USSR. The definitive host is unknown.

The sarcocysts have not been described.

Reference. Kalyakin (1975) in Kalyakin & Zasukhin (1975).

FRENKELIA SP. (ENEMAR, 1965) LEVINE & IVENS, 1987

Synonym. M-organism in *Lemmus lemmus* of Enemar (1965).

This form occurs in the brain of the lemming *Lemmus lemmus* (type intermediate host) in Europe. Its definitive host is unknown.

The brain meronts are lobulated, compartmented, and occupy a space with a diameter of 0.5—0.6 mm. The bradyzoites are elongate, fat banana-shaped, with one end rounded and the other pointed, with a nucleus at the pointed end and a globule staining with chrome alum hematoxylin at the broad end; they are 5—7 μm long. Metrocytes are also present. There is no cellular reaction in the brain.

References. Enemar (1965); Levine & Ivans (1987).

HOST GENUS *DICROSTONYX*

EIMERIA DICROSTONICIS LEVINE, 1952

This species occurs commonly in the varying lemming *Dicrostonyx groenlandicus richardsoni* (type host) and lemmings *D. torquatus* and *Lemmus sibiricus* in North America and the USSR.

The oocysts are ellipsoidal, 27—31 × 23—27 (mean 29 × 25) μm, with a two-layered wall, the outer layer rough, pitted, yellowish brown, a little more than 1 μm thick, the inner layer colorless, about 0.5 μm thick, without a micropyle, ordinarily without a residuum, with one or two polar granules. The sporocysts are ellipsoidal but slightly pointed at both ends, 13—15 × 7—9 (mean 14 × 8) μm, without or with a very small Stieda body, ordinarily without a residuum. The sporozoites lie lengthwise head to tail in the sporocysts.

References. Arnastauskene (1980); Levine (1952).

SARCOCYSTIS RAUSCHORUM CAWTHORN, GAJADHAR, & BROOKS, 1984

This species occurs in the snowy owl *Nyctea scandiaca* (definitive host) and varying lemming *Dicrostonyx richardsoni* (intermediate host) in North America.

It cannot be transmitted to the laboratory rat *Rattus norvegicus*, house mouse *Mus musculus*, white-footed mouse *Peromyscus leucopus*, red-backed vole *Clethrionomys gapperi*, or brown lemming *Lemmus sibiricus*.

Cawthorn & Brooks (1985) found that in *D. richardsoni* infected with oocysts from *N. scandiaca*, merogonous stages developed beginning 4 d after inoculation (DAI) in the cytoplasm of the hepatocytes; they were all mature by 6 DAI. They found single merozoites in the cytoplasm of polymorphonuclear neutrophils and monocytes in liver impression smears at 5 d; if more than 500 sporocysts had been given, all the lemmings died 5.5—6 DAI, probably because of loss of liver function. Sarcocysts were formed beginning 9 DAI, in the skeletal muscles, and contained bradyzoites as early as 28 DAI. They also described sporogony in *N. scandiaca*. Sporulation occurs in the lamina propria of the small intestine. Oocysts containing two sporocysts, each with four sporozoites, are formed.

References. Cawthorn, Gajadhar, & Brooks, 1984; Cawthorn & Brooks, 1985.

HOST GENUS *ELLOBIUS*

EIMERIA ELLOBII SVANBAEV, 1956

This species occurs moderately commonly to commonly in the mole lemming *Ellobius talpinus* in the USSR.

The oocysts are ovoid, greenish, 30 × 24 μm, with a smooth, two-layered wall 1.3—1.7 μm thick, without a micropyle or residuum, with a polar granule. The sporocysts are spherical, 9 μm in diameter, presumably without a Stieda body, with a residuum. The sporozoites are kidney-shaped, 6.5 × 6 μm.

Reference. Svanbaev (1956).

EIMERIA KAZAKHSTANENSIS LEVINE & IVENS, 1965

Synonym. *Eimeria volgensis* Sassuchin & Rauschenbach of Svanbaev (1956).

This species occurs commonly in the mole lemming *Ellobius talpinus* in the USSR.

The oocysts are ovoid, greenish to occasionally yellowish brown, 27.5 × 24 μm, with a smooth, colorless, one-layered wall 1.6 μm thick, with a micropyle and polar granule, without a residuum. The sporocysts are ovoid or piriform, 11—13 × 7—8 (mean 12 × 7) μm, with a residuum. The sporozoites are ovoid, 8 × 6 μm.

References. Levine & Ivens (1965); Svanbaev (1956).

EIMERIA LUTESCENAE MUSAEV & VEISOV, 1963

This species occurs in the mountain lemming *Ellobius lutescens* in the USSR.

The oocysts are ellipsoidal or ovoid, 22—27 × 18—23 (mean 25 × 21) μm, with a smooth, colorless, one-layered wall 2 μm thick, without a micropyle, residuum, or polar granule. The sporocysts are ellipsoidal or ovoid, 8—10 × 6—8 (mean 9 × 6) μm, apparently without a Stieda body, with a residuum. The sporozoites are bean-shaped. The sporulation time is 2 d at 25—30°C in 2.5% potassium bichromate solution.

Reference. Musaev & Veisov (1963).

EIMERIA TADSHIKISTANICA VEISOV, 1964

This species occurs uncommonly in the mole lemming *Ellobius talpinus* in the USSR.

The oocysts are spherical, colorless, 16—26 (mean 22) μm in diameter, with a smooth, two-layered wall, the outer layer colorless, 1—1.25 μm thick, the inner layer dark brown, 1—1.25 μm thick, without a micropyle or residuum, with a polar granule. The sporocysts are ellipsoidal or piriform, 6—10 × 4—6 (mean 9 × 6) μm, with a Stieda body and residuum. The sporozoites are comma-shaped, with a clear globule at the broad end.

Reference. Veisov (1964).

EIMERIA TALPINI LEVINE & IVENS, 1965

Synonym. *Eimeria beckeri* Yakimoff & Sokoloff of Svanbaev (1956) in *Ellobius talpinus*.

This species occurs commonly in the mole lemming *Ellobius talpinus* in the USSR.

The oocysts are spherical, 25 μm in diameter, with a smooth, colorless, two-layered wall 1.5 μm thick, without a micropyle, residuum, or polar granule. The sporocysts are ovoid, 11 × 9 μm, with a residuum. The sporozoites are 6.5 × 4 μm.

References. Levine & Ivens (1965); Svanbaev (1956).

HOST GENUS *CRICETOMYS*

EIMERIA CRICETOMYSI PRASAD, 1960

This species occurs in the giant field rat *Cricetomys gambianus* in Africa and Europe (London Zoo).

The oocysts are ovoid or ellipsoidal, 18—21 × 15—17 (mean 19 × 16) μm, with a smooth, yellow, two-layered wall, the outer layer thicker than the inner one, without a micropyle or residuum, with a polar granule. The sporocysts are ovoid, 11—13 × 6—8 μm, with a Stieda body and residuum. The sporozoites have clear globules.

Reference. Prasad (1960).

EIMERIA MORELI VASSILIADES, 1966

This species occurs in the giant field rat *Cricetomys gambianus* in Africa.

The oocysts are ellipsoidal (illustrated as slightly ovoid), 27—34 × 18—24 (mean 29 × 22) μm, with a two-layered wall 1.5—1.9 (mean 1.7) μm thick, the outer layer irregularly striated, slightly opaque, brownish, twice as thick as the inner layer, the inner layer transparent, colorless, or pale yellow, without a micropyle or residuum, with a polar granule. The sporocysts are ovoid, 15—20 × 7—8 (mean 16 × 8) μm, without an apparent Stieda body, with a residuum. The sporozoites are elongate, with a clear globule at the large end, and lie lengthwise head to tail in the sporocysts. The sporulation time is 3—6 d at 30°C in 2% potassium bichromate solution.

Reference. Vassiliades (1966).

EIMERIA SCHOUTEDENI VAN DEN BERGHE & CHARDOME, 1957

This species occurs in the hamster rat *Cricetomys dissimilis* in Africa.

The oocysts are ovoid, 14—15 × 11—13 μm, with a presumably smooth, whitish, possibly one-layered wall 0.5 μm thick, without a micropyle or polar granule, with a residuum. The sporocysts are 7 × 6 μm, illustrated without a Stieda body, with a residuum. The sporozoites are reniform, 6 × 2 μm, with coarse granulations about 0.7 μm in diameter on their surface. The sporulation time is 5—10 d in 1% chromic acid solution.

Reference. Van den Berghe & Chardome (1957).

HOST GENUS *APODEMUS*

EIMERIA AGRARII MUSAEV & VEISOV, 1965

This species occurs in the field mouse *Apodemus agrarius* in the USSR.

The oocysts are spherical or subspherical, 20—28 (mean 25) μm in diameter, with a smooth, two-layered wall 3 μm thick, the outer layer colorless, 1.5 μm thick, the inner layer dark brown, 1.5 μm thick, without a micropyle, residuum, or polar granule. The sporocysts are ellipsoidal or ovoid, 8—12 × 6—8 (mean 10 × 7) μm, illustrated without a Stieda body, with a residuum. The sporozoites are bean-shaped, with a clear globule in the broad end. The sporulation time is 2—3 d.

Reference. Musaev & Veisov 1965.

EIMERIA APIONODES PELLÉRDY, 1954

This species occurs in the epithelial cells of the small intestine mucosa of the field mouse *Apodemus flavicollis* in Europe.

The oocysts are short piriform, somewhat tapered at both ends, 17—23 × 13—18 (mean 20 × 17) μm, with a relatively delicate, pale, smooth wall, without a micropyle or polar granule, without or with a residuum. The sporocysts are 12 × 8 μm, apparently without a Stieda body, with a residuum. The sporulation time is 2—4 d at room temperature in 2% potassium bichromate solution. The prepatent period is 8—10 d.

This species cannot be transmitted to *Mus musculus, Microtus arvalis, Clethrionomys glareolus,* or *Cricetus cricetus.*

Reference. Pellérdy (1954).

EIMERIA APODEMI PELLÉRDY, 1954

This species occurs uncommonly to commonly in the small intestine of the field mice *Apodemus flavicollis* (type host), *A. sylvaticus,* and *A. agrarius* in Europe.

The oocysts are broadly ellipsoidal, often asymmetrical, 21—30 × 15—26 μm, with a two-layered wall, the outer layer brown, often detached from the lighter inner layer during concentration in glycerol, without a micropyle, residuum, or polar granule. The sporocysts are 12 × 7 μm, apparently without a Stieda body, with a finely granular residuum. The sporulation time is 3—5 d at room temperature in 2% potassium bichromate solution. The prepatent period is 6—7 d.

This species cannot be transmitted to *Mus musculus, Microtus arvalis, Clethrionomys glareolus,* or *Cricetus cricetus.*

Remarks. Lewis & Ball (1983), whose description of the oocysts differed somewhat from the above, considered *E. krijsmanni* sensu Svanbaev, 1956, *E. prasadi* Veisov, 1953, and *E. svanbaevi* Levine & Ivens, 1965 to be synonyms of this species. They accepted only *A. sylvaticus* as the host of this species and called the similar form in *A. flavicollis, Eimeria* Type B.

References. Černá & Daniel (1956); Golemanski (1979); Lewis & Ball (1983); Pellérdy (1954); Ball & Lewis (1984).

EIMERIA ARKUTINAE GOLEMANSKY, 1967

Synonym. *Eimeria keilini* Yakimoff & Gousseff of Ryšavý (1954), Černá & Daniel (1956) and Černá (1962).

This species occurs uncommonly to moderately commonly in the Old World field mice *Apodemus sylvaticus* (type host), *A. flavicollis,* and *A. agrarius* in Europe.

The oocysts are ellipsoidal, often asymmetrical, and narrowed at the poles, colorless or pale yellow, 22—32 × 15—21 μm, with a two-layered wall about 1.8 μm thick, without a micropyle or residuum, sometimes with one or two polar granules. The sporocysts are ovoid, 12 × 6.5 μm, without a Stieda body, with a residuum. The sporozoites are elongate, with one end rounded and the other pointed, have a clear globule at the broad end, and lie lengthwise head to tail in the sporocysts.

Remarks. Lewis & Ball (1983) considered this name to be a synonym of *E. divichinica.*

References. Černá (1962); Černá & Daniel (1956); Golemansky (1978, 1979); Lewis & Ball (1983); Ryšavý (1954).

EIMERIA BADAMLINICA MUSAEV & VEISOV, 1963

This species occurs in the field mouse *Apodemus sylvaticus* in the USSR.

The oocysts are spherical, colorless, 12—20 (mean 16) μm in diameter, with a smooth, one-layered wall 1 μm thick, without a micropyle or residuum, with a polar granule. The sporocysts are ellipsoidal or spherical, 4—8 × 4—6 μm. The sporozoites are comma-shaped, with a clear globule at the broad end. The sporulation time is 2—3 d at 25—30°C in 2.5% potassium bichromate solution.

Reference. Musaev & Veisov (1963).

EIMERIA DIVICHINICA MUSAEV & VEISOV, 1963

This species occurs in the field mice *Apodemus sylvaticus* and *A. flavicollis* in the USSR and Europe.

The oocysts are ovoid, rarely ellipsoidal, 16—32 × 10—26 (mean 24 × 19) μm, with a smooth, colorless, one-layered wall 1.2 μm thick, without a micropyle, residuum, or polar granule. The sporocysts are ellipsoidal, ovoid, or rarely spherical, 7—13 × 5—11 μm, with a residuum. The sporozoites are comma-shaped or piriform, with a clear globule at the broad end. The sporulation time is 2 d at 25—30°C in 2.5% potassium bichromate solution.

Remarks. Lewis & Ball (1983), whose description of the oocysts differed from the above, considered *E. arkutinae* and *E. keilini* sensu Ryšavý to be synonyms of this species. They saw a similar form in *A. flavicollis*, but called it *Eimeria* Type A.

References. Lewis & Ball (1983); Musaev & Veisov (1963); Ryšavý (1954); Ball & Lewis (1984); Glebezdin (1978).

EIMERIA GANDOBICA MUSAEV & VEISOV, 1965

This species occurs in the field mouse *Apodemus agrarius* in the USSR.

The oocysts are ellipsoidal or subspherical, colorless, 14—22 × 12—18 (mean 19 × 17) μm, with a smooth, one-layered wall 1.5—2 μm thick, without a micropyle or residuum, almost always with a polar granule. The sporocysts are ellipsoidal or ovoid, 6—10 × 4—6 (mean 8 × 5) μm, with a Stieda body and residuum. The sporozoites are bean-shaped, illustrated without a clear globule. The sporulation time is 2 d.

Reference. Musaev & Veisov (1965).

EIMERIA GOLEMANSKII LEVINE, 1985

Synonym. *Eimeria* sp. Golemanski & Yankova, 1973.

This species occurs moderately commonly in the field mice *Apodemus sylvaticus* (type host) and *A. flavicollis* in Europe.

The oocysts are ellipsoidal, sometimes ovoid, 21—28 × 15—19 (mean 25 × 17) μm, with a smooth, one-layered wall 1.8—2 μm thick, without a micropyle or residuum, with a polar granule. The sporocysts are ovoid, pointed at one end, 12—13 × 6—8 μm, with a residuum. The sporozoites are elongate, with one end rounded and the other tapered, have a clear globule at the broad end, and lie lengthwise head to tail in the sporocysts.

Remarks. The oocysts of this species differ from those of all other coccidia described from *Apodemus*. Whether is is the same as the form mentioned below as *Eimeria* sp. Ryšavý, 1954 (syn., *E. keilini* of Ryšavý [1954] and Černá & Daniel [1962]), as Golemanski & Yankova (1973) believed, is uncertain.

References. Černá & Daniel (1962); Golemanski & Yankova (1973); Levine (1985); Ryšavý (1957).

EIMERIA GOMURICA MUSAEV & VEISOV, 1963

This species occurs in the field mouse *Apodemus sylvaticus* in the USSR.

The oocysts are ovoid, colorless, 24—29 × 18—23 (mean 27 × 20.5) μm, with a smooth, one-layered wall 2 μm thick, without a micropyle or polar granule, with a residuum. The sporocysts are ovoid, 8—13 × 6—11 (mean 12 × 8) μm, apparently without a Stieda body, with a residuum. The sporozoites are bean-shaped, apparently without clear globules. The sporulation time is 3 d in 2.5% potassium bichromate solution.

Reference. Musaev & Veisov (1963).

EIMERIA GUMBASCHICA MUSAEV & VEISOV, 1963

This species occurs in the field mouse *Apodemus sylvaticus* in the USSR.

The oocysts are spherical, colorless, 14—18 (mean 17) μm in diameter, with a smooth, two-layered wall, the outer layer colorless, 0.9 μm thick, the inner layer dark yellow, 0.9 μm thick, without a micropyle, residuum, or polar granule. The sporocysts are ellipsoidal or spherical, 4 × 6 or 4—6 μm, illustrated without a Stieda body, with a residuum. The sporozoites are bean-shaped, without a clear globule. The sporulation time is 3 d at 25—30°C in 2.5% potassium bichromate solution.

Reference. Musaev & Veisov (1963).

EIMERIA HUNGARYENSIS LEVINE & IVENS, 1965

Synonym. *Eimeria muris* Galli-Valerio of Pellérdy (1954) and Černá (1962).

This species occurs moderately commonly in the small intestine mucosa of the field mice *Apodemus flavicollis* (type host), *A. sylvaticus,* and *A. agrarius* in Europe.

The oocysts in *A. flavicollis* are spherical or subspherical, 17—23 × 16—19 (mean 20 × 18) μm, with a somewhat rough, light yellowish brown, quite thick wall, without a micropyle, residuum, or polar granule. The sporocysts are pointed at one end and rounded at the other, 15 × 9 μm, apparently with a Stieda body, with a residuum. The form in *A. sylvaticus* has broadly ellipsoidal, ovoid, or spherical oocysts, 14—24 × 14—16 μm, illustrated with a one-layered wall, without a micropyle, residuum, or polar granule. The sporocysts are elongate ovoid, 9—13 × 4—8 μm, with a Stieda body at the pointed end, with a residuum. The sporulation time is 2—5 d at room temperature in 2% potassium bichromate solution. The prepatent period is 3—5 d and the patent period 8 d or more.

The mature macrogametes are 12—21 × 11—14 μm and the mature microgamonts 13—17 × 8—14 μm.

This species cannot be transmitted to *Mus musculus, Microtus arvalis, Clethrionomys glareolus,* or *Cricetus cricetus.*

Remarks. Lewis & Ball (1983) accepted only *A. sylvaticus* as the host of this species. They found a similar form in *A. flavicollis,* but called it *Eimeria* Type D.

References. Černá (1962); Golemanski (1979); Levine & Ivens (1965); Lewis & Ball (1983); Pellérdy (1954); Ball & Lewis (1984); Higgs & Nowell (1983).

EIMERIA JERFINICA MUSAEV & VEISOV, 1963

This species occurs in the field mouse *Apodemus sylvaticus* in the USSR.

The oocysts are ellipsoidal or subspherical, 20—28 × 16—24 (mean 26 × 22) μm, with a rough, yellow-brown, one-layered wall 2 μm thick, without a micropyle, residuum, or polar granule. The sporocysts are ellipsoidal, rarely spherical, 8—14 × 6—11 μm, apparently without a Stieda body, with a residuum. The sporozoites are comma-shaped, rarely piriform, with a clear globule at the broad end. The sporulation time is 3—4 d at 25—30°C in 2.5% potassium bichromate solution.

Remarks. Lewis & Ball (1983) considered this to be a synonym of *E. apodemi*.

References. Lewis & Ball (1983); Musaev & Veisov (1983).

EIMERIA KAUNENSIS ARNASTAUSKENE, KAZLAUSKAS, & MAL'DZHYUNAITE, 1978

This species occurs in the field mouse *Apodemus agrarius* in the USSR.

The oocysts are ovoid to ellipsoidal, 21.5—24.5 × 15.5—17.5 μm, with a smooth wall illustrated as one-layered, without a micropyle or residuum, with polar granules. The sporocysts are ellipsoidal, without a Stieda body, with a residuum. The sporozoites are elongate, with one end smaller than the other, have a clear globule at the broad end, and lie lengthwise head to tail in the sporocysts.

Reference. Arnastauskene, Kazlauskas, & Mal'dzhyunaite (1978).

EIMERIA MONTGOMERYAE LEWIS & BALL, 1983

This species occurs quite commonly in the field mouse *Apodemus sylvaticus* in Europe.

The oocysts are ovoid, 18—24 × 16—23 (mean 22 × 19) μm, with a light-colored, rough, two-layered wall 1.1—1.3 μm thick, without a micropyle or residuum, with one to three polar granules. The sporocysts are 12—13 × 6.5—7.5 (mean 12 × 7) μm, with a Stieda body and residuum. The prepatent period is 3.5 d.

References. Lewis & Ball (1983); Ball & Lewis (1984).

EIMERIA MURIS GALLI-VALERIO, 1932

This species occurs in the intestine of the field mouse *Apodemus sylvaticus* in Europe and the USSR.

The oocysts are ovoid or ellipsoidal, 18—29 × 14—23 μm, with a smooth, colorless, one-layered wall 1.5 μm thick, with a micropyle, without a residuum or polar granule. The sporocysts are ellipsoidal or ovoid, 6—13 × 4—9 μm, apparently without a Stieda body, with or without a residuum. The sporozoites are comma-shaped, with a clear globule at the broad end. The sporulation time is 3 d at 25—30°C in 2.5% potassium bichromate solution.

References. Galli-Valerio (1932); Musaev & Veisov (1963).

EIMERIA NAYE GALLI-VALERIO, 1940

This species occurs commonly in the intestine of the field mouse *Apodemus sylvaticus* in Europe.

The oocysts are cylindroid, with one end convex and the other flattened, 18—21 × 12—14 μm, with a micropyle at the flattened end. The sporocysts are spherical, 6 μm in diameter, with a residuum. The sporozoites are piriform. The sporulation time is 4 d on moist filter paper.

Remarks. Lewis & Ball (1983) considered this a *nomen dubium*.
References. Galli-Valerio (1940); Lewis & Ball (1983).

EIMERIA NEREENSIS GLEBEZDIN, 1973

This species occurs in the field mouse *Apodemus sylvaticus* in the USSR.

The oocysts are ovoid to subspherical, 26—33 × 23—29 (mean 29.5 × 27) µm, with a two-layered wall 2.6 µm thick, the outer layer rough, striated, the inner layer dark, without a micropyle, residuum, or polar granule. The sporocysts are ovoid, 14.5—17 × 11.5—14.5 (mean 16 × 13) µm, with a Stieda body and residuum. The sporozoites are elongate, with one end broad and the other narrow, and lie lengthwise head to tail in the sporocysts.

Reference. Glebezdin (1973).

EIMERIA PRASADI LEVINE & IVENS, 1965

Synonym. *Eimeria hindlei* Yakimoff & Gousseff of Svanbaev (1956).

This species occurs uncommonly to moderately commonly in the field mice *Apodemus sylvaticus* (type host) and *A. flavicollis* in Europe and the USSR.

The oocysts are ovoid, 26 × 20 µm, with a smooth, double-contoured wall 1.6 µm thick, without a micropyle or residuum, with or occasionally without a polar granule. The sporocysts are ovoid, 9 × 7 µm, with a residuum.

Remarks. Lewis & Ball (1983) considered this to be a synonym of *E. apodemi*.

References. Golemanski (1970); Levine & Ivens (1965); Lewis & Ball (1983); Svanbaev (1956).

EIMERIA RUGOSA PELLÉRDY, 1954

This species occurs in the intestine of the field mouse *Apodemus flavicollis* in Europe.

The oocysts are piriform, 23—27 × 15—19 (mean 24 × 16) µm, with a yellowish brown wall somewhat thicker than usual, the micropylar end appearing wrinkled under high magnification, with a micropyle and residuum, without a polar granule. The sporocysts are slender, 16 × 10 µm, with a residuum. The sporulation time is 2—4 d at room temperature in 2% potassium bichromate solution.

This species cannot be transmitted to *Mus musculus*, *Microtus arvalis*, *Clethrionomys glareolus* or *Cricetus cricetus*.

Reference. Pellérdy (1954).

EIMERIA RUSSIENSIS LEVINE & IVENS, 1965

Synonym. *Eimeria musculi* Yakimoff & Gousseff of Svanbaev (1956) from *Apodemus sylvaticus*.

This species occurs in the field mouse *Apodemus sylvaticus* in the USSR.

The oocysts are spherical, 22 µm in diameter, with a smooth, greenish, double-contoured wall 1.5—2 µm thick, without a micropyle or residuum, with or occasionally without a polar granule. The sporocysts are ovoid or spherical, 9 μm in diameter, without a residuum. The sporozoites are comma-shaped, 6 × 4 μm.

References. Levine & Ivens (1965); Svanbaev (1956).

EIMERIA SVANBAEVI LEVINE & IVENS, 1965

Synonyms. *Eimeria kriygsmanni* [sic] Yakimoff & Gousseff of Svanbaev (1956) in *Apodemus sylvaticus*; ? *Eimeria krijgsmani* [sic] of Ryšavý (1954).

This species occurs in the field mouse *Apodemus sylvaticus* in the USSR and Europe.

The oocysts are ovoid, colorless or occasionally greenish, 24—26 × 20—22 (mean 25 × 20.5) μm, with a smooth, double-contoured wall 1.6—1.8 μm thick, without a micropyle or residuum, with a polar granule. The sporocysts are ovoid, 9—13 × 7—9 (mean 10 × 8) μm, without a residuum. The sporozoites are comma-shaped.

Remarks. Lewis & Ball (1983) considered this name to be a synonym of *E. apodemi*.

References. Levine & Ivens (1965); Lewis & Ball (1983); Ryšavý (1954); Svanbaev (1956).

EIMERIA SYLVATICA PRASAD, 1960

This species occurs in the ileum epithelium of the field mouse *Apodemus sylvaticus* in Europe and the USSR.

The oocysts are ovoid or spherical, 16—21 × 10—17 μm, with a pale yellow, three-layered wall, the outer layer rough, the middle layer rather thicker than the inner layer, which is very thin, without a micropyle or residuum, with a polar granule. The sporocysts are ovoid, 11—13 × 5—7 μm, probably without a Stieda body, with a residuum. The sporulation time is 3—4 d at 25—30°C in 2.5% potassium bichromate solution. The meronts are 8—11 μm in diameter, and the macrogametes are 7—9 × 6—8 μm.

References. Musaev & Veisov (1980); Prasad (1960).

EIMERIA UPTONI LEWIS & BALL, 1983

This species occurs quite commonly in the field mouse *Apodemus sylvaticus* in Europe.

The oocysts are ovoid, 13—16 × 11—13.5 (mean 14 × 12) μm, with a smooth, colorless, one-layered wall 0.7 μm thick, without a micropyle or residuum, with one to three polar granules. The sporocysts are 8—8.5 × 4.5—6 (mean 8 × 5) μm, with a Stieda body and residuum.

Remarks. Lewis & Ball (1983) found a similar form in *A. flavicollis* but they called it *Eimeria* Type C.

Reference. Lewis & Ball (1983); Ball & Lewis (1984).

EIMERIA ZAURICA MUSAEV & VEISOV, 1963

This species occurs in the field mouse *Apodemus sylvaticus* in the USSR.

The oocysts are ellipsoidal, rarely ovoid, 20—28 × 18—23 (mean 23 × 21) μm, with a two-layered wall, the outer layer smooth, yellowish, 1.2 μm thick, the inner layer brownish, 1.8 μm thick, without a micropyle, with a residuum and polar granule. The sporocysts are ellipsoidal, 8—11 × 6—8 (mean 9 × 7) μm, illustrated without a Stieda body, with a residuum. The sporozoites are bean-shaped, illustrated without a clear globule. The sporulation time is 3—4 d at 25—30°C in 2.5% potasium bichromate solution.

Reference. Musaev & Veisov (1983).

EIMERIA SP. RYSAVY, 1954 EMEND. LEVINE & IVENS, 1965

Synonym. *Eimeria keilini* Yakimoff & Gousseff of Ryšavý (1954) and Černá & Daniel (1962).

This form occurs commonly in the field mice *Apodemus sylvaticus* (type host) and *A. flavicollis* in Europe.

The oocysts are ellipsoidal, narrow at both ends, 24—29 × 16—20 μm, without a micropyle. The sporocysts are apparently 8—9 × 6—7 μm.

Remarks. As stated by Levine & Ivens (1965), this form cannot be *E. keilini*, but its description is too incomplete to permit a name being assigned to it.

References. Černá & Daniel (1962); Levine & Ivens (1965); Ryšavý (1954).

EIMERIA SP. RYŠAVÝ, 1954 EMEND. LEVINE & IVENS, 1965

Synonym. *Eimeria falciformis* of Elton, Ford, Baker, & Gardner (1931), Rysavy (1954), and Černá & Daniel (1962).

This form occurs in the field mice *Apodemus sylvaticus* (type host) and *A. flavicollis* in Europe. Its oocyst structure is uncertain.

Remarks. As stated by Levine & Ivens (1965), this form cannot be *E. falciformis*. However, it has not been described well enough to permit a name to be assigned to it.

References. Černá & Daniel (1982); Elton, Ford, Baker, & Gardner (1931); Levine & Ivens (1965); Ryšavý (1954).

EIMERIA (?) SP. RYŠAVÝ, 1954 EMEND. LEVINE & IVENS, 1965

Synonym. *Eimeria hindlei* Yakimoff & Gousseff of Ryšavý (1954).

This form occurs in the field mouse *Apodemus sylvaticus* in Europe.

The oocysts are ovoid, 23—27 × 18—20 (mean 25 × 18) μm, with a thin, pale yellow, one-layered wall, without a micropyle. The oocysts did not sporulate.

Remarks. As Levine & Ivens (1965) pointed out, it is not certain that this form was actually *Eimeria*, since the oocysts did not sporulate.

References. Levine & Ivens (1965); Ryšavý (1954).

EIMERIA (?) SP. ČERNÁ & DANIEL, 1956 EMEND. LEVINE & IVENS, 1965

Synonym. *Eimeria* sp. Černá & Daniel, 1956.

This form occurs commonly in the intestine of the field mouse *Apodemus flavicollis* in Europe.

The oocysts are broadly ovoid to spherical, 11—14 × 10—13 μm, with a very delicate wall, without a micropyle. The oocysts did not sporulate.

Remarks. As Levine & Ivens (1965) pointed out, it is not certain that this form was actually *Eimeria*, since the oocysts did not sporulate.

References. Černá & Daniel (1956); Levine & Ivens (1965).

EIMERIA TYPE A LEWIS & BALL, 1983

This form occurs commonly in the yellow-necked mouse *Apodemus flavicollis* in Europe. The oocysts are similar to those of *E. divichinca*.

Reference. Lewis & Ball (1983).

EIMERIA TYPE B LEWIS & BALL, 1983

This form occurs commonly in the yellow-necked mouse *Apodemus flavicollis* in Europe. The oocysts are similar to those of *E. apodemi*.

Reference. Lewis & Ball (1983).

EIMERIA TYPE C LEWIS & BALL, 1983

This form occurs quite commonly in the yellow-necked mouse *Apodemus flavicollis* in Europe. The oocysts are similar to those of *E. uptoni*.

Reference. Lewis & Ball (1983).

EIMERIA TYPE D LEWIS & BALL, 1983

This form occurs moderately commonly in the yellow-necked mouse *Apodemus flavicollis* in Europe. The oocysts are similar to those of *E. hungaryensis*, except that they are 20 × 14 μm.

Reference. Lewis & Ball (1983).

ISOSPORA GOLEMANSKII LEVINE 1982

Synonym. *Isospora* sp. Golemanski & Yankova, 1973.

This species occurs moderately commonly in the field mice *Apodemus flavicollis* (type host) and *A. sylvaticus* in Europe.

The oocysts are spherical or subspherical, 23—28 × 20—28 μm, without a micropyle, residuum, or polar granule. The sporocysts are ovoid, 12—16 × 8—11 μm, with a residuum. The sporozoites are elongate, with one end rounded and the other tapered, with a clear globule in the broad end, and lie lengthwise all oriented in the same direction in the sporocysts.

References. Golemanski & Yankova (1973); Levine (1982).

ISOSPORA URALICAE SVANBAEV, 1956

Synonym. *Isospora uralica* Svanbaev, 1956 emend. Pellérdy, 1974.

This species occurs commonly in the field mouse *Apodemus sylvaticus* in the USSR.

The oocysts are ovoid, 26 × 22.5 μm, with a smooth, greenish, double-contoured wall 1.6 μm thick, without a micropyle or residuum, with a polar granule. The sporocysts are ovoid, 14×9 μm, without a Stieda body or residuum. The sporozoites are comma-shaped, 7×4 μm.

Reference. Svanbaev (1956).

SARCOCYSTIS SEBEKI (TADROS & LAARMAN, 1976) LEVINE, 1978

Synonyms. *Endorimospora sebeki* Tadros & Laarman, 1976; *Sarcocystis sebeki* (Tadros & Laarman, 1976) Tadros & Laarman, 1978.

Sarcocysts of this species occur in the muscles of the long-tailed field mouse *Apodemus sylvaticus* (type intermediate host), and oocysts and sporocysts in the feces of the tawny owl *Strix aluco* (type definitive host) in Europe.

The oocysts are 21×15 μm and the sporocysts 15×10.5 μm. Sporulation occurs in the owl intestine.

Sarcocysts in the mouse muscles are up to several centimeters long, compartmented, with a thin wall without cytophaneres, with rounded metrocytes and slender, elongate bradyzoites.

References. Levine (1978); Tadros & Laarman (1976, 1978, 1979); Tadros (1981).

SARCOCYSTIS SP. DYL'KO, 1962

Sarcocysts of this form occur in the muscles of the field mouse *Apodemus flavicollis* (type intermediate host) in the USSR. The definitive host is unknown.

Reference. Dyl'ko (1962).

SARCOCYSTIS SP. KRASNOVA, 1971

Sarcocysts of this form occur in the muscles of the field mouse *Apodemus sylvaticus* (type intermediate host) in the USSR. The definitive host is unknown.

Reference. Krasnova (1971).

FRENKELIA SP. (ŠEBEK, 1975) LEVINE & IVENS, 1987

Synonym. M-organism in *Apodemus flavicollis* of Šebek (1975).

This form occurs rarely in the brain of the field mouse *Apodemus flavicollis* (type intermediate host) in Europe. The definitive host is unknown.

References. Šebek (1975); Levine & Ivens (1987).

KLOSSIA SP. GOLEMANSKI & YANKOVA, 1973

This form occurs in the field mouse *Apodemus sylvaticus* in Europe.

The oocysts are ellipsoidal, 31—44×23—36 μm, without a micropyle or residuum, with several polar granules, with six to twelve spherical sporocysts, each with four comma-shaped sporozoites and many residual granules, without a Stieda body.

Remarks. It is probable that this is a pseudoparasite of the field mouse and a true parasite of some invertebrate that it has eaten.

Reference. Golemanski & Yankova (1973).

CRYPTOSPORIDIUM MURIS TYZZER, 1907

This species was reported by Elton et al. (1931) from the field mouse *Apodemus sylvaticus* in England. See under *Mus* for further discussion.

HOST GENUS *CHIROPODOMYS*

EIMERIA MUULI MULLIN, COLLEY, & STEVENS, 1972

This species occurs commonly in the pencil-tailed tree mouse *Chiropodomys gliroides* in Asia.

The oocysts are ellipsoidal, with a finely granular outer surface, 17—30 × 14—22 (mean 25 × 19) μm, with a striated, one-layered wall about 1 μm thick and an inner membrane, without a micropyle or residuum, with a polar granule. The sporocysts are ellipsoidal, 12—14 × 6—8 (mean 13 × 8) μm, with a Stieda body and residuum. The sporozoites are elongate, with clear globules at both the large and small ends, and lie lengthwise head to tail in the sporocysts.

Reference. Mullin, Colley, & Stevens (1972).

HOST GENUS *MICROMYS*

EIMERIA MICROMYDIS GOLEMANSKY, 1978

This species occurs commonly in the Old World harvest mouse *Micromys minutus* in Europe.

The oocysts are ellipsoidal, colorless, 20—24 × 17—19 μm, with a two-layered wall about 2 μm thick, without a micropyle or residuum, with one or two polar granules. The sporocysts are ellipsoidal, 12 × 8 μm, without a Stieda body, with a residuum. The sporozoites are elongate, with one end rounded and the other pointed, have a clear globule at the broad end, and lie lengthwise in the sporocysts.

References. Golemansky (1978, 1979).

HOST GENUS *THALLOMYS*

SARCOCYSTIS SP. VILJOEN, 1921

Sarcocysts of this form were found in the muscles of the acacia rat *Thallomys moggi* (type intermediate host) in Africa. The definitive host is unknown.

HOST GENUS *THAMNOMYS*

EIMERIA VINCKEI RODHAIN, 1954

This species occurs commonly in the cecal epithelial cells of the forest mouse *Thamnomys s. surdaster* in Africa. It is nonpathogenic.

The oocysts are cylindroid, colorless, 20—24 × 12—15 (mean 22 × 13) μm, with a thin, colorless, double-contoured wall, without a micropyle, residuum, or presumably polar granule. The sporocysts were illustrated as spherical but stated to be 7×4 µm, apparently without a Stieda body, with a residuum. Sporulation takes 7—8 d.

Each meront produces six to ten merozoites. The mature macrogametes are spherical, about 9 µm in diameter. The microgamonts are ovoid, 14×9 µm. The prepatent period is 11 d.

This species could not be transmitted to *Mus musculus* or *Rattus norvegicus*.

Reference. Rodhain (1954).

HOST GENUS *PSEUDOMYS*

SARCOCYSTIS SP. MUNDAY, MASON, HARTLEY, PRESIDENTE, & OBENDORF, 1978

Sarcocysts of this form occur moderately commonly in all striated muscles except the heart of the rat *Pseudomys higginsi* in Australia. The definitive host is unknown.

The sarcocysts are up to 200×75 µm, with thick (4—8 µm), striated walls, fine trabeculae, a primary wall folded into long and very wide protrusions up to 6 µm long and 4 µm wide, and relatively small bradyzoites.

Remarks. Munday et al. (1978) thought that this form is apparently the same as that with microscopic, thick-walled sarcocysts that they found in *Rattus lutreolus, R. norvegicus, R. rattus,* and *Mastacomys fuscus.*

Reference. Munday, Mason, Hartley, Presidente, & Obendorf (1978).

HOST GENUS *MASTACOMYS*

SARCOCYSTIS SP. MUNDAY, MASON, HARTLEY, PRESIDENTE, & OBENDORF, 1978

Sarcocysts of this form occur commonly in all striated muscles except the heart of the rat *Mastacomys fuscus* in Australia.

The sarcocysts are described above under *Pseudomys.*

Remarks. Munday et al. (1978) thought that this form is apparently the same as that with microscopic, thick-walled sarcocysts that they found in *Rattus lutreolus, R. norvegicus, R. rattus,* and *Pseudomys higginsi.*

Reference. Munday et al. (1978).

HOST GENUS *LOPHUROMYS*

EIMERIA AFRICANA LEVINE, BRAY, IVENS, & GUNDERS, 1959

This species occurs in the harsh-furred mouse *Lophuromys s. sikapusi* in Africa.

The oocysts are subspherical to ellipsoidal, 14—21 × 11—16 (mean 17 × 14) µm, with a smooth, colorless to pale yellowish, one-layered wall about 0.9 µm thick that collapses rather quickly in Sheather's sugar solution, without a micropyle or residuum, with a polar granule. The sporocysts are ovoid, 8—11 × 6 (mean 10×6) μm, very thin-walled, with a Stieda body and usually a residuum. The sporozoites are crowded and curled in the sporocysts, have a large clear globule at the broad end and a small one at the other end, and lie either lengthwise or at the ends of the sporocysts. The sporulation time is 12 h at room temperature in Liberia.

Reference. Levine, Bray, Ivens, & Gunders (1959).

EIMERIA HARBELENSIS LEVINE, BRAY, IVENS, & GUNDERS, 1959

This species occurs in the harsh-furred mouse *Lophuromys s. sikapusi* in Africa.

The oocysts are ovoid, 32 × 23 μm, with a smooth, two-layered wall, the outer layer 1 μm thick at the thick end of the inner layer, maintaining this thickness for a little more than $^2/_3$ the length of the oocyst and then narrowing rapidly, forming a shoulder, and seeming to disappear anteriorly, the inner layer ellipsoidal, 1.5 μm thick at one end but becoming progressively thinner along the sides until it reaches a thickness of about 0.9 μm at the other end, with a broad, dimplelike indentation 6 μm in diameter at the large end which may be a micropyle, without a residuum or polar granule. The sporocysts are 14 × 8 μm, football-shaped, rather thin-walled, with a Stieda body and residuum. The sporozoites are very granular, have a clear globule at one end and lie lengthwise in the sporocysts.

Reference. Levine, Bray, Ivens, & Gunders (1959).

EIMERIA KRUIDENIERI LEVINE, BRAY, IVENS, & GUNDERS, 1959

This species occurs in the epithelial cells of the jejunum of the harshfurred mouse *Lophuromys s. sikapusi* in Africa.

The oocysts are somewhat ellipsoidal, but narrowed and flattened at both ends, sometimes asymmetrical, 27—31 × 19—21 (mean 29 × 19) μm, with a roughened, sometimes pitted, yellowish brown, one-layered wall about 1.3 μm thick at the sides and narrowing to 0.6—0.7 μm at both ends, without a micropyle or residuum, with a polar granule. The sporocysts are elongate ovoid, rounded at one end and truncate at the other, 15 × 8 μm, without a Stieda body, with or without a residuum. The sporozoites are somewhat granular, with a large clear globule at one end. The sporulation time is 3 d at room temperature in Liberia.

Reference. Levine, Bray, Ivens, & Gunders (1959).

EIMERIA LIBERIENSIS LEVINE BRAY IVENS, & GUNDERS, 1959

This species occurs in the harsh-furred mouse *Lophuromys s. sikapusi* in Africa.

The oocysts are ellipsoidal, 20—27 × 14—20 (mean 24 × 17) μm, with a smooth, pale yellowish, one-layered wall about 0.9 μm thick, without a micropyle or residuum, with a polar granule. The sporocysts are ellipsoidal, 10×5 μm, with a thin wall, without a Stieda body, with a residuum. The sporozoites lack clear globules and lie slantwise or curled longitudinally in the sporocysts.

Reference. Levine, Bray, Ivens, & Gunders (1959).

EIMERIA LOPHUROMYSIS LEVINE, BRAY, IVENS & GUNDERS, 1959

This species occurs in the harsh-furred mouse *Lophuromys* s. *sikapusi* in Africa.

The oocysts are ellipsoidal to ovoid, 19—23 × 13—15 (mean 21 × 14) μm, with a smooth, yellowish to brownish yellow, one-layered wall about 0.8 μm thick, without a micropyle or residuum, with or without a polar granule. The sporocysts are elongate ellipsoidal to slightly ovoid, 11 × 6—7 μm, with a very thin wall, with a very small Stieda body, without a residuum. The sporozoites are yellowish, have a large clear globule at the large end and a small one at the other end, and lie lengthwise in the sporocysts.

Reference. Levine, Bray, Ivens, & Gunders (1959).

EIMERIA SIKAPUSII LEVINE, BRAY, IVENS, & GUNDERS, 1959

This species occurs in the harsh-furred mouse *Lophuromys* s. *sikapusi* in Africa.

The oocysts are subspherical to ellipsoidal, 20—23 × 15—18 (mean 21 × 17) μm, with a smooth to slightly pitted, pale yellowish, one-layered wall about 1 μm thick which often collapses rather quickly in Sheather's sugar solution, without a micropyle or residuum, with one to four polar granules. The sporocysts are football-shaped, 10—12 × 7—8 μm, without a Stieda body, with a residuum. The sporozoites are clear, lack clear globules, and lie more or less lengthwise or at the ends of the sporocysts. Sporulation occurs outside the host.

Reference. Levine, Bray, Ivens, & Gunders (1959).

SARCOCYSTIS SP. MANDOUR, BIRD, & MORRIS, 1965

Sarcocysts of this species occur in the skeletal and heart muscles of *Lophuromys flavopunctatus* (type intermediate host) in Africa. The definitive host is unknown. The sarcocysts have a thin, smooth wall containing minute papillae; they are compartmented and contain bradyzoites about 7 μm long.

Reference. Mandour, Bird, & Morris (1965).

HOST GENUS ACOMYS

KLOSSIELLA SP. MESHORER, 1970

This form occurs in the spiny mouse *Acomys cahirinus* in Asia (Israel). Gamonts are in the kidney tubules and sporocysts in the kidney lumen.

Reference. Meshorer (1970).

HOST GENUS *RHABDOMYS*

EIMERIA CHINCHILLAE DE VOS & VAN DER WESTHUIZEN, 1968
See under *Chinchilla laniger*.

EIMERIA PRETORIENSIS DE VOS & DOBSON, 1970
This species occurs uncommonly in the small intestine, especially the ileum, of the 4-striped grass mouse *Rhabdomys pumilio* in Africa.

The oocysts are broadly ellipsoidal, 18—27×16—22 (mean 24×19) μm, with a two-layered wall about 1.5 μm thick, the outer layer the thicker, rough, light brown, the inner layer colorless, without a micropyle, with a residuum and polar granule. The sporocysts are broadly ellipsoidal, tapering slightly toward one end, 8—12 × 5—8 (mean 10 × 7) μm, with a Stieda body at the small end, with a residuum. The sporozoites are difficult to see, illustrated as sausage-shaped, without a clear globule. The sporulation time is 5 d at 28°C in 2% potassium bichromate solution.

This species will not infect *Chinchilla laniger*.

Reference. De Vos & Dobson (1970).

EIMERIA PUMILIOI DE VOS & DOBSON, 1970
This species occurs commonly, mainly in the cecum and colon, to a lesser extent the small intestine, of the 4-striped grass mouse *Rhabdomys pumilio* in Africa.

The oocysts are predominantly ellipsoidal but also subspherical or cylindrical, 15—21 × 10—14 (mean 18 × 12) μm, with a smooth, colorless, apparently one-layered wall about 1 μm thick, without a micropyle or residuum, with a polar granule. The sporocysts are usually broadly ellipsoidal, 7—10 × 4—7 (mean 9 × 6) μm, with a Stieda body and residuum. The sporozoites are about 8 μm long, with one end pointed and the other broad and rounded, have a clear globule at the broad end, and lie lengthwise head to tail in the sporocysts. The sporulation time is 2 d at 28°C in 2% potassium bichromate solution. The prepatent period is 3 d.

This species will not infect *Chinchilla laniger*.

Reference. De Vos & Dobson (1970).

EIMERIA RHABDOMYIS DE VOS & DOBSON, 1970
This species occurs uncommonly in the cecum and colon of the 4-striped grass mouse *Rhabdomys pumilio* in Africa.

The oocysts are spherical to subspherical, 10—14×9—13 (mean 13×12) μm, with a smooth, colorless, apparently one-layered wall 0.7 μm thick, without a micropyle or residuum, usually with a polar granule. The sporocysts are ellipsoidal to ovoid, 6—8 × 4—6 (mean 7 × 5) μm, with a Stieda body and residuum. The sporulation time is 2 d at 28°C in 2% potassium bichromate solution. The prepatent period is 3 days.

This species cannot be transmitted to *Chinchilla laniger*.

Reference. De Vos & Dobson (1970).

HOST GENUS *LEMNISCOMYS*

EIMERIA LEMNISCOMYSIS LEVINE, BRAY, IVENS, & GUNDERS, 1959

This species occurs in the jejunum and ileum of the striped grass mouse *Lemniscomys s. striatus* in Africa.

The oocysts are broadly spindle-shaped with somewhat flattened ends, 27—30 × 18—19 (mean 28 × 19) μm, with a moderately rough and pitted, brownish yellow, one-layered wall about 1.2 μm thick at the sides and 0.8 μm thick at the ends, lined by a thin membrane, without a micropyle or residuum, with a polar granule. The sporocysts are elongate ovoid, pointed at one end, 16 × 8 μm, with or without a small Stieda body, with a residuum. The sporozoites lie more or less diagonally in the sporocysts. The oocysts collapse rather quickly in Sheather's sugar solution. The sporulation time is 3 d at room temperature in Liberia.

Reference. Levine, Bray, Ivens, & Gunders (1959).

EIMERIA PUTEVELATA BRAY, 1958

This species occurs in the ileum of the striped grass mouse *Lemniscomys s. striatus* in Africa.

The oocysts are ovoid, 22—30 × 17—22 (mean 26.5 × 20) μm, with a two-layered wall, the outer layer thick, yellow, covered with small, well-defined pits, the inner layer thin, without a micropyle or residuum, occasionally with polar granules. The sporocysts are ovoid, 10—13 × 8—10 μm, with a very small Stieda body, usually with a residuum. The sporozoites are banana-shaped, 10 × 3 μm. The sporulation time is 10 d.

Reference. Bray (1958).

HOST GENUS *ARVICANTHIS*

EIMERIA ARVICANTHIS VAN DEN BERGHE & CHARDOME, 1956

This species occurs in the grass mouse *Arvicanthis abyssinicus rubescens* in Africa.

The oocysts were described as ovoid but illustrated as ellipsoidal, 23—24 × 10—14 μm, with a thin, rose-colored wall, presumably without a micropyle, with a residuum composed of some rare granules. The sporocysts are ellipsoidal, 10—11 × 7 μm, without a Stieda body, with a residuum. The sporozoites are 7 × 3 μm, with one end broader than the other. The sporulation time is 3 d in 1% chromic acid solution.

Reference. Van den Berghe & Chardome (1956).

HOST GENUS *DASYMYS*

EIMERIA DASYMYSIS LEVINE, BRAY, IVENS, & GUNDERS, 1959

This species occurs in the epithelial cells of the lower jejunum of the shaggy-haired rat *Dasymys incomptus rufulus* in Africa.

The oocysts are subspherical, ellipsoidal, sometimes slightly ovoid, 17—23 × 15—21 (mean 20 × 17) µm, with a smooth, colorless to pale yellowish, one-layered wall about 0.8 µm thick, without a micropyle or residuum, with a polar granule. The sporocysts are slightly ovoid, 10—11 × 6 µm, with a very thin wall, tiny Stieda body, and residuum. The sporozoites are more or less curled up in the sporocysts. The sporulation time is 3 d at room temperature in Liberia.

Reference. Levine, Bray Ivens, & Gunders (1959).

HOST GENUS *BANDICOTA*

EIMERIA BANDICOTA BANDYOPADHYAY & DASGUPTA, 1982

This species occurs uncommonly in the intestine of the bandicoot rat *Bandicota bengalensis* in India.

The oocysts are subspherical, 24—28.5 × 21.5—23 (mean 26 × 22.5) µm, with a two-layered wall, the outer layer very thin and yellowish brown, the inner layer 1.5 µm thick, said to have a micropyle (illustrated without one), without a residuum, said not to have a polar granule but illustrated with one. The sporocysts are ovoid, 10.5—13.5 × 6—9 (mean 12 × 8) µm, with a Stieda body and residuum. The sporozoites are 9—12 × 3.5—4.5 (mean 10 × 4) µm, illustrated without clear globules, and lie lengthwise head to tail in the sporocysts. The sporulation time is 36—48 h at room temperature in 2.5% potassium bichromate solution.

Reference. Bandyopadhyay & Dasgupta (1982).

SARCOCYSTIS SP. SINNIAH, 1979

Sarcocysts of this species were found in the muscles of the bandicoot rat *Bandicota indica* in Malaysia. They were not described.

Reference. Sinniah (1979).

HOST GENUS *NESOKIA*

EIMERIA GOUSSEFFI GLEBEZDIN, 1978

This species occurs commonly in the short-tailed bandicoot rat *Nesokia indica* in the USSR (Turkmenia).

The oocysts were described as oval but illustrated as ellipsoidal, 14—25 × 11—20 (mean 17 × 14) µm, with a smooth, one-layered wall, without a micropyle, residuum, or polar granule. The sporocysts are elongate ovoid, 8—11 × 6—9 (mean 9 × 6) µm, without a Stieda body or residuum. The sporozoites are elongate, with a clear globule at the broad end, and lie lengthwise head to tail in the sporocysts.

Reference. Glebezdin (1978).

EIMERIA NESOKIAI MIRZA, 1975

This species occurs in the ileum of the short-tailed bandicoot rat *Nesokia indica* in Asia (Iraq) and Africa (Egypt).

The oocysts are subspherical or broadly ellipsoidal, 17—30 × 17—23 (mean 24 × 20) μm, with a two-layered wall, the outer layer slightly pitted, yellowish, about 1.5 μm thick, the inner layer dark brown, about 1 μm thick, without a micropyle, with a residuum and polar granule. The sporocysts are lemon-shaped, with a Stieda body and residuum. The sporozoites are elongate, with one end narrower than the other, lack a clear globule, and lie lengthwise head to tail in the sporocysts. The sporulation time is 4—5 d at 22-24°C in 2.5% potassium bichromate solution.

There are two asexual generations. Mature first generation meronts are 10—14 × 9—12 μm and contain 14—20 fusiform merozoites 9—13 × 3—4 μm. Second generation meronts are 11—18 × 10—16 μm and contain 16—26 fusiform merozoites 8—16 × 2—3 μm. Microgamonts are ellipsoidal or cylindrical, 14—21 × 11—19 μm and produce many microgametes. Macrogametes are ovoid or cylindrical, 10—17 × 10—14 μm. The prepatent period is 6—8 d.

The infection may decrease the appetite and body weight of infected 2-month-old *N. indica*, but does no permanent damage.

References. Jawdat & Al-Jafary (1979); Mirza (1970, 1975).

EIMERIA (?) SP. (YAKIMOFF, GOUSSEFF, & SUZ'KO, 1945) LEVINE & IVENS, 1965

This form occurs in the feces of the scaly-toothed rat *Nesokia indica huttoni* in the USSR.

The oocysts are ovoid, colorless, 22—31 × 15—23 (mean 26 × 20) μm, without a micropyle. They failed to sporulate.

References. Levine & Ivens (1965); Yakimoff, Gousseff, & Suz'ko (1945).

WENYONELLA BAGHDADENSIS MIRZA & AL-RAWAS, 1978

This species occurs in the bandicoot rat *Nesokia indica* in Asia (Iraq).

The oocysts are subspherical to broadly ellipsoidal, 18—22 × 15—18 μm, with a two-layered wall, the outer layer mammillated, sometimes appearing striated, without a micropyle, micropylar cap, residuum, or polar granule. The sporocysts are ovoid, 9—15 × 6—10 (mean 12 × 8) μm, with a Stieda body and residuum. Two of the sporozoites in each sporocyst are elongate, with granular cytoplasm and a clear globule; the other two sporozoites are bean-shaped, without granules or a clear globule. Sporulation occurs outside the host.

Reference. Mirza & Al-Rawas (1978)

HOST GENUS *PRAOMYS*

EIMERIA MASTOMYIS DE VOS & DOBSON, 1970

This species occurs moderately commonly in the multimammate rat *Praomys* (syns., *Rattus, Mastomys*) *natalensis* in Africa.

The oocysts are broadly ellipsoidal, 24—32 × 20—23 (mean 27 × 21) μm, with a finely punctate, two-layered wall about 1.5 μm thick, the outer layer light brown and radially striated, the inner layer colorless, without a micropyle or residuum, usually with a polar granule. The sporocysts are ellipsoidal, 13—15 ×7—9 (mean 14 × 8) μm, with a Stieda body and residuum. The sporozoites are virtually impossible to see, but have a large clear globule.

Reference. De Vos & Dobson (1970).

EIMERIA PRAOMYSIS LEVINE, BRAY, IVENS, & GUNDERS, 1959

This species occurs in the rat *Praomys* (syn., *Rattus*) *tullbergi rostratus* in Africa.

The oocysts are subspherical to ellipsoidal, 17—24 × 16—23 (mean 21 × 19) μm, with a smooth to somewhat rough, pale yellowish to brownish, one-layered wall about 1 μm thick, without a micropyle or residuum, with one or two polar granules. The sporocysts are ellipsoidal to slightly ovoid, 10—12 × 6—7 (mean 11 × 6.5) μm, with a Stieda body and residuum. The sporozoites lie more or less lengthwise in the sporocysts. The sporulation time is 6 d at room temperature in Liberia.

Reference. Levine, Bray, Ivens, & Gunders (1959).

EIMERIA THEILERI DE VOS & DOBSON, 1970

Synonym. *Eimeria* sp. Fantham, 1926.

This species occurs quite commonly in the small intestine of the multimammate rat *Praomys* (syn., *Rattus*) (*Mastomys*) *natalensis* in Africa.

The oocysts are subspherical to ellipsoidal, 16—25 × 14—20 (mean 20 × 17) μm, with a smooth, colorless, apparently one-layered wall 1 μm thick, without a micropyle or residuum, often with a polar granule. The sporocysts are ellipsoidal, 9—12 × 5—8 (mean 11 × 6) μm, with a Stieda body, sub-Stiedal body, and residuum. The sporozoites are elongate, 8 μm long, rounded at one end and tapered toward the other, have a clear globule at the broad end, and lie lengthwise head to tail in the sporocysts. The sporulation time is 2 d at 28°C in 2% potassium bichromate solution.

Chinchilla laniger could not be infected with this species.

Reference. DeVos & Dobson (1970).

KLOSSIELLA MABOKENSIS BOULARD & LANDAU, 1971

This species occurs in the kidneys, intestine, liver, lungs, and spleen of the rat *Praomys* (syn., *Rattus*) *jacksoni* (type host) and (experimentally) laboratory mouse *Mus musculus* in Africa.

The oocysts contain 9—20 (mean 12) sporocysts, each with 18—24 (mean 20) sporozoites.

The life cycle involves (1) an initial merogony in the intestine, (2) a first generation of renal meronts in the glomeruli, and (3) a second generation of renal and extrarenal merogony that appears to be the origin of the gamonts. The forms attained were small cysts in the liver and lungs of heavily infected white mice; they looked like cysts formed by endogeny.

The initial merogony occurs in the intestine (and not in the kidney, liver, spleen, or lung) during the first month after inoculation. The parasites develop in the core of the villi. At 3 weeks the intestinal meronts are 15—30 μm in diameter and contain 30—100 nuclei. The first renal meronts appear at 1 month. They are in the Bowman's capsules of the glomeruli, average 19×14 μm, and contain 45—70 nuclei.

The second generation meronts in the kidney are seen 30—60 d after inoculation. They average 48×28 μm and contain 200 nuclei. During the remainder of the parasitic period, the kidney meronts vary in size, and it was impossible to tell whether marked asynchrony or a succession of generations was responsible. They were found both in the glomeruli and other reticuloendothelial cells of the kidney.

Extrarenal meronts in the liver, lung, and spleen appear at the same time as the second generation kidney meronts and are structurally identical.

The first gamonts appear in the kidneys of infected mice at 2 months, and the first mature sporocysts about 15 d later. Sporogony rapidly becomes asynchronous and all the stages can coexist in the same animal. The gamonts are in the cells of the proximal convoluted tubules. Males and females develop in the same cell. The macrogametes are spherical or ovoid. The microgamonts are small, elongate, with clear cytoplasm, and form four intracellular, nonflagellate round gametes. The parasitized cells and their nuclei are hypertrophied. The cells are pushed into the lumen of the renal tubule, remaining attached to it by a pedicle.

Rounded intracellular cysts about 15 μm in diameter are present in the reticuloendothelial system of the liver and lungs 200 d after inoculation. They have a thick capsule and contain about 25 banana-shaped bodies, thus resembling mature sporocysts in the lumen of the renal tubules.

The prepatent period is 2 months before the first gamonts appear in the kidney tubules and 2.5 months before sporocysts appear in the urine. The patent period is indeterminately long.

Three laboratory rats *Rattus norvegicus* could not be infected.

References. Boulard & Landau (1971); Boulard (1973).

HOST GENUS *MAXOMYS*

BESNOITIA SPP. MATUSCHKA & HÄFNER, 1984

Matuschka & Häfner (1984) found that snakes of the genus *Bitis* are definitive hosts of *Besnoitia* spp. in Africa and that rodents of the genus *Maxomys* and other genera are experimental intermediate hosts. The rodents had macroscopic *Besnoitia* cysts up to 2 mm in diameter in their connective tissue. These authors thought that rodents might be reservoir hosts of *Besnoitia* of cattle.

Reference. Matuschka & Häfner (1984).

HOST GENUS *RATTUS*

EIMERIA ALISCHERICA MUSAEV & VEISOV, 1965

This species occurs in the Norway rat *Rattus norvegicus* in the USSR.

The oocysts are ovoid or ellipsoidal, yellowish, 28—36 × 16—26 (mean 34 × 23) μm, with a rough, striated, one-layered wall 2.5 mm thick containing small granules, without a micropyle or polar granule, with a residuum. The sporocysts are ovoid or piriform, with a Stieda body and residuum. The sporozoites are piriform or bean-shaped, illustrated without a clear globule. The sporulation time is 4—5 d.

Reference. Musaev & Veisov (1965).

EIMERIA BYCHOWSKYI MUSAEV & VEISOV, 1965 EMEND. PELLÉRDY, 1974

Synonym. *Eimeria bychowsky* Musaev & Veisov, 1965.

This species occurs in the Norway rat *Rattus norvegicus* in the USSR.

The oocysts are ovoid or ellipsoidal, colorless, 20—28 × 14—20 (mean 25 × 18) μm, with a smooth, two-layered wall 3 μm thick, the outer layer 1 μm thick, the inner layer 2 μm thick, without a micropyle, with a residuum and polar granule. The sporocysts are ovoid, rarely ellipsoidal, 6—12 × 4—8 (mean 9 × 5) μm, illustrated without a Stieda body, with a residuum. The sporozoites are comma-shaped, with a clear globule in the broad end. The sporulation time is 3—4 d.

References. Musaev & Veisov (1965); Pellérdy (1974).

EIMERIA CONTORTA HABERKORN, 1971

This is a mixture of *Eimeria nieschulzi* of *Rattus norvegicus* and *E. falciformis* of *Mus musculus*.

References. Cerná (1975); Haberkorn (1971); Müller (1975); Müller, Hammond, & Scholtyseck (1973); Stockdale, Tiffin, Kozub, & Chobotar (1979).

EIMERIA EDWARDSI COLLEY & MULLIN, 1971

This species occurs in Edwards' rat *Rattus edwardsi* in Asia.

The oocysts are ovoid, 25—32 × 21—23 (mean 29 × 22) μm, with a two-layered wall, the outer layer smooth, yellowish brown, 1.5 μm thick, the inner layer light brown, about 0.5 μm thick, with a micropyle, without a residuum or polar granule. The sporocysts are ellipsoidal to slightly ovoid, 13—18 × 5—10 (mean 14.5 × 6.5) μm, with or without a Stieda body, with a residuum. The sporozoites are elongate, with one end broader than the other, without a clear globule, and lie lengthwise head to tail in the sporocysts.

Reference. Colley & Mullin (1971).

EIMERIA ELERYBECKERI LEVINE, 1984

Synonyms. *Eimeria levinei* Krishnamurthy & Kshirsagar, 1980; *E. lavinei* Krishnamurthy & Kshirsagar, 1980 *lapsus calami*.

This species occurs moderately commonly in the black rat *Rattus r. rattus* in Asia (India).

The oocysts are ellipsoidal ("ovoid"), 36—40 × 26—30 (mean 39 × 28) μm, with a one-layered, smooth, colorless wall 1.6 μm thick, without a micropyle or residuum, with a polar granule. The sporocysts are ovoid or elongate, with a very small Stieda body and prominent residuum. The sporozoites are elongate or kidney-shaped, with a clear globule at one end, and lie lengthwise head to tail in the sporocysts. The sporulation time is about 80—86 h at 28—30°C in 2.5% potassium bichromate solution.

References. Krishnamurthy & Kshirsagar (1980); Levine (1984).

EIMERIA HASEI YAKIMOFF & GOUSSEFF, 1936

This species occurs uncommonly in the black rat *Rattus rattus* in the USSR.

The oocysts are ovoid, ellipsoidal, or spherical, 12—24 × 12—24 μm, with a smooth wall illustrated as one-layered, without a micropyle or residuum, with a polar granule. The sporocysts are 8.5 × 5 μm, with a residuum. The sporulation time is 3 d at 28°C in 2% potassium bichromate solution.

Reference. Yakimoff & Gousseff (1936).

EIMERIA MIYAIRII OHIRA, 1912

Synonyms. *Eimieria miyiarii* [sic] of Ryšavý (1954); *E. carinii* Pinto, 1928.

This species occurs rather uncommonly in the epithelial cells of the villi and occasionally the glands of Lieberkühn of the small intestine of the Norway rat *Rattus norvegicus* (type host), the Malaysian woodrat *R. tiomanicus*, Whitehead's rat *R. whiteheadi*, and probably the black rat *R. rattus* throughout the world.

The oocysts in *R. norvegicus* are spherical to subspherical, 17—29 × 16—26 (mean 24 × 22) μm, with a two-layered wall, the outer layer moderately rough, yellowish, radially striated, about 1.3 μm thick, the inner layer brownish, about 0.4 μm thick, without a micropyle or residuum, with a polar granule. The sporocysts are ovoid, 11—14 × 7—9 (mean 12 × 8) μm, with a wall about 0.2 μm thick, without a Stieda body or rarely with a tiny one, without a sub-Stiedal body, with a residuum. The sporozoites are comma-shaped, with a clear globule in the broad end, and lie lengthwise head to tail in the sporocysts. Sporulation takes 4—6 d.

The oocysts in *R. whiteheadi* are ellipsoidal to spherical, 17—22 × 16—21 (mean 20 × 18) μm, with a rough, striated, two-layered wall 1.5 μm thick (of which the outer layer is difficult to distinguish and might come off after storage in potassium bichromate solution), without a micropyle or residuum, with a polar granule. The sporocysts are ellipsoidal, 9—13 × 6—8 (mean 10 × 7) μm, with a Stieda body and residuum. The sporocysts are elongate, lack clear globules, and lie lengthwise head to tail in the sporocysts.

The oocysts in *R. tiomanicus* are 23—28 × 20—23 (mean 25 × 22) μm; the sporocysts are 13—14 × 8—9 (mean 13 × 8) μm; the other structures are the same as those described for the oocysts in *R. whiteheadi*.

Following inoculation (Roudabush, 1937), the sporozoites are most abundant in the small intestine in about 12 h. They are banana-shaped, 12—17 × 2—3 (mean 14.5 × 2—3) μm, with a clear globule at each end. The nucleus has a marginal chromatic ring and a central karyosome, and lies anterior to the middle of the sporozoite.

The sporozoite enters an intestinal cell and rounds up to form a first generation meront. This produces 12—24 first generation merozoites 6—7 × 1—2 (mean 7 × 1) μm, and a residuum; these merozoites have a central nucleus with a central karyosome, but no clear globules. They leave the host cell in 2 d, enter new host cells, and form second generation meronts. Each of these produces 8—16 second generation merozoites 8—11 × 1—2 (mean 9 × 1) μm plus a residuum on the 3rd day. Their nucleus is in the posterior $1/_4$ of the body. The second generation merozoites enter new host cells and form third generation meronts. These produce 20—24 third generation merozoites 4—5 × 1—2 (mean 4 × 1) μm with a central nucleus on the 4th day.

The merozoites (Andreassen & Behnke, 1968) are banana-shaped and have a plasma membrane and an inner, thicker, two-layered membrane. There are 26 subpellicular microtubules, a polar ring and possibly a posterior ring, a conoid, two rhoptries, micronemes, a micropore about the middle of the body, Golgi apparatus, endoplasmic reticulum, and three different kinds of granules.

The third generation merozoites enter new host cells (Roudabush, 1937) and form macrogametes and microgamonts. The latter form many biflagellate microgametes 3 × 0.75 μm. Fertilization presumably occurs while the macrogametes are still in their host cells. Mature oocysts appear 5.5 d after inoculation. The prepatent period is 6—8 d and the patent period 5—7 d.

E. miyairii can produce 25,360 to 73,728 (mean 38,016) oocysts per oocyst fed (Roudabush, 1937).

Mus musculus, Sylvilagus floridanus, Cavia porcellus, and *Spermophilus tridecemlineatus* cannot be infected.

References. Andreassen & Behnke (1968); Becker (1933); Becker & Burroughs (1933); Firlotte (1948); Levine & Husar (1979); Levine & Ivens (1965); Matubayasi (1938); Mullin, Colley, & Stevens (1972); Mullin, Colley, & Welch (1975); Ohira, (1912); Pinto (1928); Roudabush (1937); Ryšavý (1954); Owen (1983).

EIMERIA NIESCHULZI DIEBEN, 1924

Synonyms. *Eimeria falciformis* (Eimer) Schneider of some *auctores*; *E. miyairii* Ohira of Pérard (1926), Becker, Hall, & Hager (1932), Becker (1934), Yakimoff & Gousseff (1935), Matubayasi (1938) and Beltran & Perez (1950); *E. miyiairii* [sic] of Yakimoff & Gousseff (1936), *E. halli* Yakimoff, 1935.

This species occurs commonly in the small intestine of the Norway rat *Rattus norvegicus* (type host), chestnut rat *R. fulvescens,* Hawaiian rat *R. hawaiiensis,* Mueller's rat *R. muelleri,* black rat *R. rattus,* and red spiny rat *R. surifer* throughout the world. It is one of the most popular coccidia for laboratory research. It may produce severe diarrhea or even death in young rats less than 6 months old. Up to 2000 oocysts daily for 5 d produce immunity but no clinical evidence of severe disease. However, 30,000—100,000 oocysts will produce severe diarrhea on about the 7th day and may cause death.

The oocysts from *R. norvegicus* are ellipsoidal to ovoid, tapering at both ends, 16—26 × 13—21 μm, with a smooth or rough, colorless to yellowish, two-layered wall about 1.1 μm thick, without a noticeable micropyle or residuum, with a polar granule. The sporocysts are elongate ovoid, 11—12 × 7 μm, with a thin wall, very small Stieda body, and compact residuum. The sporozoites are elongate, with a clear globule at the broad end, and lie lengthwise head to tail in the sporocysts. They have a three-layered pellicle, two polar rings, a conoid, numerous rhoptries, many micronemes, about 25 subpellicular microtubules, ribosomes, and more than one type of granule.

The oocysts in *R. surifer* are ellipsoidal, tapering toward each end, 18—26 × 15—18 (mean 23 × 17) μm, with a smooth, brownish yellow wall about 1.5 μm thick with a thin membrane visible only when the oocyst is crushed, without a residuum, with a polar granule. The sporocysts are ellipsoidal, 10—12 × 6—8 (mean 11 × 7) μm, with a small Stieda body and compact residuum. The sporozoites are elongate, comma-shaped, with a clear globule at the large end, and lie lengthwise head to tail in the sporocysts.

The oocysts in *R. fulvescens* are 21—28 × 15—20 (mean 24 × 18) μm, and the sporocysts are 10—12 × 6—8 (mean 11 × 7) μm.

The sporulation time is 2—3.5 d. Sporogony has been described in detail (Dieben, 1924).

The endogenous cycle in the rat includes four asexual generations. Sporozoites can be found in the lumen of the small intestine 3—4 h after inoculation, and may remain there as long as 4 d. They are 10—12 × 1—2 (mean 11 × 2) μm. They pass through the cells of the villi, enter the lacteals, pass down them, and round up to enter cells in the crypts of Lieberkuehn, where they become first generation meronts. These can be seen 31 h after inoculation. They form 20—36 first generation merozoites 7—11 × 1—2 μm, which enter new epithelial cells, turn into second generation meronts, and produce 8—14 merozoites 13—16 × 1—2.5 μm which enter new epithelial cells and become third generation meronts. The latter mature in about 2—5 d and produce about 8—20 third generation merozoites about 17—27 × 1—3 μm. These again enter new epithelial cells, round up, and produce about 36—60 fourth generation merozoites about 4—10 μm long. The merozoites have a polar ring, about 25 subpellicular microtubules, a conoid, 2 rhoptries, numerous micronemes, and (first generation merozoites only) 2 paranuclear bodies.

Young gamonts can be seen as early as 5.5 d after inoculation. Both the macrogametes and microgamonts have micropores. The microgametes are about 4 μm long and have two or three flagella. Oocysts can be found in the intestinal lumen 6—12 d after inoculation. The prepatent period is 6—8 d and the patent period 4—8 d or as long as 14 d.

A single oocyst can theoretically produce 460,800—4,819,200 oocysts. The more oocysts fed, the smaller the number of oocysts produced by each one. In addition, starved rats produce fewer oocysts than normal ones.

Bristol, Pinon, & Mayberry (1983) found that *E. nieschulzi* infections increased the longevity of *Nippostrongylus muris* (5% level) in rats.

This species cannot be transmitted to *Mus musculus, Cavia porcellus,* or *Oryctolagus cuniculus.*

References. Becker (1934, 1939, 1941, 1942); Becker & Dilworth (1941); Becker, Hall, & Hager (1932); Becker, Manresa, & Smith (1943); Becker & Smith (1942); Beltrán & Pérez (1950); Bonfante, Faust, & Giraldo (1961); Chbouki & Dubremtz (1982); Colley (1967, 1967a, 1968); Dieben (1924); Dubremetz, Colwell, & Mahrt, (1975); Dubremetz & Torpier (1978); Duszynski (1972); Duszynski & Conder (1977); Duszynski, Ramaswamy, & Castro (1982); Duszynski, Roy, & Castro (1978); Duszynski, Roy, Stewert, & Castro (1978); Hall (1934); Landers (1960); Levine & Ivens (1965); Liburd (1973); Liburd, Pabst, & Armstrong (1972); Marchiondo, Duszynski, & Speer (1978); Frandsen (1983); Marquardt (1966, 1966a); Matubayasi (1938); Mayberry (1973); Mayberry & Marquardt (1974); McQuistion & Schurr (1978); Mullin & Colley (1972); Mullin, Colley, & Welch (1975); Mullin, Colley, & Stevens (1972); Pérard (1926); Romero (1979); Rommel & Heydorn (1971); Rose, Hesketh, & Ogilvie (1979, 1980); Rose, Ogilvie, Hesketh, & Festing (1979); Rose, Peppard, & Hobbs (1984); Roudabush (1937); Sheppard (1974); Sibert & Speer (1980); Yakimoff (1935); Yakimoff & Gousseff (1935); Frandsen (1983); Castro & Duszynski (1983, 1984); Owen (1983); Bristol, Pinon, & Mayberry (1983); Marquardt, Osman, & Muller (1983, 1984); Frank (1978); Osman & Marquardt (1984); Broaddus, Mayberry, Bristol, & Upton (1987).

EIMERIA NOCHTI YAKIMOFF & GOUSSEFF, 1936

This species occurs uncommonly in the black rat *Rattus rattus* (type host) and possibly Norway rat *Rattus norvegicus* in the USSR and Europe.

The oocysts are ovoid, 14—25 × 12—22 μm, with a smooth, double-contoured wall, without a micropyle, residuum, or polar granule. The sporocysts lack a residuum.

References. Ryšavý (1954); Yakimoff & Gousseff (1936).

EIMERIA RATTI YAKIMOFF & GOUSSEFF, 1936

This species occurs uncommonly in the black rat *Rattus rattus* in the USSR.

The oocysts are cylindrical to ovoid, 16—28 × 15—16 (mean 23 × 15) μm, with a smooth, double-contoured wall, without a micropyle or residuum, with a polar granule. The sporocysts have a residuum.

This name may be a synonym of *E. separata.* Only ten oocysts were measured.

Reference. Yakimoff & Gousseff (1936).

EIMERIA SABANI COLLEY & MULLIN, 1971

This species occurs quite commonly in the long-tailed giant rat (noisy rat) *Rattus sabanus* (type host) and Whitehead's rat *R. whiteheadi* in Asia.

The oocysts are ellipsoidal, 22—32 × 18—26 (mean 28.5 × 22) μm, with a smooth, two-layered wall, the outer layer greenish yellow, about 1.5 μm thick, the inner layer dark brown, about 0.5 μm thick, without a micropyle or residuum, with a polar granule. The sporocysts are ellipsoidal to slightly ovoid, 10—15 × 7—9 (mean 12 × 8) μm, with or without a Stieda body, with a residuum. The sporozoites are elongate, comma-shaped, without clear globules, and lie lengthwise head to tail in the sporocysts.

Reference. Colley & Mullin (1971).

EIMERIA SEPARATA BECKER & HALL, 1931

Synonym. *Eimeria separatica* Becker & Hall, 1931 of Litvenkova (1969) *lapsus calami.*

This species occurs quite commonly to commonly in the epithelial cells of the cecum and colon of the Norway rat *Rattus norvegicus* (Type host), dark-tailed tree rat *R. cremoriventer, R. defua,* chestnut rat *R. fulvescens,* Hawaiian rat *R. hawaiiensis,* Mueller's rat *R. muelleri,* Malaysian woodrat *R. tiomanicus,* probably black rat *R. rattus,* and (experimentally) certain strains of the house mouse *Mus musculus* throughout the world. It is nonpathogenic or mildly pathogenic.

The oocysts in *R. norvegicus* are predominantly ellipsoidal, sometimes subspherical or ovoid, 10—19 × 9—17 μm, with a smooth, colorless to pale yellowish, one-layered wall about 0.6 μm thick, without a micropyle or residuum, with one to three polar granules. The sporocysts are ellipsoidal, 8—10 × 5—6 (mean 9 × 5) μm, with a very thin wall, tiny Stieda body, and small, compact (sometimes not discernible) residuum. The sporozoites are elongate, often have a clear globule at one end, and lie lengthwise head to tail in the sporocysts. The sporulation time is less than 36 h to 3 d at room temperature.

The oocysts in *R. defua* are subspherical, ellipsoidal, or slightly ovoid, 16—21 × 15—17 (mean 18 × 16) μm, with a smooth, colorless to pale yellowish or greenish, one-layered wall about 1 μm thick, without a micropyle or residuum, with a polar granule. The oocyst wall collapses rather quickly in Sheather's sugar solution. The sporocysts are almost ellipsoidal, 11 × 7 μm with a thin wall, small Stieda body, and small residuum. The sporozoites seem to have a clear globule at one end and lie lengthwise in the sporocysts but are curled at the ends.

The oocysts in *R. hawaiiensis* are ellipsoidal to subspherical, 12—16 × 10—13 (mean 13.5 × 12) μm, with a smooth, pale yellowish, one-layered wall 0.6—0.8 μm thick, without a micropyle or residuum, with one or two polar granules. The sporocysts are almost ellipsoidal, 8—9 × 5 (mean 9 × 5) μm, with a small, somewhat flat Stieda body, with or without a residuum. The sporozoites are elongate, have a clear globule at the large end, and lie lengthwise head to tail in the sporocysts.

The oocysts in *R. muelleri* are ellipsoidal, 15—20 × 11—16 (mean 17 × 13) μm, with a smooth, one-layered wall about 1 μm thick, without a residuum, with a polar granule. The sporocysts are ellipsoidal, 9—10 × 5—6 (mean 9 × 6) μm, with a small Stieda body and residuum. The sporozoites are elongate, apparently lack a clear globule, and lie lengthwise head to tail in the sporocysts.

The oocysts in *R. fulvescens* are 14—21 × 10—17 (mean 18 × 14) μm. The sporocysts are 6—9 × 4—6 (mean 7 × 5) μm.

The oocysts in *R. tiomanicus* are 14—17 × 10—12 (mean 15 × 11) μm. The sporocysts are 6—8 × 4—5 (mean 7 × 5) μm.

Sporocysts can be found in the intestinal lumen 6 h to 3 d after inoculation; they are 8—10 × 2—3 (mean 9.5 × 2) μm, have a central nucleus and two clear globules. They enter intestinal cells and round up to form first generation meronts. Each of these forms 6—12 (mean 8) first generation merozoites 11—13 × 2—3 (mean 12 × 2) μm with a central nucleus, without clear globules. They enter new host cells in about a day, round up, and form second generation meronts. Each of these produces 4—6 (mean 5.5) second generation merozoites 6—9 × 2—3 (mean 8 × 2) μm, which break out of the host cell 2 d after inoculation, enter new host cells, and round up to form third generation meronts. Each of these forms 2—6 (mean 4) third generation merozoites 13—15 × 2—3 (mean 14 × 3) μm by the 3rd day after inoculation. They have no residuum.

The third generation merozoites enter new host cells, round up, and turn into macrogametes or microgamonts. The prepatent period is 3—6 d and the patent period is 3 d. *E. separata* can produce 384—3,456 macrogametes or microgamonts per oocyst fed.

E. Separata does not infect *Cavia porcellus, Sylvilagus floridanus,* or *Spermophilus tridecemlineatus* and can infect only certain strains of *Mus musculus.*

References. Becker (1933, 1934); Becker & Hall (1931, 1933); Becker, Hall, & Hager (1932); Beltran & Perez (1950); Duszynski (1971, 1972); Levine, Bray, Ivens, & Gunders (1959); Levine & Ivens (1965); Litvenkova (1969); Matubayasi (1938); Mayberry & Marquardt (1973); Mayberry, Marquardt, Nash, & Plan (1982); Mayberry, Plan, Nash, & Marquardt (1975); Mullin & Colley (1972); Mullin, Colley & Stevens (1972); Mullin, Colley, & Welch (1975); Romero (1979); Roudabush (1937).

EIMERIA SURIFER COLLEY & MULLIN, 1971

This species occurs commonly in the red spiny rat *Rattus surifer* in Asia.

The oocysts are ellipsoidal, 30—38 × 22—28 (mean 35 × 25) μm, with a three-layered wall, the outer layer slightly rough, pitted, striated, greenish yellow, about 1.5 μm thick, the middle and inner layers colorless, about 0.5 μm thick each, without a micropyle or residuum, with a polar granule. The sporocysts are ellipsoidal, 15—17 × 9—10 (mean 15 × 9.5) μm, with a Stieda body, sub-Stiedal body, and residuum. The sporozoites are elongate, with one end narrower than the other, without clear globules, and lie lengthwise head to tail in the sporocysts.

Reference. Colley & Mullin (1971).

EIMERIA TIKUSI COLLEY & MULLIN, 1971

This species occurs in Edwards' rat *Rattus edwardsi* in Asia.

The oocysts are ellipsoidal, 27—34 × 21—28 (mean 30 × 24) μm, with a smooth, greenish yellow, one-layered wall about 1.5 μm thick, without a micropyle or residuum, with a polar granule. The sporocysts are ovoid, 13—16 × 9—11 (mean 14 × 10) μm, with a Stieda body and residuum. The sporozoites are elongate, comma-shaped, without clear globules, and lie lengthwise head to tail in the sporocysts.

Reference. Colley & Mullin (1971).

EIMERIA TUBERCULATA **KRISHNAMURTHY & KSHIRSAGAR, 1980**

This species occurs rarely in the black rat *Rattus r. rattus* in Asia.

The oocysts are subspherical, 37—43 × 32—38 (mean 40 × 36) μm, with a three-layered wall 3.2 μm thick, the outer layer thin and colorless, the middle layer tuberculate, darker, the inner layer yellowish brown, 1 μm thick, without a micropyle, residuum, or polar granule. The sporocysts are ovoid, without a Stieda body, with several residual granules. The sporozoites are elongate, banana-shaped, have a clear globule at one end, and lie lengthwise in the sporocysts. The sporulation time is 80—86 hours at 28—30°C in 2.5% potassium bichromate solution.

Reference. Krishnamurthy & Kshirsagar (1980).

EIMERIA TURKESTANICA **VEISOV, 1964**

This species occurs moderately commonly in the Turkestan rat *Rattus turkestanicus* in the USSR.

The oocysts are ovoid or ellipsoidal, 22—28 × 16—20 (mean 26 × 19) μm, with a two-layered wall 2.5 μm thick, the outer layer smooth, yellowish, 1.25 μm thick, the inner layer dark brown, 1.25 μm thick, without a micropyle or residuum, with a polar granule. The sporocysts are ellipsoidal or piriform, 8—12 × 6—8 (mean 11 × 7) μm, with a Stieda body and residuum. The sporozoites are piriform or comma-shaped.

Reference. Veisov (1964).

ISOSPORA AURANGABADENSIS **KSHIRSAGAR, 1980**

This species occurs in the black rat *Rattus r. rattus* in India.

The oocysts are spherical or subspherical, 32—44 × 32—40 (mean 35 × 34) μm, with a yellowish brown, one-layered wall 1.5—2 μm thick, without a micropyle or residuum, with a polar granule. The sporocysts are ovoid, globose, or elongate, with a small Stieda body and a compact residuum. The sporozoites are elongate, with one end pointed and the other rounded, with a clear globule at the broad end. The sporulation time is 2—3 d. Kshirsagar (1980) found this species in 0.11% of 900 rats in Maharashtra State.

Reference. Kshirsagar (1980).

ISOSPORA KRISHNAMURTHYI **KSHIRSAGAR, 1980**

This species occurs in the black rat *Rattus r. rattus* in India.

The oocysts are spherical or slightly ovoid, 36—48 × 35—40 (mean 42 × 37) μm, with a pale yellowish, one-layered wall 1.8 μm thick, without a micropyle or residuum, with a polar granule. The sporocysts are ovoid, 21—26 × 17—22 (mean 24 × 19) μm, with a prominent Stieda body and a residuum composed of a few coarse, dispersed granules. The sporozoites are elongate, without a clear globule. The sporulation time is 6 h. Kshirsagar (1980) found this species in 0.11% of 900 rats in Maharashtra State.

Reference. Kshirsagar (1980).

ISOSPORA RATTI LEVINE & IVENS, 1965

This species occurs moderately commonly in the Norway rat *Rattus norvegicus* in North America.

The oocysts are subspherical, 22—24 × 20—21 μm, with a smooth, pale tan to tan, one-layered wall about 1 μm thick, without a micropyle or residuum, with two polar granules. The sporocysts are symmetrical, broadly ovoid, 16 × 11 μm, with a Stieda body and residuum. The sporozoites are colorless, not arranged in any particular order in the sporocysts.

Reference. Levine & Ivens (1965).

WENYONELLA LEVINEI BADYOPADHYAY, RAY, & DAS GUPTA, 1986

This species was found in the feces of house rat *Rattus rattus arboreus* in Asia (West Bengal, India).

The oocysts are ellipsoidal to slightly ovoid, 16.5—22 × 13.5—15 (mean 20 × 14) μm, with a smooth, colorless, two-layered wall 1 μm thick (the layers of equal thickness), without a micropyle, residuum, or polar granule. The sporocysts are ovoid, 7.5—10 × 5—6 (mean 9 × 5) μm, without a Stieda body, with a granular, diffuse residuum. The sporozoites are ellipsoidal, 4.5—5 × 3—4.5 (mean 5 × 3) μm, sometimes lying in pairs. The sporulation time is 36 h at room temperature (35—37°C).

Reference. Badyopadhyay, Ray, & Das Gupta (1986).

SARCOCYSTIS CYMRUENSIS ASHFORD, 1978

This species occurs in the Norway rat *Rattus norvegicus* (type intermediate host) and domestic cat *Felis catus* (type definitive host). The black rat *Rattus rattus* is an experimental intermediate host. Its sarcocysts are in the rat skeletal muscles (but not in the heart or tongue) and its sporocysts are in the cat feces.

Oocysts have not been seen. The sporocysts are ovoid, with one side notably flattened, without a Stieda body, with a residuum. Sporulation occurs in the cat intestine.

Premuscle meronts, if any, are unknown. The sarcocysts are 3—13 (mean 6.5) mm long at 6 months and up to 5 cm long at 9 months. They are not clearly septate, and are thin-walled, with both metrocytes (at first) and merozoites.

Mus musculus cannot be infected with sporocysts from the cat, and *Canis familiaris* and *Mustela putorius* cannot be infected with sarcocysts from the rat.

Reference. Ashford (1978).

SARCOCYSTIS MURINOTECHIS MUNDAY & MASON, 1980

This species occurs in the tiger snake *Notechis ater* (type definitive host) and laboratory rat *Rattus norvegicus* (type intermediate host) in Australia.

In addition, Munday & Mason (1980) believed without proof but on the basis of the similarity of their sarcocysts that the following Tasmanian rodents were also intermediate hosts: eastern swamp rat *Rattus lutreolus*, southern bush rat *R. fuscipes*, black rat *R. rattus*, long-tailed rat *Pseudomys higginsi,* and broad-toothed rat *Mastacomys fuscus.* The sporocysts are in the feces of the snake. Early meronts are in the heart and other tissues; sarcocysts are in the tongue, diaphragm, and skeletal muscles of the rat.

This species does not appear to be pathogenic for the tiger snake, but may kill rats 7—13 d after inoculation.

Sporocysts are found either in pairs or singly in the feces of the snake. They are 10.5—11.5 × 7—9 (mean 11 × 7) μm, without a Stieda body, with numerous residual granules. Sporulation occurs in the snake.

Meronts 9—29 × 9—11.5 μm are in the myocardium, skeletal muscles, lungs, and kidney glomeruli of rats that die 7—13 d after inoculation. At 12 weeks, thick-walled sarcocysts are in the tongue, diaphragm, and skeletal muscles; they are 300 × 60—90 μm or larger, and have short, broad protrusions 6 × 4 μm with many invaginations but without necks.

Mus musculus probably cannot be infected with sporocysts from the tiger snake.

Reference. Munday & Mason (1980).

SARCOCYSTIS MURIS? ŠEBEK, 1975

Sarcocysts of this form have been found rarely in the muscles of the Norway rat *Rattus norvegicus* in Europe.

Reference. Šebek (1975).

SARCOCYSTIS SINGAPORENSIS ZAMAN & COLLEY, 1976

Synonym. *Sarcocystis orientalis* Zaman & Colley, 1975.

This species occurs in the reticulated python *Python reticulatus* (type definitive host), Timor python *P. timorensis*, black-headed python *Aspidites melanocephalus*, rock python *P. sebae*, Norway rat *Rattus norvegicus* (type intermediate host), *R. r. diardii, R. exulans, R. jalorensis,* and *Bandicota indica,* and other rats in Asia. Oocysts, macrogametes, and microgamonts are in the epithelial cells of the snake small intestine, and sarcocysts in the rat muscles. This species is pathogenic for Norway rats. It may kill them in 5—12 d; signs of acute respiratory distress, massive pleural effusions, and sometimes partial posterior paralysis occur.

The oocysts are subspherical or ovoid, 14—17 × 10—12 (mean 15 × 11) μm, with a smooth, colorless, one-layered wall. The sporocysts are ellipsoidal, 8—

11 × 7—10 µm, without a Stieda body, with a residuum. The sporozoites are slightly curved, 6—7 × 2 µm. Sporulation occurs in the host's body.

Sarcocysts are in the skeletal (but not cardiac) muscles. They are spindle-shaped, up to 1 mm long and (in contracted fibers) 120 µm wide; they attain a diameter of 20 µm and contain only metrocytes at 4 weeks; when mature they are septate, with a thick, striated wall with villi up to 5 × 1.5 µm on a base about 1 µm thick, all of whose surfaces bear tiny invaginations but which do not contain fibrils. Their bradyzoites are banana-shaped, up to 6 × 1.5 µm. The metrocytes of mature sarcocysts are several times larger than the bradyzoites and are attached to the primary sarcocyst wall. The bradyzoites in the sarcocysts have micronemes, rhoptries, a conoid, and a micropore.

The prepatent period is apparently 7—12 d and the patent period apparently 73—117 d.

Häfner & Frank (1984) said that *S. singaporensis* has the following hosts: **Definitive Hosts.** *Python reticulatus, P. timorensis*, black-headed python *Aspidites melanocephalus*, rock python *Python sebae*.

Intermediate Hosts. *Rattus norvegicus, R. rattus*, ricefield rat *R. argiventer*, Malaysian field rat *R. tiomanicus*, Polynesian rat *R. exulans*, lesser ricefield rat *R. losea*, long-haired rat *R. villosissimus, R. colletti*, greater bandicoot-rat *Bandicota indica*, lesser bandicoot-rat *B. bengalensis, B. savilei*.

Mus musculus, Mesocricetus auratus, Cavia porcellus, Meriones unguiculatus, Praomys natalensis, Microtus arvalis, Oryctolagus cuniculus, Felis catus, Canis domesticus, Gallus gallus, and *Columba livia* cannot be infected with sporocysts from the snake.

O'Donoghue, Watts, & Dixon (1987) found what they considered to be *S. singaporensis* in the rodents *Maxomys bartelsii, M. musschenbroekii, M. dominator, Rattus xanthurias, Rattus fratorum,* and *R. exulans* from North Sulawesi and West Java; these had a thick, radially striated sarcocyst wall, whereas what they called *S. sulawesiensis* n. sp. in various rodents had a thin sarcocyst wall containing numerous hair-like protrusions (see *S. sulawesiensis* below).

References. Beaver & Maleckar (1981); Brehm (1979); Brehm & Frank (1980); Häfner & Frank (1984); Kan (1976); Kan & Dissanaike (1977); O'Donoghue, Watts, & Dixon (1987); Zaman (1975, 1976); Zaman & Colley (1975, 1976).

SARCOCYSTIS SULAWESIENSIS O'DONOGHUE, WATTS, & DIXON, 1987

This species was reported from the rats *Bunomys chrysocomus, B. fratorum,* and *Paruromys dominator* (Nowak & Paradiso, 1987, assigned all three species to the genus *Rattus*, subgenus *Bullimus*) in North Sulawesi and West Java, Indonesia. The definitive host is unknown. The sarcocysts (muscle meronts) have a thin wall containing numerous hair-like protrusions, in contrast to *S. singaporensis*, whose sarcocysts have a thick, radially striated wall.

Reference. O'Donoghue, Watts, & Dixon (1987).

SARCOCYSTIS VILLIVILLOSI BEAVER & MALECKAR, 1981

This species occurs in the reticulated python *Python reticulatus* (type definitive host) and Norway rat *Rattus norvegicus* (type intermediate host) in Asia. Sporocysts are in the feces of the snake and sarcocysts in the muscles of the rat.

Sarcocysts are in the skeletal but not cardiac muscles of the rat; they are present one month after inoculation. They are elongate fusiform, with tapered ends, septate, up to about 1 mm × 100 μm. Metrocytes are ordinarily present only during the first 6 weeks. The sarcocyst wall is about 2 μm or less thick, coarsely striated, with villi 1.6 × 0.5 μm bearing many microvilli which also arise from the primary sarcocyst wall and tiny pinocytotic-like vessels between them. The bradyzoites from formalin-fixed sarcocysts are about 5 × 1 μm.

Mus musculus and *Macaca mulatta* cannot be infected with sporocysts from the snake.

Häfner & Frank (1984) reported the following definitive and intermediate hosts for *S. villivillosi*:

Definitive Hosts. *Python reticulatus, P. timorensis, P. sebae, Aspidites melanocephalus.*

Intermediate Hosts. *Rattus norvegicus, R. rattus, R. argiventer, R. tiomanicus, R. exulans, R. losea, R. villosissimus, R. colletti, Bandicota indica, B. bengalensis, B. savilei.*

References. Beaver & Maleckar (1981); Häfner & Frank (1984).

SARCOCYSTIS ZAMANI BEAVER & MALECKAR, 1981

This species occurs in the reticulated python *Python reticulatus* (type definitive host), Norway rat *Rattus norvegicus* (type intermediate host) and apparently also *R. rattus diardii, R. exulans, R. annadelei,* and *Bandicota indica.* Sporocysts are in the feces of the snake and sarcocysts in the skeletal (but not the cardiac) muscles of the rat.

The sarcocysts are well differentiated 6 weeks after inoculation of sporocysts, and some persist for at least 2 years. They are broadly fusiform with rounded ends, 1—2 mm by 300—500 μm, and are compartmented when mature. They contain only metrocytes at 6 weeks, both metrocytes and bradzyoites at 8 and 9 weeks, and bradyzoites alone when fully mature. Their wall is relatively thin and generally smooth, with projections of the primary sarcocyst wall forming thin, branched, anastomosing structures up to 2 μm long; these villi are barely visible by light microscopy. The bradyzoites are banana-shaped, 10—12 μm long in the living sarcocyst and 7—8 × 1—1.5 μm in the formalin-fixed sarcocyst.

Mus musculus and *Macaca mulatta* cannot be infected with this species.

References. Beaver & Maleckar (1981); Kan (1979).

SARCOCYSTIS SP. POPE, BICKS, & COOK, 1957

Sarcocysts of this form were found in the muscles of the rat *Rattus assimilis* (type intermediate host) in Australia. The definitive host is unknown.

Reference. Pope, Bicks, & Cook (1957).

SARCOCYSTIS SP. RZEPCZYK, 1974

Sarcocysts of this form were found in the rat *Rattus fuscipes* (type intermediate host) in Australia, transmitted experimentally to the carpet python *Morelia spilotes varigata* (type definitive host), and from it to the laboratory rat *R. norvegicus*. Oocysts and sporocysts are in the python feces, gamonts in its duodenal wall, and sarcocysts in the rat muscles.

The oocysts are thin-walled, presumably ellipsoidal, 14 × 10 μm. The sporocysts are 10 × 7 μm, colorless, without a Stieda body, with a residuum.

Rzepczyk & Scholtyseck (1976) found two species. In one the sarcocysts were 0.59—1.05 mm long, with striated walls 8.5—11 μm thick. In the other the sarcocysts were 0.16—1.09 mm long, with thin, nonstriated walls. Both contained both metrocytes and bradyzoites. Rzepczyk (1974) saw meronts which appeared to be in the capillaries of the skeletal muscles in a baby *R. norvegicus* which had died after being fed sporocysts from a python. These meronts were 13—34 × 7—16 μm in sections, and some contained merozoites about 5 × 2 μm.

References. Rzepczyk (1974); Rzepczyk & Scholtyseck (1976).

SARCOCYSTIS SP. MEHLHORN, HARTLEY, & HEYDORN, 1976

Sarcocysts of this form were found in the muscles of the rat *Rattus lutreolus* (type intermediate host) in Australia. The definitive host is unknown.

Reference. Mehlhorn, Hartley, & Heydorn (1976).

SARCOCYSTIS SP. ŠEBEK, 1963

Sarcocysts of this form were found in the muscles of the Norway rat *Rattus norvegicus* (type intermediate host) in Europe. The bradyzoites are 9—10 μm long. The definitive host is unknown.

Reference. Šebek (1963).

SARCOCYSTIS SP. KAN & DISSANAIKE, 1977

Sarcocysts of this form were found in the muscles of the Malaysian house rat *Rattus rattus diardii* (type intermediate host) in Asia. The definitive host is unknown. Kan & Dissanaike (1977) said that the sarcocysts' fine structure resembled that of *S. singaporensis* and also to a somewhat lesser extent the Type A *Sarcocystis* sp. described from *R. fuscipes* by Rzepczyk & Scholtyseck (1976).

References. Kan & Dissanaike (1977); Rzepczyk & Scholtyseck (1976).

SARCOCYSTIS SP. MARKUS, 1982

This form was found in the black rat *Rattus rattus* in South Africa. Markus (1982) said that its definitive host is the cat and that it infected all laboratory-bred *R. rattus* and some *R. norvegicus* but not *Mus musculus*.

SARCOCYSTIS SP. MIR, CHHABRA, BHARDWAJ, & GAUTAM, 1982

This form was found in the muscles of 12.5% of 186 wild rats *Rattus rattus* in India.

Reference. Mir, Chhabra, Bhardwaj, & Gautam (1982).

SARCOCYSTIS I (S. MURIS-LIKE) LAI, 1977

Sarcocysts of this form were found in Asia in the rats *Rattus exulans* and *R. jalorensis*. The definitive host is unknown.

The sarcocysts are thin-walled, 3—8 mm × 161—290 μm and compartmented. The bradyzoites are 13 × 4 μm.

Reference. Lai (1977).

SARCOCYSTIS II (S. ORIENTALIS-LIKE) LAI, 1977

Sarcocysts of this form were found in the muscles of the rats *Rattus exulans*, *R. annandeli, R. jalorensis,* and *R. rattus diardii* in Asia. The definitive host is unknown. The sarcocysts are 0.4—0.6 mm × 64—115 μm, not compartmented, with a wall 5—7 μm thick, the outer wall with villi 4—6 μm long, the inner wall striated. The bradyzoites are 9 × 2 μm.

Reference. Lai (1977).

SARCOCYSTIS III LAI, 1977

Sarcocysts of this form were found in the muscles of the rats *Rattus annandeli* and *R. rattus diardii* in Asia. The sarcocysts are 0.7—2.3 mm × 223—775 μm, with a thin wall. It is not known whether they have cytophaneres or trabeculae. The bradyzoites are 9 × 2 μm.

Reference. Lai (1977).

SARCOCYSTIS IV LAI, 1977

Sarcocysts of this form occur in the muscles of the rats *Rattus annandeli, R. jalorensis,* and *R. rattus diardii* in Asia. The sarcocysts are 0.4—1.5 mm × 32—102 μm, with a thin wall, compartmented, with bradyzoites 8 × 2 μm.

Reference. Lai (1977).

SARCOCYSTIS V LAI, 1977

Sarcocysts of this form occur in the muscles of the rat *Rattus exulans* in Asia. The sarcocysts are up to 185 × 59 μm, with a bead-like wall 1 μm thick or with striations 1 μm apart, non-compartmented, with bradyzoites 8 × 2 μm.

Reference. Lai (1977).

SARCOCYSTIS SP. (?) MUNDAY, MASON, HARTLEY, PRESIDENTE, & OBENDORF, 1978

Sarcocysts of this form occur in the skeletal and heart muscles of the rats *Rattus fuscipes, R. lutreolus,* and *R. rattus* in Australia. The sarcocysts have thin walls and small bradyzoites. The longest sarcocyst they saw was 900 μm long, and the diameters varied between 18 and 70 μm.

Reference. Munday, Mason, Hartley, Presidente, & Obendorf (1978).

SARCOCYSTIS SP. (?) MUNDAY, MASON, HARTLEY, PRESIDENTE, & OBENDORF, 1978

Sarcocysts of this form were found in all striated muscles except the heart of the rats *Rattus lutreolus*, *R. norvegicus*, and *R. rattus* in Australia. The sarcocysts are microscopic, up to 200 × 75 μm, with thick (4—8 μm), striated walls, fine trabeculae, relatively small bradyzoites, and a primary sarcocyst wall folded into long and very wide protrusions up to 6 μm long and 4 μm wide. Munday & Mason (1980) apparently believed (without proof) that this form was *S. murinotechis*. They could not infect kittens *Felis domestica*, quolls *Dasyurus viverrinus*, or a masked owl *Tyto novaehollandiae* by feeding sarcocysts in naturally-infected eastern swamp rats *Rattus lutreolus*.

References. Munday & Mason (1980); Munday, Mason, Hartley, Presidente, & Obendorf (1978).

SARCOCYSTIS SP. MUNDAY, MASON, HARTLEY, PRESIDENTE, & OBENDORF, 1978

Sarcocysts of this form were found in all striated muscles except the heart of the rats *Rattus norvegicus* and *R. rattus* in Australia. The sarcocysts are microscopic, 54—576 × 27—160 μm, with a thin wall and large bradyzoites. Some have fine trabeculae, and some of the smaller sarcocysts have metrocytes.

Reference. Munday, Mason, Hartley, Presidente, & Obendorf (1978).

SARCOCYSTIS SP. MUNDAY, MASON, HARTLEY, PRESIDENTE, & OBENDORF, 1978

Sarcocysts of this form were found in the muscles of the native rat *Pseudomys higginsi* in Australia. They are macroscopic, up to 2 mm long and 70 μm wide, with a thick (5—10 μm), striated wall, with trabeculae. The bradyzoites are 7 × 2 μm. Munday & Mason (1980) apparently believed (without proof) that this form was *S. murinotechis*. They were unable to infect kittens *Felis domestica*, quolls *Dasyurus viverrinus*, or a masked owl *Tyto novohollandiae* by feeding sarcocysts in naturally infected long-tailed rats *Pseudomys higginsi*.

References. Munday & Mason (1980); Munday, Mason, Hartley, Presidente, & Obendorf (1978).

SARCOCYSTIS SP. COLLINS, 1981

Sarcocysts of this form were found in the leg muscles of the black rat *Rattus rattus* on Stewart Island, New Zealand.

Reference. Collins (1981).

SARCOCYSTIS SP. RUIZ & FRENKEL, 1980

Sarcocysts of this form were found in the striated muscles of the Norway rat *Rattus norvegicus* in Costa Rica. They were not described.

Reference. Ruiz & Frenkel (1980).

SARCOCYSTIS SPP. CROSS, FRESH, JONES, & GUNAWAN, 1973

Sarcocysts of these forms were found in the diaphragms of rats *Rattus rattus diardi*, *R. exulans*, and *R. argentiventer* in Java. They were not described.

Reference. Cross, Fresh, Jones, & Gunawan (1973)

SARCOCYSTIS SPP. SINNIAH, 1979

Sarcocysts of these forms were found in the rats *Rattus annadalei, R. argentiventer, R. rattus diardi, R. exulans, R. tiomanicus, R. norvegicus, R. sabanus* and *Bandicota indica* in Malaysia. They were not described.

Reference. Sinniah (1979).

SARCOCYSTIS SPP. BROWN, CARNEY, VAN PEENEN, & SUDOMO, 1974

Sarcocysts of these forms were found in the rats *Rattus exulans* and *R. polelae* in Indonesia.

Reference. Brown, Carney, Van Peenen, & Sudomo (1974).

SARCOCYSTIS SP. MUNDAY, 1983

Munday (1983) found oocysts of *Sarcocystis* sp. in the feces of the masked owl *Tyto novaehollandiae* in Australia, which produced sarcocysts in the diaphragm of rats *Rattus norvegicus* but not mice *Mus musculus*. He said that the organism was not *S. azevedoi, S. cymruensis, S. murinotechis, S. oryzomyos, S. proechimyos, S. singaporensis,* or various unnamed *Sarcocystis* species, but resembled *S. booliati* in the structure of the sarcocyst wall and bradyzoites. However, he was not sure that this was what it was. Its oocysts were 15—18 × 12—13 (mean 17 × 12.5) μm and its free sporocysts were 11—13 × 7—9 (mean 12 × 8) μm and contained a residuum but no Stieda body.

Reference. Munday (1983).

FRENKELIA SP. HAYDEN, KING, & MURTHY, 1976

This form was found in the brain of the white rat *Rattus norvegicus* in North America. The brain meronts were lobulated, 470—1000 μm in diameter.

Reference. Hayden, King, & Murthy (1976).

BESNOITIA WALLACEI (TADROS & LAARMAN, 1976) DUBEY, 1977

Synonym. *Isospora wallacei* Tadros & Laarman, 1976.

This species occurs in the domestic cat *Felis catus* (type definitive host) and (experimentally) laboratory rat *Rattus norvegicus* (type intermediate host), Polynesian rat *R. exulans,* and house mouse *Mus musculus* in Hawaii and Australia. Meronts and merozoites occur in the small intestine, lamina propria, adjacent endothelial cells, or intimal cells, and liver of rats. Gamonts, gametes, zygotes, and oocysts are formed in the goblet cells of the small intestine of the cat. Meronts and bradyzoites occur in the heart, tongue muscles, mesentery, small intestine wall, and to a lesser extent in the liver and lungs of rats and mice.

The unsporulated oocysts are subspherical, 16—19 × 10—16 μm, with a two-layered, smooth, light brown to pink wall 0.5 μm thick. The sporulated oocysts are ellipsoidal to subspherical, 15—18 × 12—16 μm. The sporocysts are without a Stieda body, with a residuum. The sporozoites are about 10 × 2 μm within the sporocysts and 8 × 2.5 μm outside them when stained with Giemsa; they lie

lengthwise in the sporocysts. The oocyst wall may collapse around the sporocysts upon sporulation. The sporulation time is 64—96 h at 24°C in 1—2%
sulfuric acid solution.

Tachyzoites apparently have not been seen following the feeding of sporulated oocysts to rodents, but they may be present, since a single transfer of 6-d
peritoneal washings from one *Rattus exulans* to another was successful (Frenkel,
1975). Last generation meronts can be found with a dissecting microscope in the
tissues of both rats and mice 30—60 d after the inoculation of oocysts, and can
be seen with the naked eye in mice 40—60 d after inoculation of oocysts. They
degenerate in the mouse by 70 d, but persist in *Rattus norvegicus* at least 393 d
and in *R. exulans* for 64 d. These meronts consist of a host cell (probably a
fibroblast) forming a thick wall with multiple, hypertrophied nuclei and containing some hundreds or thousands of bradyzoites. These meronts are approxi-
mately spherical, up to 200 μm in diameter in sections, with a wall up to 30 μm
thick. They can be found in the heart, tongue muscles, and lungs of experimentally infected mice.

After newborn kittens have ingested meronts containing merozoites in mouse
tissues, one generation of asexual reproduction occurs in the lamina propria of
the small intestine. Meronts 500—800 μm in diameter develop asynchronously
between days 4 and 14. They generally extend into a blood vessel, appearing to
be in endothelial or intimal cells. They also occur in the liver. Macrogametes
10—13 μm in diameter can be found in the goblet cells in the small intestine of
these kittens 13—16 d after ingestion of mouse meronts; they are usually
between the vacuole and the nucleus. Microgamonts up to 11 μm in diameter are
also present. The prepatent period is 12—15 d, and the patent period is 5—12 d.

There is no evidence that this species is pathogenic, either for cats or rodents.

Serologic cross-reactions occur in the indirect fluorescent antibody test
between this species and *B. jellisoni, Toxoplasma gondii,* and *Sarcocystis muris,*
and between this species and *B. jellisoni* with the dye test, but not between this
species and *T. gondii* with the dye test. Mice previously infected with *B. wallacei*
are immune against reinfection and also against *B. jellisoni.*

This species can be transmitted from the cat to *Rattus norvegicus, R. exulans,*
and *Mus musculus,* but not to *R. rattus, Mesocricetus auratus,* or *Cavia
porcellus.* Ooocysts cannot be produced in the bobcat, cougar, fox, dog, coyote,
raccoon, skunk, owl, hawk, or a boid, a colubrid, and three viperid snakes (the
scientific names of these reptiles were not given) (Wallace & Frenkel, 1975).

Remarks. This organism was isolated by G. D. Wallace from a stray cat on
Oahu, Hawaii, and found by Mason in a wild *R. norvegicus* in Australia.

References. Dubey (1977); Frenkel (1955, 1975, 1977); Mason (1980);
Tadros & Laarman (1976); Wallace (1975); Wallace & Frenkel (1975).

BESNOITIA JELLISONI FRENKEL, 1955

The laboratory rat *Rattus norvegicus* is an experimental intermediate host of
this species.

References. See under *Peromyscus.*

BESNOITIA SP. McKENNA & CHARLESTON, 1980

McKenna & Charleston (1980) found *Besnoitia* sp. in a feral cat *Felis domestica* in New Zealand. They transmitted it to the laboratory rat *Rattus norvegicus*, and laboratory rabbit *Oryctolagus cuniculus*, laboratory mouse *Mus musculus*, but not to the guinea pig *Cavia porcellus*. They then transmitted it from the rat to kittens. They said that this form differed from *B. besnoiti, B. darlingi, B. wallacei,* and a cat-rat form described in Japan by Ito, Tsunoda, & Shimura (1978). (Ito & Shimura [1986] said that this Japanese species was actually *B. wallacei.*)

Meronts killed rats 33 and 78 d after infection from the original cat. They were in the walls of the ileum and cecum, and to a lesser extent in the skeletal muscles, heart, and brain. They were up to 260 μm in diameter, with a wall 7—11 μm thick containing several host cell nuclei and thousands of crescentic bradyzoites 10—11 × 2—3 (mean 11 × 3) μm when unfixed; the bradyzoites had pointed ends.

References. Ito & Shimura, 1986; Ito, Tsunoda, & Shimura, 1978; McKenna & Charleston, 1980a.

KLOSSIA SP. MULLIN & COLLEY, 1972

This form was found in Whitehead's rat *Rattus whiteheadi* in Asia.

The oocysts are spherical to subspherical, 33—48 × 22—40 (mean 41 × 34) μm, with a smooth, thin, one-layered wall, without a micropyle, with a residuum, apparently without a polar granule. Four sporulated oocysts contained 12—20 spherical sporocysts without a Stieda body, with a residuum; the number of sporozoites per sporocyst could not be determined.

This may be a pseudoparasite of the rat, since an adult *R. whiteheadi* could not be infected with oocysts of this species.

Reference. Mullin & Colley (1972).

BESNOITIA WALLACEI (TADROS & LAARMAN, 1976) DUBEY, 1977

Synonyms. *Besnoitia* sp. Wallace & Frenkel, 1975; *Isospora wallacei* Tadros & Laarman, 1976.

The definitive host of this species is the cat *Felis catus*. The type intermediate host is the Norway rat *Rattus norvegicus*. Other (experimental) intermediate hosts are the house mouse *Mus musculus*, Polynesian rat *Rattus exulans*, Mongolian gerbil *Meriones unguiculatus*, vole *Microtus montebelli*, hamster *Mesocricetus auratus,* and rabbit *Oryctolagus cuniculus*. *B. wallacei* has been found in Oceania (Hawaii), Australia, and Japan. It occurs in the intestinal wall of both intermediate and definitive hosts and the heart and tongue muscles of the rat.

The intestinal meronts are spherical, 90—210 μm in diameter, with a wall 1—30 μm thick; they probably occur in fibroblasts. The intestinal merozoites are crescentic, 10—13 × 2—3 μm. The muscle meronts are 108—162 × 90—162 μm at 55—56 d and have a fibrillar wall. The oocysts are ellipsoidal, 15—19 × 10—16 μm, with a two-layered wall about 0.5 μm thick, without a micropyle, residuum, or polar granule. The sporocysts are ellipsoidal, 10—11 × 7—8 (mean 11 × 8) μm, without a Stieda body, with a residuum. The sporozoites are elongate, 10 × 2 μm within the sporocysts or 8 × 2.5 μm outside the sporocysts when stained with Giemsa.

Wallace & Frenkel (1975) were unable to produce a meront infection in the golden hamster, but Ito & Shimura (1986) said that this species is slightly infective for the hamster. Wallace & Frenkel (1975) could not produce oocysts in the bobcat, cougar, fox, dog, coyote, raccoon, skunk, owl, hawk, a boid snake, a colubrid snake, or three viperid snakes. (They did not give the scientific names of these animals.) Fayer & Frenkel (1979) could not infect calves with oocysts from calves.

Ito & Shimura (1986) said that the large type of *Isospora bigemina* from the cat is actually *B. wallacei*.

References. Dubey (1977); Ito & Shimura (1986); Ito, Tsunoda, & Shimura (1978); Mason (1980); Tadros & Laarman (1976) Wallace & Frenkel (1975).

KLOSSIELLA (?) SP. HARTIG & HEBOLD, 1970

This form was found in the kidney tubules of albino Sprague-Dawley laboratory rats *Rattus norvegicus* in Germany.

Reference. Hartig & Hebold (1970).

HOST GENUS *MUS*

CARYOSPORA BIGENETICA WACHA & CHRISTIANSEN, 1982

Wacha & Christiansen (1982) found that *Caryospora bigenetica* of the timber rattlesnake *Crotalus horridus* could be transmitted experimentally to the laboratory mouse *Mus musculus*, in which it could be found in the connective tissue, dermis, and hypodermis. They considered the genus *Caryospora* to be heteroxenous, with two hosts (primary and secondary) "with gamogony occurring in each; development is intestinal in the primary host, resulting in formation of oocysts; it is extraintestinal in the secondary host, resulting in formation of caryocysts. Primary hosts are mainly birds and reptiles, secondary hosts are mammals, where known."

The mouse is actually a transport host; *Caryospora* (or at least *C. bigenetica*) can be considered heteroxenous only if the existence of a transport host is considered to be acceptable as part of the definition of heteroxeny.

How is *C. bigenetica* actually transmitted? By oocysts from snake to snake? Can it be transmitted from mouse to snake? Is a mammalian intermediate host necessary?

Reference. Wacha & Christiansen (1982).

CARYOSPORA BUBONIS CAWTHORN & STOCKDALE, 1981

This species was found in the feces of the great horned owl *Bubo virginianus*.

Stockdale & Cawthorn (1981) and Cawthorn & Stockdale (1982) found that it may have either a direct or indirect life cycle. In the latter case the white mouse *Mus musculus* can be the intermediate (transport) host.

References. Cawthorn & Stockdale (1981, 1982); Stockdale & Cawthorn (1981).

CARYOSPORA SIMPLEX LÉGER, 1904

Upton, Current, Ernst, & Barnard (1984) transmitted this species from the Ottoman viper *Vipera x. xanthina* to the laboratory mouse *Mus musculus* and cotton rat *Sigmodon hispidus* by oral inoculation of sporulated oocysts or sporocysts. The protozoon produced microgametes, macrogametes and sporulated oocysts in the connective tissue of the cheek, tongue, and nose of the rodents. They also transmitted the protozoon to mice by intraperitoneal injection of the sporozoites. They saw unsporulated oocysts in the orally inoculated mice 10—12 d after inoculation, and sporulated oocysts 12 d after inoculation. They said that this was "the first report of a heteroxenous coccidium with both sexual and asexual development in the primary (predator) and secondary (prey) hosts."

Clearly, other species of *Caryospora* may turn out to be transmitted directly from predator host to predator host by means of sporulated oocysts or via a prey transport host. We need to learn the situation in the other 17 named species of *Caryospora*.

References. Léger (1904); Upton, Current, Ernst, & Barnard (1984).

EIMERIA ARASINAENSIS MUSAEV & VEISOV, 1965

This species occurs in the house mouse *Mus musculus* in the USSR.

The oocysts are ellipsoidal, colorless, 12—24 × 10—20 (mean 19 × 16) µm, with a one-layered, colorless wall 1.5 µm thick, with a micropyle and small micropylar cap, without a residuum, usually with a polar granule. The sporocysts are comma-shaped or piriform, with a clear globule in the large end. The sporulation time is 2—3 d.

Reference. Musaev & Veisov (1965).

EIMERIA BAGHDADENSIS MIRZA, 1975

This species occurs uncommonly in the house mouse *Mus musculus* in Asia (Iraq).

The oocysts are ellipsoidal but narrowed and flattened at both ends, 20—24 × 17—20 (mean 22 × 18) µm, with a smooth, yellow to yellowish green, one-layered wall about 1 µm thick, without a micropyle or residuum, with a polar granule. The sporocysts are elongate ovoid, 10—13 × 6—8 (mean 12 × 7) µm, with a thin wall, prominent Stieda body, and residuum. The sporozoites are elongate, with one end narrower than the other, lack clear globules, and lie lengthwise head to tail in the sporocysts.

References. Mirza (1970, 1975).

EIMERIA CONTORTA HABERKORN, 1971

This is a mixture of *E. nieschulzi* of *Rattus norvegicus* and *E. falciformis* of *Mus musculus*.

EIMERIA FALCIFORMIS (EIMER, 1870) SCHNEIDER, 1875

Synonyms. *Gregarina falciformis* Eimer, 1870; *G. muris Rivolta*, 1878; *Coccidium falciforme* (Eimer, 1870) Schuberg, 1892; *Pfeifferia schubergi* (Labbé, 1896) Labbé, 1899; *Eimeria schubergi* (Labbé, 1896) Doflein, 1916; *E. pragensis* Černá & Senaud, 1969; *E. falciformis* var. *pragensis* Černá, Sénaud, Mehlhorn, & Scholtyseck, 1975; *Eimeria*-like species Clarke, 1895.

This species occurs in the epithelial cells of the cecum and upper colon (and occasionally the lower ileum) of the laboratory or house mouse *Mus musculus* throughout the world. It is uncommon in the U.S., but common in other countries. A great deal of research has been done on this species.

The oocysts are ovoid, ellipsoidal, subspherical, or spherical, 14—27 × 11—24 µm, with a smooth, apparently colorless, one- or two-layered wall, without a micropyle, with or without a residuum and polar granule. The sporocysts are ovoid, ellipsoidal, or spherical, 10—12 × 6—8 µm, with a Stieda body and residuum. The sporozoites are elongate, with one end broader than the other, have a clear globule at the broad end, and lie lengthwise in the sporocysts. The sporulation time is 3—6 d.

There are ordinarily four sexual generations. Mature first generation meronts occur 2 d after inoculation; they are 18 × 12 µm and contain 10—12 merozoites 13 × 2 µm. Mature second generation meronts occur 3 d after inoculation; they are 10 × 9 µm and contain eight to ten merozoites 5 × 1.6 µm. Mature third generation meronts occur 5 d after inoculation; they are 17.5 × 10 µm and contain 10—20 merozoites 12 × 2 µm. Mature fourth generation meronts occur 6 d after inoculation; they are 18 × 15 µm and contain 18—50 merozoites 4.5 × 1 µm. Meronts of all generations also contain a residuum (Mesfin & Bellamy, 1978; Mehlhorn, Sénaud, & Scholtyseck, 1973). The merozoites all have a typical pellicle with two polar rings and a single micropore. They also have a conoid, two or more rhoptries, many micronemes, and 26 subpellicular microtubules. The rhoptries and micronemes are joined by cross-connections to form a single unit (Scholtyseck & Mehlhorn, 1970). The merozoites are not sexually predetermined, since both macrogametes and microgamonts are produced after infection with a single merozoite (Haberkorn, 1970). Oral, subcutaneous, intramuscular, intravenous, and intraperitoneal exposures produce identical infections (Haberkorn, 1970).

Using a pure strain obtained from a single oocyst in a specific-pathogen-free mouse, Owen (1974) found three asexual generations and meronts of two types. A large meront about 15 × 8.5 µm contained 12—20 merozoites and a small one about 3 × 4 µm contained 3—7 merozoites.

Sexual stages may be produced by any merozoite generation, but mostly by the second and third generations. They may appear 3—8 d after inoculation. The macrogametes are about 16 × 14.5 μm; they have several micropores and many wall-forming bodies I and II ("plastic granules"). The microgamonts are about 7—9 × 7—15 μm; they have several micropores and form relatively few microgametes about 3—6 μm long and a residuum. Each microgamete has a perforatorium, a mitochondrion, a longitudinal filamentous structure of unknown function behind the perforatorium at the anterior end of the nucleus, and presumably two or three flagella. The prepatent period is 3—7 d, and the patent period is 6—16 d.

The number of oocysts produced per oocyst fed decreases progressively with increasing doses of oocysts. When 10 oocysts were fed, Tilahun & Stockdale (1981) found that 289,000 oocysts were produced per oocyst fed.

Light infections have little effect, but heavy ones may cause diarrhea and even death. A catarrhal enteritis may occur, together with desquamation of the intestinal epithelium and hemorrhage. The apical area of the absorptive enterocytes increases, and the goblet cells increase in number at first and then eventually disappear (Speer & Pollari, 1984). The feces may contain blood. Mortality is highest 4—10 d after inoculation. Different strains vary in pathogenicity. Germ-free mice are apparently immune (Owen, 1974). Day-old conventional mice are much more resistant than older ones. Susceptibility to infection rises slowly with age until weaning, when there is a marked increase (Schrecke & Durr, 1970).

There are slight cross-reactions between *E. falciformis, E. ferrisi, E. papillata* and *E. vermiformis* in the indirect immunofluorescence test (IFAT) using sporozoites as antigen, and strong IFAT reactions against the homologous species (Tilahun & Stockdale, 1982). IgA is the most important antibody in local immunity (Nash, Callis, & Speer, 1985).

The Norway rat and dog cannot be infected with this species

References. Becker (1934); Bonfante, Faust & Giraldo (1961); Černá (1970); Černá & Sénaud (1969, 1970); Černá, Sénaud, Mehlhorn, & Scholtyseck (1974); Clarke (1895); Cordero del Campillo (1959); Doflein (1916); Eimer (1870); Golemanski (1979), Gousseff & Suz'ko (1945); Haberkorn (1970); Koffman (1936); Labbé (1896, 1899); Levine & Ivens (1965); Mehlhorn, Sénaud, & Scholtyseck (1973); Mehlhorn & Scholtyseck (1974); Mesfin & Bellamy (1978, 1979, 1979a, 1979b); Mesfin, Bellamy, & Stockdale (1978); Mirza (1970); Nieschulz & Bos (1931); Nöller (1920); Owen (1974); Pellérdy, Haberkorn, Mehlhorn, & Scholtyseck (1971); Pérard (1926); Reimer (1923); Rivolta (1878); Schneider (1875); Scholtyseck & Ghaffar (1981); Scholtyseck & Mehlhorn (1970); Scholtyseck, Mehlhorn, & Haberkorn (1971); Scholtyseck, Mehlhorn, & Hammond (1971, 1972); Scholtyseck, Mehlhorn, & Sénaud (1972); Scholtyseck, Pellérdy, Mehlhorn, & Haberkorn (1973) Rose & Hesketh (1987); Schrecke & Dürr (1970); Schuberg (1892); Sénaud & Černá (1968, 1970); Tilahun & Stockdale (1981, 1982); Wenyon (1926); Ball & Lewis (1984); Kyle & Speer (1984); Speer & Pollari (1984); Whitmire & Speer (1984); Douglass & Speer (1985); Nash, Callis, & Speer (1985); Owen (1983); Whitmire & Speer (1986)

EIMERIA FERRISI LEVINE & IVENS, 1965

This species occurs moderately commonly in the epithelial cells of the villi (and sometimes the lamina propria) of the cecum and colon, and sometimes in the epithelial cells of the crypts of Lieberkuehn of the house mouse *Mus musculus* in North America

The oocysts are ellipsoidal to subspherical, spherical or rarely ovoid, 12—22 × 11—18 (mean 17—18 × 14—15) μm, with a smooth, colorless to light brown, one-layered wall 0.9—1 μm thick, often lined by a dark membrane, without a micropyle or residuum, with one to three polar granules. The sporocysts are elongate ovoid or ellipsoidal, 8—11 × 5—7 μm, with a small Stieda body, with or without a residuum, with a thin wall. The sporozoites are elongate, have a clear globule at the small end, and lie lengthwise head to tail in the sporocysts. Sporulation occurs outside the host body.

There are three meront generations. The first generation meronts are mature 1 d after inoculation; they are 7—14 × 6—13 (mean 11 × 10) μm and contain 7—14 (mean 10) merozoites. The second generation meronts are seen 1.5—3 d after inoculation; they are 5—13 × 6—12 (mean 10 × 8) μm and contain 6—25 (mean 18) merozoites. The third generation meronts are seen 3 d after inoculation; they are 12—15 × 9—13 (mean 14 × 11) μm and contain 5—16 (mean 12.5) merozoites (Ankrom, Chobotar, & Ernst, 1975). Fully developed merozoites have a three-layered pellicle, 32 subpellicular microtubules which originate at the anterior polar ring, 2 polar rings, a conoid, 2 rhoptries, micronemes, Golgi apparatus, endoplasmic reticulum, mitochondria, polysaccharide granules, and a posterior ring (Chobotar, Scholtyseck, Sénaud, & Ernst, 1975).

Macrogametes are mature 3 d after inoculation; they are 5—14 × 6—13 (mean 11 × 10) μm, have a crescent-shaped body in the parasitophorous vacuole, and three limiting membranes. Microgamonts are mature 4 d after inoculation; they are 7—14 × 6—10 (mean 11 × 8) μm and have many microgametes in long, narrow whorls on their periphery or in whorls at the surface of two to five compartments. The prepatent period is 3 d and the patent period 3—4 d.

The number of oocysts produced per oocyst fed decreases progressively with increasing doses of oocysts. With 10 oocysts, 160,000 oocysts were produced per oocyst fed (Tilahun & Stockdale, 1981).

This species is pathogenic for mice and may kill them. There are strain differences in resistance in mice. Resistance is cell-mediated, thymus-dependent, and due to a dominant gene (Klesius & Hinds, 1979).

Infection of mice with this species produces partial immunity to subsequent challenge (Blagburn, Chobotar, & Smith, 1979). Transfer factor from calves resistant to *E. bovis*, given either by mouth or by systemic injection, immunizes mice against *E. ferrisi* (Klesius et al., 1979, 1979a).

There are slight IFAT cross-reactions between *E. ferrisi* and *E. falciformis*, *E. papillata* and *E. vermiformis* (Tilahun & Stockdale, 1982). However, Douglas & Speer (1985) found no cross-reaction between *E. ferrisi* and *E. falciformis*.

References. Ankrom, Chobotar, & Ernst (1975); Blagburn, Chobotar, &

Smith (1979); Chobotar, Scholtyseck, Sénaud & Ernst (1975); Klesius, Elston, Chambers, & Fudenberg (1979); Klesius & Hinds (1979); Klesius, Qualls, Elston, & Fudenberg (1979); Levine & Ivens (1965); Tilahun & Stockdale (1981, 1982); Douglas & Speer (1985).

EIMERIA HANSONORUM LEVINE & IVENS, 1965

This species occurs uncommonly in the house mouse *Mus musculus* in North America.

The oocysts are subspherical, 15—22 × 13—19 (mean 18 × 16) μm, with a smooth, pale yellowish, one-layered wall 0.8 μm thick, without a micropyle or residuum, with a polar granule. The sporocysts are ovoid, about 9 × 7 μm, thick-walled, with a broad, thick Stieda body and a residuum. The sporozoites are pale and lie lengthwise in the sporocysts. Sporulation occurs outside the host's body.

Reference. Levine & Ivens (1965).

EIMERIA HINDLEI YAKIMOFF & GOUSSEFF, 1938

This species occurs moderately commonly in the house *Mus musculus* in the USSR and Europe.

The oocysts are ovoid, presumably 22—27 × 18—21 μm, with a smooth wall, without a micropyle or residuum, with a polar granule. The sporocysts are presumably ovoid, 9 × 6 μm, possibly with a residuum. The sporozoites have a clear globule at one end. The sporulation time is presumably 3 d.

Remarks. Veisov (1973) concluded on the basis of his own studies that Yakimoff & Gousseff's *E. hindlei* was actually *E. krijgsmanni*. Whatever it was, their description was so incomplete that *E. hindlei* will probably never be recognized with certainty again.

References. Golemanski (1979); Levine & Ivens (1965); Musaev & Veisov (1965); Svanbaev (1956); Veisov (1964, 1973); Yakimoff & Gousseff (1938).

EIMERIA KEILINI YAKIMOFF & GOUSSEFF, 1938

This species occurs moderately commonly in the house mouse *Mus musculus* in the USSR.

The oocysts are yellowish, pointed at both ends, 24—32 × 18—21 (mean 29 × 19) μm, with a smooth wall, without a micropyle, residuum, or polar granule. The sporocysts are 12 × 6 μm, possibly without a Stieda body.

Remarks. The above description is so deficient that it is doubtful whether this species will ever be recognized with certainty again.

Reference. Yakimoff & Gousseff (1938).

EIMERIA KRIJGSMANNI YAKIMOFF & GOUSSEFF, 1938

This species occurs moderately commonly to commonly in the house mouse *Mus musculus* in the USSR. It is possibly in the lower small intestine.

The oocysts were described by Yakimoff & Gousseff (1938) as oval but illustrated as ellipsoidal, 18—23 × 13—16 (mean 22 × 15) μm, with a smooth, colorless or yellowish wall, without a micropyle or residuum, with a polar granule. The sporocysts were ovoid, 12 × 9 μm. Musaev & Veisov (1965) described the oocysts that they called those of this species as ellipsoidal or ovoid, colorless, 12—32 × 10—26 (mean 19 × 15) μm, with a smooth, one-layered, colorless wall 1—1.5 μm thick without a micropyle or residuum, with a polar granule. The sporocysts were ellipsoidal or ovoid, 6—14 × 4—10 (mean 9 × 7) μm, illustrated without a Stieda body but with a residuum. The sporozoites were comma-shaped, piriform, or lemon-shaped, illustrated without a clear globule. The sporulation time is presumably 2—7 d.

There are presumably two asexual generations. The prepatent period is presumably 4—5 d and the patent period presumably 9—10 d (Veisov, 1973).

This species may be weakly pathogenic.

Remarks. Černá (1962) considered *E krijgsmanni* to be a synonym of *E. schueffneri*. Veisov (1973) thought that *E. hindlei* was actually *E. krijgsmanni*. However, the original description was so deficient that it is doubtful whether this species will ever be recognized with certainty.

References. Černá (1982); Chobotar, Sénaud, & Scholtyseck (1978); Musaev & Veisov (1965); Svanbaev (1956); Veisov (1964, 1973); Yakimoff & Gousseff (1938).

EIMERIA MUSCULI YAKIMOFF & GOUSSEFF, 1938

This species occurs moderately commonly in the house mouse *Mus musculus* in the USSR.

The oocysts are spherical, 21—26 μm in diameter, smooth, greenish, without a micropyle, residuum, or polar granule. The sporocysts are broadly ovoid.

Remarks. The description of this species is so deficient that it is doubtful if this species will ever be recognized with certainty again.

This was probably a misidentification. Svanbaev (1979) reported this species from *Microtus arvalis* in Kazakhstan.

Reference. Yakimoff & Gousseff (1938); Svanbaev (1979).

EIMERIA MUSCULOIDEI LEVINE, BRAY, IVENS, & GUNDERS, 1959

This species occurs in the upper ileum of the mouse *Mus* (*Leggada*) *musculoides* (type host) in Africa and possibly in the house mouse *M. musculus* in Asia.

The oocysts are subspherical to ellipsoidal, 17—22 × 15—19 (mean 20 × 17) μm, with a pale yellowish to yellowish brown, smooth, one-layered wall about 1 μm thick, without a micropyle or residuum, with one to several polar granules. The sporocysts are lemon-shaped, 10—12 × 7 (mean 10 × 7) μm, without a Stieda body or with a small, rather flat one, with a residuum. The sporozoites lie more or less lengthwise in the sporocysts. The sporulation time is 2—4 d at room temperature in Liberia.

The mature meronts average 10 × 8 μm and have 28—36 merozoites.

References. Levine, Bray, Ivens, & Gunders (1959); Mirza (1970).

EIMERIA PAPILLATA ERNST, CHOBOTAR, & HAMMOND, 1971

This species occurs uncommonly to moderately commonly in the house mouse *Mus musculus* in North America and Europe.

The oocysts are spherical, subspherical, or ellipsoidal, 18—26 × 16—24 (mean 22 × 19) μm, with a one-layered yellowish brown wall about 1.2 μm thick, appearing rough and striated because of many short, papilla-like projections on its surface, without a micropyle or residuum, with one to three polar granules. The sporocysts are ovoid, 10—13 × 6—9 (mean 11 × 8) μm, with a small Stieda body, a lentiform sub-Stiedal body about 1.5 × 2 μm, and a residuum. The sporozoites have a large clear globule.

The merozoites have the usual apical complex of organelles. These disappear when the macrogametes are formed. The prepatent period is 3 d and the patent period 8 d. The number of oocysts produced per oocyst fed decreases progressively with increasing doses of oocysts. With 10 oocysts, 221,000 oocysts are produced per oocyst fed (Tilahun & Stockdale, 1981).

There are slight cross reactions against *Eimeria falciformis, E. ferrisi,* and *E. vermiformis* with the IFAT (Tilahun & Stockdale, 1982).

References. Chobotar, Sénaud, Ernst, & Scholtyseck (1980); Chobotar, Sénaud, & Scholtyseck (1978); Ernst, Chobotar, & Hammond (1971); Golemanski (1979); Tilahun & Stockdale (1981, 1982); Danforth, Augustine, & Chobotar (1984); Danforth, Chobotar, & Entzeroth (1984).

EIMERIA PARAGACHAICA MUSAEV & VEISOV, 1965

This species occurs in the house mouse *Mus musculus* in the USSR.

The oocysts are ellipsoidal or ovoid, 24—32 × 18—24 (mean 28 × 22) μm, with a colorless, one-layered wall 1.5—2 μm thick, with a micropyle, micropylar cap, residuum, and polar granule. The sporocysts are piriform, 10—14 × 6—9 (mean 13 × 8) μm, with a Stieda body and residuum. The sporozoites are comma- or lemon-shaped, with a clear globule at the large end. The sporulation time is 3—4 d.

References. Glebezdin (1974a); Musaev & Veisov (1965).

EIMERIA SCHUEFFNERI YAKIMOFF & GOUSSEFF, 1938

This species occurs moderately commonly in the house mouse *Mus musculus* in the USSR.

The oocysts are cylindrical (illustrated with rounded ends), 18—26 × 15—16 μm, with a smooth, colorless wall, without a micropyle, residuum, or polar granule. The sporocysts are ovoid.

Remarks. The description of this species is so deficient that it is doubtful if it will ever be recognized with certainty again.

Reference. Yakimoff & Gousseff (1938).

EIMERIA TENELLA (RAILLIET & LUCET, 1891) FANTHAM, 1909

Naciri & Yvoré (1982) found that house mice *Mus musculus* could act as transport hosts of this chicken species. The mice did not become ill, and no oocysts were produced in them.

Reference. Naciri & Yvoré (1982).

EIMERIA VERMIFORMIS ERNST, CHOBOTAR, & HAMMOND, 1971

This species occurs quite commonly, mostly in the epithelial cells of the lower $^2/_3$ of the small intestine but occasionally in the cecum and colon and (in heavy infections) in the lamina propria of the lower small intestine of the house mouse *Mus musculus* in North America.

The oocysts are subspherical to broadly ellipsoidal, sometimes tapering toward each end, 18—26 × 15—21 (mean 23 × 18) μm, with a 2-layered wall, the outer layer lightly pitted, brownish yellow, about 0.5 μm thick, the inner layer colorless to light yellow, about 1 μm thick, without a micropyle or residuum, with one to three polar granules. The sporocysts are ovoid, 11—14 × 6—10 μm, with a small Stieda body and a residuum. The sporozoites are elongate, have a clear globule at the large end, and lie lengthwise in the sporocysts.

There are three meront generations. The first generation meronts mature at 40 h, are 16—27 × 13—19 (mean 21 × 16) μm and contain about 22 merozoites. The second generation meronts are mature 4 d after inoculation; they are 16—25 × 9—16 (mean 20 × 13) μm and contain 28—50 (mean 39) vermiform merozoites 15—18 × 2 (mean 17 × 2) μm in fresh smears; the merozoites usually lie lengthwise within the meronts but occasionally have a spiral arrangement; their nuclei are in their anterior third. Mature third generation meronts are first seen 5 d after inoculation. They are 8—18 × 7—14 (mean 13 × 11) μm and contain about 10—20 short, blunt merozoites 5—7 × 2 (mean 6 × 2) μm.

Immature gamonts are present 5 d after inoculation. The mature macrogametes are 12—19 × 11—16 (mean 16 × 13) μm. The mature microgamonts are 17—32 × 12—25 (mean 23.5 × 19) μm and contain many microgametes about 2—3 × 0.5 μm and a diffuse mass of residual material. The prepatent period is 6—7 d and the patent period 7—22+ d. The number of oocysts produced per oocyst fed decreases progressively with increasing doses of oocysts. With 10 oocysts, 187,000 oocysts are produced per oocyst fed (Tilahun & Stockdale, 1981).

Slight cross reactions occur against *E. falciformis*, *E. ferrisi*, and *E. papillata* in the IFAT.

Different strains of mice differ in susceptibility (Rose, Owen, & Hesketh, 1984).

Normal weanling laboratory rats *Rattus norvegicus* cannot be infected, but they can be infected if fed 0.5—1 mg dexamethasone daily for 6—7 d before feeding them sporulated oocysts (Todd, Lepp, & Trayser, 1971; Todd & Lepp, 1972).

E. vermiformis has been cultivated from sporozoite to mature first generation meront in bovine kidney cells (Kelley & Youssef, 1977).

References. Blagburn, Adams, & Todd (1982, 1984); Blagburn & Todd

(1986); Ernst, Chobotar, & Hammond (1971); Kelley & Youssef (1977); Tilahun & Stockdale (1981, 1982); Todd & Lepp (1971, 1972); Todd, Lepp, & Trayser (1971); Adams & Todd (1983); Owen (1983); Rose & Hesketh (1987); Rose, Owen, & Hesketh (1984).

EIMERIA SP. MUSAEV & VEISOV, 1965

This form was found in the house mouse *Mus musculus* in the USSR.

The oocysts are broadly ovoid, almost spherical, 16—22 × 14—18 (mean 21 × 17) μm, with a smooth, colorless, two-layered wall 2.5 μm thick, the layers of equal thickness, with a residuum, without a polar granule. The sporocysts are "oval," 6—10 × 4—7 (mean 8 × 6) μm, with a residuum. The sporozoites are bean-shaped. The sporulation time is 3—4 d.

Reference. Musaev & Veisov (1965).

EIMERIA SP. VEISOV, 1973

This form occurs in the large intestine of the house mouse *Mus musculus* in the USSR.

The oocysts are ellipsoidal or ovoid, 12—23 × 10—20 (mean 18 × 14) μm, with a double-contoured wall and a "refractile globule." The sporulation time is 130-192 h at 27—28°C.

There are three asexual generations in the mouse, but they have not been described. The prepatent period is 106—120 h and the patent period is 230—240 h.

This species is pathogenic, especially for month-old mice.

Reference. Veisov (1973).

ISOSPORA CANIS NEMESÉRI, 1959

The type definitive host of this species is the dog *Canis familiaris*. However, Dubey (1975) found that the house mouse *Mus musculus* is a transport host. Hypnozoites are present in its lymph nodes. Other information on this species was given by Levine & Ivens (1981).

References. Dubey (1975); Heine (1981); Levine & Ivens (1981); Markus (1983); Nemeséri (1959).

ISOSPORA FELIS WENYON, 1923

The type definitive host of this species is the cat *Felis catus*. However, the house mouse *Mus musculus*, laboratory rat *Rattus norvegicus*, and golden hamster *Mesocricetus auratus* have been found to be experimental transport hosts. Hypnozoites are in their lymph nodes. Other information on this species was given by Levine & Ivens (1981).

Reference. Levine & Ivens (1981); Heine (1981); Wenyon (1923).

ISOSPORA OHIOENSIS DUBEY, 1975

The type definitive host of this species is the dog *Canis familiaris*; other definitive hosts are the coyote *C. latrans*, dingo *C. dingo* (?), fox *Vulpes vulpes* (?), and perhaps the raccoon dog *Nyctereutes procyonoides*. However, the house mouse *Mus musculus* has been found to be an experimental transport host. Hypnozoites are in its lymph nodes. Other information on this species was given by Levine & Ivens (1981).

References. Levine & Ivens (1981); Dubey (1975).

ISOSPORA RIVOLTA (GRASSI, 1879) WENYON, 1923

The type definitive host of this species is the cat *Felis catus*. However, the house mouse *Mus musculus*, laboratory rat *Rattus norvegicus*, golden hamster *Mesocricetus auratus*, domestic dog *Canis familiaris*, and chicken *Gallus gallus* have been found by Dubey (1977) to be experimental transport hosts. Hypnozoites are in their lymph nodes. Other information on this species was given by Levine & Ivens (1981).

References. Arcay (1981); Brösigke, Heine, & Boch (1981); Dubey (1977); Levine & Ivens (1981); Grassi (1879); Wenyon (1923).

ISOSPORA VULPINA NIESCHULZ & BOS, 1933

The type definitive host of this species is the fox *Vulpes vulpes*. However, the house mouse *Mus musculus* has been found to be an experimental transport host. Hypnozoites are in its lymph nodes. Other information on this species was given by Levine & Ivens (1981).

Reference. Gdebezbin (1978); Levine & Ivens (1981); Nieschulz & Bos (1933).

SARCOCYSTIS CROTALI ENTZEROTH, CHOBOTAR, & SCHOLTYSECK, 1985

This species occurs in the rattlesnake *Crotalus s. scutulatus* (definitive host) and house mouse *Mus musculus* (intermediate host).

The sarcocysts in the mouse are septate and have an unstriated, thin, highly folded primary wall and prominent ground substance. The sporulated oocysts from the snake are 17—19 × 10—12 (mean 18 × 11) μm; they are ellipsoidal, with a thin, smooth, colorless, one-layered wall. The sporocysts are ellipsoidal, 10—12 × 8—10 (mean 11 × 9) μm, with a smooth wall, without a Stieda body, with a residuum. The sporozoites are banana-shaped, 7—8 × 2—3 (mean 7 × 2) μm, and have a prominent clear globule near the rounded end.

Reference. Entzeroth, Chobotar, & Scholtyseck (1985).

SARCOCYSTIS DISPERSA ČERNÁ, KOLÁŘOVÁ, & ŠULC, 1978

Synonyms. *Sarcocystis* sp. Černá, 1976 in *Tyto alba*; *Sarcocystis* sp. Munday, 1977, in *Tyto novaehollandiae*.

This species occurs in the barn owl *Tyto alba* (type definitive host), masked owl *T. novaehollandiae*, and long-eared owl *Asio otus* (both definitive hosts) in Europe and Australia. The house mouse *Mus musculus* (type intermediate host) is an experimental intermediate host. *Microtus nivalis* cannot be infected. Early meronts are in the mouse hepatocytes and terminal meronts (sarcocysts) in the mouse muscles. Oocysts are in the owl feces.

The oocysts are 17—22 × 10—14 μm, without a micropyle, residuum, or polar granule. The sporocysts are ovoid, 11—14 × 8—12 μm, without a Stieda body, with a residuum. Sporulation occurs in the owl intestine.

Pre-muscle meronts can be found in the mouse hepatocytes 4—7 d after inoculation. They are ovoid, 19—14 μm in diameter, without a parasitophorous vacuole, and form tachyzoites 5—7 × 2—3 μm by synchronous endodyogeny. These enter the bloodstream, where they multiply, probably by endodyogeny. They are free in the blood or in the macrophages 7—8 d after inoculation. They then enter the striated muscles and form sarcocysts. These are elongate, compartmented, up to 1.8 mm × 30—60 μm, with a delicate wall. They contain ovoid metrocytes 4 × 3 μm at first and then banana-shaped bradyzoites 5—9 × 2—4 μm. The tachyzoites in the liver have a conoid, polar rings, a three-membraned wall, a micropore, an apical ring from which 22 subpellicular microtubules arise, micronemes, rhoptries, a central nucleus, Golgi apparatus, abundant ergastoplasm in the cytoplasm formed by saccules and vesicles, and little polysaccharide reserves. The metrocytes in the sarcocysts are globular, with a typical apical complex, conoid, three-membraned wall, and one or several micropores. The metrocytes divide by endodyogeny to form bradyzoites. These have a three-membraned wall, one micropore, an apical complex composed of a conoid, two polar rings, an apical ring from which 22 subpellicular microtubules arise, often a posterior ring, a few rhoptries, many micronemes, a dictyosome near the globular nucleus, mitochondria with pediculated crests, ergastoplasmic saccules, some lipid globules, and polysaccharide granules. They multiply by endogenesis.

The IFAT can be used for diagnosis, and the eluate from dried blood samples of mice is a satisfactory source of antibody.

Microtus arvalis cannot be infected.

References. Černá (1976, 1977, 1983); Černá & Kolářová (1978); Černá, Kolářová, & Šulc (1978); Černá & Sénaud (1977); Černá & Vánová (1979); Gut (1982); Kolářova (1986); Munday (1977); Sénaud & Černá (1978); Tadros (1981); Červá & Černá (1982).

SARCOCYSTIS MURIS (BLANCHARD, 1885) LABBÉ, 1899

Synonyms. *Miescheria muris* Blanchard, 1885; *Coccidium bigeminum* var. *cati* Railliet & Lucet, 1891; *Sarcocystis musculi* Blanchard, 1885 of Kalyakin & Zasukhin (1975) *lapsus calami*; *Endorimospora muris* (Blanchard, 1885) Tadros & Laarman, 1976.

This species occurs in the domestic cat *Felis catus* (type definitive host) and house mouse *Mus musculus* (type intermediate host) throughout the world. *Microtus pennsylvanicus* and *Meriones unguiculatus* cannot be infected (Woodmansee & Powell, 1984). Oocysts and sporocysts are in the cat feces and meronts in the mouse hepatocytes at first and then in the mouse muscles.

158 *The Coccidian Parasites of Rodents*

The sporocysts in cat feces are ellipsoidal, 8—12 × 7.5—9 (mean 11 × 9) μm, without a Stieda body, with a residuum. The sporozoites are slightly curved, about 10 × 2 μm, with a central nucleus. Sporulation occurs in the cat intestine.

Pre-muscle meronts are in the hepatocytes. Their tachyzoites are arranged in a rosette around a residuum. Terminal meronts (sarcocysts) in the muscles are up to 6 mm long, compartmented, with a thick wall, and contain metrocytes and bradyzoites measuring 14—16 × 4—6 μm in Giemsa-stained impression smears.

The laboratory rat *Rattus norvegicus* cannot be infected, nor can the ox *Bos taurus*, either as an intermediate, transport, or definitive host.

For further information on this species, see Levine & Ivens (1981).

References. Abbas & Powell (1983); Blanchard (1885); Entzeroth (1984); Fayer & Frenkel (1979); Kalyakin & Zasukhin (1975); Labbé (1899); Levine & Ivens (1981); McKenna & Charleston (1980), Mehlhorn & Frenkel (1980); O'Donoghue & Weyreter (1983, 1984); Railliet & Lucet (1891); Ruiz & Frenkel (1976; 1980a); Šebek (1975); Selá-Perez, Martinez, Arias, & Ares (1982); Sheffield, Frenkel, & Ruiz (1977); Tadros & Laarman (1976); Woodmansee (1983); Woodmansee & Powell (1984).

SARCOCYSTIS SCOTTI LEVINE & TADROS, 1980

Synonym. *Sarcocystis* sp. Tadros & Laarman, 1980.

This species occurs in the tawny owl *Strix aluco* (Definitive host) and house mouse *Mus musculus* (Intermediate host).

References. Tadros & Laarman (1980); Levine & Tadros (1980); Tadros (1981).

SARCOCYSTIS SP. MUNDAY, MASON, HARTLEY, PRESIDENTE, & OBENDORF, 1978

This species occurs rarely in the skeletal muscles of the house mouse *Mus musculus* (type intermediate host) in Australia. The definitive host is unknown.

The sarcocysts are thin-walled, 40—60 μm in diameter, without visible trabeculae, and contain large bradyzoites.

Reference. Munday, Mason, Hartley, Presidente, & Obendorf (1978).

SARCOCYSTIS SP. MUNDAY, MASON, HARTLEY, PRESIDENTE, & OBENDORF, 1978

This species occurs rarely in the skeletal muscles of the house mouse *Mus musculus* (type intermediate host) in Australia. The definitive host is unknown.

The sarcocysts are thin-walled, up to 750 μm long and 35—60 μm in diameter, and have relatively small bradyzoites.

Reference. Munday, Mason, Hartley, Presidente, & Obendorf (1978).

TOXOPLASMA GONDII (NICOLLE & MANCEAUX, 1908) NICOLLE & MANCEAUX, 1909

Synonyms. See under *Ctenodactylus* for a list of the 18 synonyms.

The type definitive host of this species is the domestic cat *Felis catus*. Intermediate hosts are the house mouse *Mus musculus* and about 200 species of mammals and birds, including many rodents and man. *T. gondii* is common and worldwide in distribution. The oocysts are in felid feces, and the meronts in the brain, peritoneal exudate (mononuclear leukocytes), intestine, and other sites of the intermediate hosts.

The oocysts are spherical to ellipsoidal, 11—14×9—11 (mean 12.5×11) µm, without a micropyle, residuum, or polar granule. The sporocysts are ellipsoidal, about 8.5 × 6 µm, without a Stieda body, with a residuum. The sporozoites are elongate, wider at one end than the other, and about 8 × 2 µm. Sporulation occurs outside the host's body.

The tachyzoites (pre-brain merozoites) and bradyzoites (brain merozoites) are elongate, wider at one end than at the other, 4—8 × 2—4 µm. There are no metrocytes.

A great deal of research has been done on this species. The mouse is a favorite laboratory host because it is readily infected with merozoites. Several books have been written on *T. gondii*, and further information can be obtained from them or any textbook of human, veterinary, or general parasitology.

References. Levine (1985); Levine & Ivens (1981); etc. See under *Ctenodactylus*.

TOXOPLASMA HAMMONDI (FRENKEL, 1974) LEVINE, 1977

Synonyms. *Hammondia hammondi* Frenkel, 1974; *H. hammondi* Frenkel & Dubey, 1975; *Isospora (Toxoplasma) datusi* Overdulve, 1978.

This species occurs in the domestic cat *Felis catus* (type definitive host) and house *Mus musculus* (type intermediate host); other intermediate hosts are the laboratory rat *Rattus norvegicus*, golden hamster *Mesocricetus auratus*, guinea pig *Cavia porcellus*, deermouse *Peromyscus* sp., *Mastomys* sp., and marmoset *Saguinus nigricollis*. All the intermediate hosts are experimental. *T. hammondi* has been found in Hawaii, Ohio, and Europe. The oocysts are in cat feces and the meronts in the striated muscles, small intestine, lymph nodes, brain, and probably elsewhere of the intermediate hosts.

The oocysts are subspherical to spherical, 10—13 × 11—13 (mean 11 × 11) µm, with a colorless, two-layered wall about 0.5 µm thick, without a micropyle, residuum, or polar granule. The sporocysts are ellipsoidal, 8—11 × 6—8 (mean 10 × 6.5) µm, without a Stieda body, with a residuum. The sporozoites are elongate, curved, about 7 × 2 µm, with a nucleus near the center. Sporulation occurs outside the host's body.

The terminal (muscle) meronts are elongate, 100—340×40—95 µm, uncompartmented, without cytophaneres, with a smooth, intensely folded primary wall beneath which is a zone of granular ground substance and no secondary wall. The bradyzoites are about 4—7 × 2 µm. There are no metrocytes.

Remarks. This species is closely related serologically to *Toxoplasma gondii* (see, for instance, Araujo, Dubey, & Remington 1984).

References. Araujo, Dubey, & Remington (1984); Fayer & Frenkel (1979); Frenkel (1974); Frenkel & Dubey (1975); Gjerde (1983); Hiepe, Nickel, Jungman, Hansel, & Unger (1980); Levine (1977); Mehlhorn & Frenkel (1980); Overdulve (1978); Tadros & Laarman (1976).

TOXOPLASMA PARDALIS (HENDRICKS, ERNST, COURTNEY & SPEER, 1979) LEVINE & IVENS, 1981

Synonyms. *Hammondia pardalis* Hendricks, Ernst, Courtney, & Speer, 1979; *Isospora* sp. Long & Speer, 1977.

This species was found in the ocelot *Felis pardalis* (type definitive host) in Panama and transmitted experimentally to the house mouse *Mus musculus*. Hendricks et al. (1979) then transmitted it from the mouse to the jaguarundi *F. yagouroundi*. It occurs presumably in the intestine of the definitive hosts, and the mesenteric lymph nodes, lungs, and intestinal mucosa of the mouse.

The oocysts are ovoid, 36—46 × 25—35 (mean 41 × 28.5) μm, with a one-layered wall 1.8 μm thick lined by a dark membrane, with a micropyle (?) or operculum (?) at the small end and a similar structure at the broad end, usually without a residuum, with a polar granule. The sporocysts are ellipsoidal, 19—25 × 14—19 (mean 22 × 16) μm, without a Stieda body or sub-Stiedal body, with one to several small, residual granules. The sporozoites are sausage-shaped, without clear globules.

References. Hendricks, Ernst, Courtney, & Speer (1979); Levine & Ivens (1981); Long & Speer (1977).

BESNOITIA BESNOITI (MAROTEL, 1913) HENRY, 1913

Synonyms. *Sarcocystis besnoiti* Marotel, 1913; *Gastrocystis robini* Brumpt, 1913; *G. besnoiti* (Marotel, 1912) Brumpt, 1913; *Globidium besnoiti* (Marotel, 1912) Wenyon, 1926; *Isospora besnoiti* (Marotel, 1912) Tadros & Laarman, 1976.

The type definitive host of this species is the domestic cat *Felis catus*; the wild cat *F. libyca* is also a definitive host. The type intermediate host is the ox *Bos taurus*. Other intermediate hosts include the following rodents (all experimental): house mouse *Mus musculus*, guinea pig *Cavia porcellus*, gerbil *Meriones tristrami shawi*, golden hamster *Mesocricetus auratus*, ground squirrel *Spermophilus fulvus*, and marmot *Marmota* sp. *B. besnoiti* occurs throughout the world. The oocysts are in the feces of the cat. The meronts are in the subcutis, cutis, conjunctiva, and other locations of the intermediate hosts.

The oocysts are ovoid, 14—16 × 12—14 μm, similar to those of *Toxoplasma gondii*.

The meronts are more or less spherical, usually visible to the naked eye, aseptate, with a two-layered wall, the outer layer thick and homogeneous or concentrically laminated, the inner layer thin, containing several flattened giant host nuclei. The bradyzoites are crescent- or banana-shaped, with one end pointed and the other rounded. There are no metrocytes.

Peteshev, Galuzo, & Polomoshnov (1974) were unable to produce oocysts in the house mouse, hedgehog, marmot, or various other mammals.

Neuman, Nobel, & Perelman (1979) and Neuman & Nobel (1981) found that this species might or might not cause central nervous system signs and death in experimentally infected rabbits, golden hamsters, guinea pigs, gerbils, and white mice.

Shkap, Pipano, & Greenblatt (1983) inoculated *B. benoiti* (from green monkey kidney cell cultures) intraperitoneally or subcutaneously into gerbils *Meriones tristrami*. Those inoculated intraperitoneally with 10 parasites survived, while those inoculated intraperitoneally with 10 million died within 6—8 d. None of the gerbils inoculated subcutaneously died, but they had nodules at the site of injection. The survivors had become immunized.

Samish et al. (1982) cultivated *B. besnoiti* in cell cultures of the ticks *Rhipicephalus appendiculatus, Boophilus microplus,* and *Dermacentor variabilis.*

See Levine (1985) and Levine & Ivens (1981) for further information.

References. Brumpt (1913); Henry (1913); Levine (1985); Levine & Ivens (1981); Marotel (1913); Neuman & Nobel (1981); Neuman, Nobel, & Perelman (1979); Peteshev, Galuzo, & Polomoshnov (1974); Samish et al. (1982); Shkap, Pipano, & Greenblatt (1983); Wenyon (1926).

BESNOITIA DARLINGI (BRUMPT, 1913) MANDOUR, 1965

Synonyms. *Fibrocystis darlingi* (Brumpt, 1913) Babudieri, 1932; *Sarcocystis darlingi* Brumpt, 1913; *Besnoitia panamensis* Schneider, 1965; *B. sauriana* Garnham, 1966.

The type definitive host of this species is the domestic cat *Felis catus*. The intermediate hosts are various marsupials, other mammals, and lizards, including the following rodents (all experimental): house mouse *Mus musculus*, squirrels *Sciurus granatensis* and *S. variegatoides*, golden hamster *Mesocricetus auratus*. The oocysts are in cat feces, and the meronts in the lung, kidney, brain, etc. of the intermediate hosts. *B. darlingi* has been found in Central and North America.

The oocysts are subspherical, 11—13 × 11—13 (mean 12 × 12) μm, without a micropyle, residuum, or polar granule. The sporocysts are ellipsoidal, 6—9 × 5—6 (mean 8 × 5) μm, without a Stieda body, with a residuum. The sporozoites are about 5 × 2 μm. Sporulation occurs outside the host.

The tachyzoites in peritoneal exudate are crescentic, spindle-shaped, piriform, or ovoid, 7—11 × 2—3 μm. The organ meronts are ovoid or spherical, may appear lobed, 66—300 × 62—156 μm, with a three-layered wall, the outer layer 2.5—3 μm thick, the middle layer containing large host cell nuclei, and the inner layer a thin membrane. There are thousands of bradyzoites but no metrocytes in the organ meronts.

References. Babudieri (1932); Brumpt (1913); Garnham (1966); Mandour (1965); Schneider (1965); Smith & Frenkel (1977).

BESNOITIA WALLACEI (TADROS & LAARMAN, 1976) DUBEY, 1977

Synonyms. *Isospora wallacei* Tadros & Laarman, 1976; *Besnoitia* sp. Wallace & Frenkel, 1975.

The type definitive host of this species is the domestic cat *Felis catus*. The house mouse *Mus musculus* and laboratory rat *Rattus norvegicus* are experimental intermediate hosts. Oocysts are in the feces of the definitive host, and meronts in the intestinal wall of the intermediate hosts. This species occurs in Hawaii and Australia. For further information, see under *Rattus norvegicus*.

References. See under *Rattus norvegicus*.

BESNOITIA JELLISONI FRENKEL, 1955

The house mouse *Mus musculus* is an experimental intermediate host of this species.

References. See under *Peromyscus*.

BESNOITIA WALLACEI (TADROS & LAARMAN, 1976) DUBEY, 1977

The house mouse *Mus musculus* is an experimental intermediate host of this species.

References. See Under *Rattus*.

BESNOITIA SP. McKENNA & CHARLESTON, 1980

This cat-rat species is transmissible to *Mus musculus* as an intermediate host.

Reference. See under *Rattus*.

BESNOITIA SPP. MATUSCHKA & HÄFNER, 1984

Matuschka & Häfner (1984) found that the snakes *Bitis arietans, B. caudalis, B. gabonica,* and *B. nasicornis* in Africa were all the definitive hosts of *Besnoitia* spp. and that rodents of the genera *Mus, Mesocricetus, Phodopus, Gerbillus, Meriones,* and *Mastomys* were intermediate hosts. The rodents had macroscopic *Besnoitia* cysts up to 2 mm in diameter in their connective tissue. They thought that rodents might be reservoir hosts of *Besnoitia* of cattle.

Reference. Matuschka & Häfner (1984).

CRYPTOSPORIDIUM MURIS TYZZER, 1907

This species is known to occur in the house mouse *Mus musculus* and ox *Bos bovis*, presumably throughout the world. It is either attached to the surface of the epithelial cells or in their brush border in the small and large intestines. It may cause diarrhea in both immunologically compromised and uncompromised hosts.

The oocysts are fully sporulated in the feces, spherical, subspherical, or ovoid, $7—8 \times 5—6.5$ (mean 7×6) μm, with a one-layered, smooth wall 0.5 μm thick, a faint, longitudinal suture extending from one pole down each side for about $^1/_3$ to $^1/_2$ the oocyst length, without micropyle or polar granule, with a large residuum and four vermiform sporozoites $10—12.5 \times 1$ (mean 11×1) μm, lying lengthwise and head-to-head in each oocyst, with a clear globule at the large end.

Remarks. Until recently, it was thought that the species of *Cryptosporidium* in man and many other mammals was *C. muris*. However, Upton & Current (1985) found oocysts of both *C. muris* and *C. parvum* in the feces of naturally infected calves, and determined that the oocysts of the two species were markedly different in size. They concluded that there are only two valid species of *Cryptosporidium* in mammals, and that all previous reports of cryptosporidiosis in mammals have been of a species indistinguishable from *C. parvum*.

References. Brändler (1982); Current (1985); Tyzzer (1907, 1910); Upton & Current (1985) .

CRYPTOSPORIDIUM PARVUM TYZZER, 1912

Synonyms. *Cryptosporidium agni* Barker & Carbonell, 1974; *C. bovis* Barker & Carbonell, 1974; *C. cuniculus* Inman & Takeuchi, 1979; *C. felis* Iseki, 1979; *C. garnhami* Bird, 1981, *C. rhesi* Levine, 1981; *C. wrairi* Vetterling, Jervis, Merrill, & Sprinz, 1971.

This species occurs in the house mouse *Mus musculus* (type host), the rodents *Apodemus sylvaticus*, *Clethrionomys glareolus*, *Sciurus carolinensis*, *Cavia porcellus*, *Mesocricetus auratus,* and *Rattus norvegicus,* and at least 15 other mammalian species of 7 orders throughout the world. It is either attached to the surface of the epithelial cells or in their brush border in the small and large intestines or stomach. It may cause diarrhea in both immunologically compromised and uncompromised hosts.

The oocysts are spherical, subspherical, ellipsoidal, or ovoid, 4.5—5 × 4—5 (mean 5 × 4.5) μm, with a smooth, colorless, one-layered wall about 0.4 μm thick, with a faint, subterminal suture at one end of the oocyst that extends diagonally across $^1/_3$ to $^1/_2$ of the oocyst's width, without a micropyle or polar granule, with a residuum and four elongate sporozoites lying head-to-head in the oocyst, with or without clear globules, with a posterior nucleus. Most authors say that there are no sporocysts, but Brändler (1982) said that there are and that the oocyst wall is five-layered and the sporocyst wall two-layered. Sporulation occurs in the host's intestinal wall. Brändler (1982) found sporocysts and some oocysts in the rectal contents of experimentally infected mice 4 d after inoculation.

Free sporozoites may be found in the stomach or intestine. They are boomerang-shaped, with a thin, pointed anterior end with a rod-shaped nucleus near it. They can be distinguished from merozoites by their greater length and by the shape of their nuclei.

The meronts, like the oocysts, microgamonts, and macrogametes, have an attachment organelle. There are apparently two generations. The first generation meronts contain eight crescentic merozoites each, and the second generation meronts contain four merozoites each. A globule of residual material is left near the attachment organelle. The merozoites are 5—8 μm long. They are present 2 d after inoculation, have an apical complex, a mitochondrion, and 28 subpellicular microtubules.

Gamogony begins 12 h after inoculation. The microgamonts are similar in structure and mode of development to meronts, but are smaller and contain a relatively large amount of chromatin. Their maximum size is 5 × 3.5 μm. Each forms 16 nonflagellate microgametes which can be found 3 d after inoculation.

The macrogametes are spherical to ellipsoidal, with an attachment organelle at one end that makes them appear flask-shaped. After fertilization, their limiting membrane becomes thicker and forms a dense oocyst wall.

This species has been transmitted from man to the sheep, pig, Norway rat, house mouse, guinea pig, chicken, and ox, from the ox to the sheep, goat, rabbit, Norway rat, house mouse, pig, golden hamster, and guinea pig, and from the red deer to the house mouse.

Pathogenicity. *C. parvum* may cause diarrhea in both immunologically compromised and uncompromised hosts.

Remarks. See Levine (1984) for a review. Until recently, it was not realized that *C. parvum* may affect many different animal species. Now, however, it is being often found in man, and its zoonotic potential has been realized.

References. Barker & Carbonell (1974); Bird (1981); Brändler (1982); Current (1985); Ernest, Blagburn, Lindsay, & Current (1986); Goebel & Braendler (1982); Inman & Takeuchi (1979); Iseki (1979); Levine (1981, 1984); Tyzzer (1912); Tzipori, Angus, Campbell, & Gray (1980); Tzipori & Campbell (1981); Upton & Current (1985); Vetterling, Jervis, Merrill, & Sprinz (1971); etc.

KLOSSIELLA MABOKENSIS BOULARD & LANDAU, 1971

See under the type host, *Praomys jacksoni. Mus musculus* is an experimental host.

KLOSSIELLA MURIS SMITH & JOHNSON, 1902

This species occurs in the house mouse *Mus musculus* throughout the world. Merogony is in the epithelial cells of the capillaries and arterioles in the kidneys, lungs, spleen, and other organs. Gamogony and sporogony occur in the epithelial cells of the convoluted tubules of the kidneys. *K. muris* has been found in 10—100% of the mice in some laboratory colonies, but its true prevalence is unknown.

It is not certain whether true oocysts occur in this genus. The sporocysts develop from a sporont within a vacuole in the host cell. They are surrounded by a membrane, but this may be a host cell structure. Sporulation occurs within the host. The sporont divides by multiple fission to form 12—16 sporoblasts and a residual body. Within each sporoblast about 25—34 banana-shaped sporozoites and a sporocyst residuum are formed by multiple fission. The resulting sporocysts are subspherical to spherical, about 16 × 13 μm, and have a very thin wall. The sporozoites lie side by side in the sporocysts. The sporocysts are released into the lumen of the kidney tubules by rupture of the host cell, and pass out in the urine. Infection presumably takes place by ingestion of sporocysts, and the sporozoites presumably pass by way of the blood stream. Merogony takes place in the kidney glomeruli and also to a lesser extent in the lungs, spleen, brain, suprarenal glands, thyroid gland, lymph nodes, and pituitary gland. The meronts are primarily in the endothelial cells of the capillaries. In each meront, 40—60 more or less falciform merozoites 7×2 µm are formed. These break out of the meront, and those produced in Bowman's capsule in the kidneys pass into the lumen of the kidney tubules.

These merozoites enter the epithelial cells of the convoluted tubules, where they become gamonts and where gamogony and sporogony take place. A macrogamete and microgamont are found together within a vacuole in the host cell. The microgamont divides to form two microgametes. Fertilization takes place, and the fertilized macrogamete (zygote) becomes a sporont or "mother sporoblast" which may reach a diameter of 40 µm. The prepatent period is unknown. It is presumably long, since sporocysts are not shed by mice less than 6 months old.

In heavy infections the kidneys have minute grayish spots over their entire surface. These spots are foci of necrobiotic changes in the cortex, followed by marked cell proliferation. A perivascular, follicular, lymphocytic infiltration in the region of the medullary cortex is of diagnostic significance. The epithelium of the infected kidney tubules is destroyed, and the tubules come to look like elongated, contorted bags filled with sporocysts. There is no inflammatory reaction. Fatal infections have not been reported, although infections may decrease metabolic capability and endurance time.

References. Bogovsky (1956); Bonciu, Dincolesco, & Petrovici (1958); Bonne (1925); Elaut (1932); Gard (1945); Otto (1945); Rosenmann & Morrison (1975); Šebek (1975); Smith & Johnson (1902); von Sternberg (1929); Stevenson (1915); Sureau (1963); Twort & Twort (1923); Wilson & Edrissian (1974).

HOST GENUS *HYDROMYS*

KLOSSIELLA HYDROMYOS WINTER & WATT, 1971

This species occurs in the Australian water rat *Hydromys chrysogaster* in Australia. Sporogony occurs in the kidney tubules, and the meronts are mainly in the kidney glomeruli and to a lesser extent in the kidney blood vessels.

The oocysts develop in the kidney tubules, forming up to 13 sporocysts each. The sporocysts contain up to 30 sporozoites each. The meronts are up to 35 µm in diameter and contain numerous very small nuclei. Each microgamont produces one to four microgametes.

Reference. Winter & Watt (1971).

SARCOCYSTIS SP. MUNDAY, MASON, HARTLEY, PRESIDENTE, & OBENDORF, 1978

This form occurs uncommonly in the muscles of the Australian water rat *Hydromys chrysogaster* in Australia. The sarcocysts are thin-walled and contain small bradyzoites.

Reference. Munday, Mason, Hartley, Presidente, & Obendorf, (1978).

HOST FAMILY GLIRIDAE

HOST GENUS *GLIS*

EIMERIA GLIRIS MUSAEV & VEISOV, 1961

This species occurs commonly in the common dormouse *Glis* (syn., *Myoxus*) *glis* in the USSR.

The oocysts are ovoid, 14—23 × 12—17 (mean 21 × 16) μm, with a smooth, two-layered wall, the outer layer yellowish, 1.25 μm thick, the inner layer dark brown, 1.25 μm thick, with a micropyle and residuum, with a polar granule. The sporocysts are ovoid, 6—10 × 4—8 (mean 9 × 7) μm, without a Stieda body, with a residuum. The sporozoites are piriform. The sporulation time is 3 d at 25—30°C in 2.5% potassium bichromate solution.

Reference. Musaev & Veisov (1961).

HOST GENUS *ELIOMYS*

EIMERIA MYOXI GALLI-VALERIO, 1940

This species occurs in the garden dormouse *Eliomys* (syn., *Myoxus*) *quercinus* in Europe.

The oocysts are ovoid, 18 × 15 μm, with a wall relatively thin at the micropylar end, with a poorly visible micropyle, presumably without a residuum or polar granule. The sporocysts are ovoid, 7.5 × 6 μm, presumably without a residuum. The sporozoites are piriform. The sporulation time is 7 d on moist filter paper.

Reference. Galli-Valerio (1940).

HOST GENUS *DRYOMYS*

EIMERIA ABDILDAEVI DZERZHINSKII, 1982

This species occurs commonly in the forest dormouse *Dryomys nitedula* in the USSR.

The oocysts were described as oval but illustrated as ellipsoidal, with a flat micropylar end, 16—22 × 14—18.5 (mean 20 × 17) μm, with a smooth, one-layered wall 1 μm thick, with a micropyle and residuum, apparently without a polar granule. The sporocysts were described as oval but illustrated as ellipsoidal, 9 × 6 μm, without a Stieda body or residuum.

Reference. Dzerzhinskii (1982).

EIMERIA ABUSALIMOVI MUSAEV & VEISOV, 1965

This species occurs commonly in the forest dormouse *Dryomys nitedula* in the USSR and Europe.

The oocysts are ovoid or ellipsoidal, 26—32 × 18—26 μm, with a rough, yellow-brown, one-layered wall 3 μm thick, without a micropyle, residuum, or polar granule. The sporocysts are ovoid, 10—15 × 6—10 μm, with a Stieda body and residuum. The sporozoites are comma-shaped or piriform, illustrated without a clear globule. The sporulation time is 4 d.

References. Golemanski (1979); Musaev & Veisov (1965).

EIMERIA ASADOVI MUSAEV & VEISOV, 1965

This species occurs in the forest dormouse *Dryomys nitedula* in the USSR.

The oocysts are ovoid or ellipsoidal, 18—32 × 14—26 (mean 25 × 22) µm, with a two-layered wall, the outer layer smooth, colorless, 1—1.5 µm thick, the inner layer yellow-brown, 1—1.5 µm thick, without a micropyle, residuum, or polar granule. The sporocysts are spherical, ellipsoidal, or rarely ovoid, 6—12 × 4—8 µm, apparently without a Stieda body, with a residuum. The sporozoites are comma-shaped or piriform, illustrated without a clear globule. The sporulation time is 3—4 d.

Reference. Musaev & Veisov (1965).

EIMERIA DYROMIDIS ZOLOTAREV, 1935

This species occurs commonly in the forest dormouse *Dryomys nitedula* in the USSR and Europe.

The oocysts are ovoid, rarely spherical, 16—30 × 13—25 µm, with a smooth, double-contoured wall apparently composed of one layer 1 µm thick, without a micropyle or residuum, with or without a polar granule (depending on the author). The sporocysts are ovoid, 7—14 × 6—9 µm, apparently without a Stieda body, with a residuum. The sporozoites are piriform, with or without a clear globule (depending on the author). The sporulation time is 2 d at 25—30°C in 2.5% potassium bichromate solution.

References. Golemanski (1979); Musaev & Veisov (1959a); Zolotarev (1935).

EIMERIA NACHITSCHEVANICA MUSAEV & VEISOV, 1959

This species occurs in the forest dormouse *Dryomys nitedula* in the USSR.

The oocysts are ovoid, 20—26 × 17—20 (mean 24 × 19) µm, with a smooth, colorless, apparently one-layered wall 1.3 µm thick, with a prominent micropyle, with a micropylar cap, without a residuum or polar granule. The sporocysts are ovoid, 10—13 × 6—9 (mean 12 × 7) µm, with a Stieda body and residuum. The sporozoites are comma-shaped, with a clear globule at the broad end. The sporulation time is 3 d at 25—30°C in 2.5% potassium bichromate solution.

Reference. Musaev & Veisov (1959).

EIMERIA NITEDULAE MUSAEV & VEISOV, 1965

This species occurs in the forest dormouse *Dryomys nitedula* in the USSR.

The oocysts are ellipsoidal or subspherical, 22—32 × 18—28 (mean 27 × 24) µm, with a smooth, colorless, one-layered wall 2 µm thick, without a micropyle or polar granule, with a residuum. The sporocysts are ellipsoidal or spherical, 8— 12 × 6—10 (mean 10 × 9) µm, apparently without a Stieda body, with a residuum.

The sporozoites are comma-shaped, with a clear globule at the broad end. The sporulation time is 2—3 d.

Reference. Musaev & Veisov (1965).

ISOSPORA DYROMIDIS GLEBEZDIN, 1974

This species occurs in the forest dormouse *Dryomys nitedula* in the USSR.

The oocysts are ellipsoidal, 23—26 × 20—23 μm, with a smooth, colorless, one-layered wall, without a micropyle, residuum, or polar granule. The sporocysts are ellipsoidal, 12—15 × 9—12 (mean 14 × 11) μm, without a Stieda body; Glebezdin (1974) said that a residuum was not seen, but residual material is illustrated in his drawing.

Reference. Glebezdin (1974).

KLOSSIA MUSABAEVAE DZERZHINSKII, 1982

This species occurs uncommonly in the forest dormouse *Dryomys nitedula* in the USSR.

The oocysts were described as oval but illustrated as ellipsoidal, 34—39 × 28.5—33 (mean 36.5 × 30) μm, with a two-layered wall 2 μm thick, without a micropyle, residuum, or polar granule, containing eight to ten spherical sporocysts 8 μm in diameter, each with four bean-shaped sporozoites and many residual granules.

Reference. Dzerzhinskii (1982).

HOST FAMILY ZAPODIDAE

HOST GENUS *SICISTA*

EIMERIA JAKUNINI DZERZHINSKII, 1981

This species occurs in the birch mouse *Sicista tianschanica* in the USSR.

The oocysts are ovoid, 21—25 × 16—19 (mean 22 × 18) μm, with a smooth, one-layered wall 1 μm thick, with a micropyle, without a residuum or polar granule. The sporocysts are ovoid, 8.5 × 6 μm, without a Stieda body or residuum. The sporulation time is 2—3 d.

Reference. Dzerzhinskii (1981).

EIMERIA SICISTAE LEVINE & IVENS, 1987

Synonyms. *Eimeria tianschanica* Dzerzhinskii & Svanbaev, 1980 *nomen nudum*; *E. tianschanica* of Dzerzhinskii (1981).

This species occurs in the birch mouse *Sicista tianschanica* in the USSR.

The oocysts were described as oval but illustrated as ellipsoidal, 30—36.5 × 21—26.5 (mean 34.5 × 24) μm, with a striated, greenish, apparently one-layered wall 2 μm thick, without a micropyle or polar granule, with a residuum. The sporocysts were described as oval but illustrated as ellipsoidal, apparently

169

without a Stieda body, with a residuum. The sporozoites are bean-shaped, 10 ×
3 μm. The sporulation time is 2—3 d.

References. Dzerzhinskii (1981); Dzerzhinskii & Svanbaev (1980); Levine
& Ivens, (1981).

HOST GENUS *ZAPUS*

EIMERIA HUDSONI DUSZYNSKI, EASTHAM, & YATES, 1982

This species occurs in the jumping mouse *Zapus hudsonius luteus* in North
America.

The oocysts are ellipsoidal, 18—23 × 13—16 (mean 21 × 14) μm, with a two-
layered wall 1 μm thick, the outer layer smooth, colorless, 0.5 μm thick, the inner
layer colorless, about 0.5 μm thick, with a micropyle and residuum, without a
polar granule. The sporocysts are ovoid, 8—11 × 5—7 (mean 10.5 × 6) μm, with
a Stieda body and residuum, without a sub-Stiedal body. The sporozoites have
a large clear globule near the posterior end and a nucleus in the anterior half.

Reference. Duszynski, Eastham, & Yates (1982).

EIMERIA ZAPI GERARD, CHOBOTAR & ERNST, 1977

This species occurs quite commonly to commonly in the jumping mice *Zapus
hudsonius* (type host), *Z. trinotatus,* and *Z. princeps* in North America.

The oocysts are subspherical, 19—24 × 17—23 (mean 22 × 20) μm, with a
rough, pitted, clear, colorless to amber, one-layered wall 1.5 μm thick, without
a micropyle or residuum, with a polar granule. The sporocysts are ovoid, 12—
18 × 7—12 (mean 16 × 10) μm, with a Stieda body, sub-Stiedal body, and
residuum. The sporozoites are elongate, have one or two clear globules, and lie
lengthwise head to tail in the sporocysts. The sporulation time is 4—8 d at 22°C
in 2.5% potassium bichromate solution.

Reference. Gerard, Chobotar, & Ernst (1977); Duszynski, Eastham, & Yates
(1982).

HOST FAMILY DIPODIDAE

HOST GENUS *JACULUS*

ISOSPORA DAWADIMIENSIS KASIM & AL SHAWA, 1985

This species occurs quite commonly in the jerboa *Jaculus jaculus* in Saudi
Arabia.

The oocysts are ovoid or nearly spherical (illustrated as ellipsoidal), 22—26.5
× 20.5—22 (mean 24 × 21) μm, with a smooth, light brown or pale green, one-
layered wall 1 μm thick, without a micropyle, residuum, or polar granule. The
sporocysts are ellipsoidal, 12—16.5 × 9—10.5 (mean 15 × 10) μm, with a one-
layered wall, without a Stieda or sub-Stiedal body, with a residuum. The

sporozoites are 8—11 × 2—3 (mean 10 × 3) μm, sausage-shaped, slightly curved, tapered at one end, illustrated without clear globules. The sporulation time is 1.5 d at 30—32°C.

Reference. Kasim & Al Shawa (1985).

KLOSSIELLA SP. (HARTMANN & SCHILLING, 1917) NÖLLER, 1921

Synonyms. *Hepatozoon jaculi* (?) Hartmann & Schilling, 1917; *Klossiella* sp. Nöller in Reichenow, 1921 and Wenyon, 1926.

This form occurs in the endothelial cells of the capillaries of the internal organs of the desert jerboa *Jaculus jaculus* in Africa.

Remarks. Hartmann & Schilling (1917) described schizogonic forms in the endothelial cells of the capillaries of the lung, liver, spleen, and other internal organs of the jerboa which they believed to be part of the life cycle of *Hepatozoon balfouri* (syn., *H. jaculi*). However, Nöller examined their preparations and informed Reichenow (1921) that they were developmental stages of a *Klossiella*. Wenyon (1926) also saw this parasite in the jerboa from the same country (the Sudan) in 1906 and agreed that it was a *Klossiella*. The only forms he saw appeared to be schizogonic stages in the kidney tubules.

References. Hartmann & Schilling (1917); Reichenow (1921); Wenyon (1926).

HOST GENUS *ALLACTAGA*

EIMERIA ALLACTAGAE IWANOFF-GOBZEM, 1934 EMEND. SVANBAEV, 1956

Synonym. *Eimeria alactagae* Iwanoff-Gobzem, 1934.

This species occurs, presumably commonly, in the great jerboa *Allactaga major* (type host) and small, 5-toed jerboa *A. elater* in the USSR.

The oocysts are spherical or ovoid, 14—26 × 12—26 μm, with a smooth, colorless, one-layered or "double-contoured" wall 1—1.7 μm thick, without a micropyle or residuum, ordinarily with a polar granule. The sporocysts are ovoid, ellipsoidal, or piriform, 6—12 × 4—9 μm, with a Stieda body, with or without a clear globule. The sporulation time is 2 d at 25—30°C in 2% potassium bichromate solution.

References. Glebezdin (1970); Iwanoff-Gobzem (1934); Musaev & Veisov (1963a, 1965); Svanbaev (1956).

EIMERIA DAMIRCHINICA MUSAEV & VEISOV, 1965

Synonym. *Eimeria caucasica* Musaev & Veisov, 1963.

This species occurs in the small, 5-toed jerboa *Allactaga elater* in the USSR.

The oocysts are ellipsoidal, sometimes spherical, 14—25 × 12—22 μm, with a two-layered wall 2 μm thick, the outer layer smooth, colorless, the inner layer light brown or yellowish brown, without a micropyle, with a residuum and polar granule. The sporocysts are ellipsoidal or piriform, 6—10 × 4—7 μm, without a Stieda body, with a residuum. The sporozoites are bean- or comma-shaped, apparently without clear globules. The sporulation time is 3 d at 25—30°C in 2.5% potassium bichromate solution.

References. Glebezdin (1970); Musaev & Veisov (1963, 1965).

EIMERIA ELATER MUSAEV & VEISOV, 1963

This species occurs in the small, 5-toed jerboa *Allactaga elater* in the USSR.

The oocysts are ellipsoidal, 20—29 × 16—25 µm, with a smooth, colorless, one-layered wall 1.5—2 µm thick, without a micropyle or polar granule, with a residuum. The sporocysts are piriform, rarely ovoid, 8—15 × 6—12 µm, with a Stieda body and residuum. The sporozoites are comma-shaped, apparently without clear globules, and lie lengthwise head to tail in the sporocysts. The sporulation time is 4—5 d at 25—30°C in 2.5% potassium bichromate solution.

References. Glebezdin (1970, 1978); Musaev & Veisov (1963a).

EIMERIA JOYEUXI YAKIMOFF & GOUSSEFF, 1936

This species occurs, apparently commonly, in the great jerboa *Allactaga major* (syn., *A. jaculus*) in the USSR.

The oocysts are subspherical or spherical, 22—29 × 21—27 µm, with a smooth, yellow, yellow-brown, or orange-brown wall 1.3—1.7 µm thick, without a micropyle or polar granule, with a residuum. The sporocysts are ovoid, ellipsoidal, subspherical, or spherical, 8—14 × 7—11 µm, without a residuum. The sporozoites are comma-shaped, 7—8 × 2—3 µm, and lie lengthwise in the sporocysts.

References. Svanbaev (1962); Yakimoff & Gousseff (1936).

EIMERIA LAVIERI YAKIMOFF & GOUSSEFF, 1936

This species occurs in the great jerboa *Allactaga maior* (syn., *A. jaculus*) in the USSR.

The oocysts are spherical or subspherical, 17—27 × 17—19 µm, with a smooth, yellow, yellow-brown, or yellowish orange wall 1.1—1.2 µm thick, without a micropyle, residuum, or polar granule. The sporocysts are subspherical or ellipsoidal, 8—9 × 7—8 µm, without a Stieda body, with a residuum. The sporozoites are comma-shaped and lie lengthwise in the sporocysts.

References. Svanbaev (1962); Yakimoff & Gousseff (1936).

EIMERIA PAVLOVSKYI MACHUL'SKII, 1949

This species occurs in the Mongolian 5-toed jerboa *Allactaga saltator mongolica* in the USSR.

The oocysts are ellipsoidal, rarely spherical, 17—25 × 15—17 (mean 23 × 16) µm, with a rose-colored, one-layered wall 1 µm thick, without a micropyle, residuum, or polar granule. The sporocysts are ellipsoidal, 8—9 × 6 µm, without a Stieda body, with a residuum. The sporozoites are comma-shaped, with a clear globule.

References. Machul'skii (1949); Musaev & Veisov (1965).

EIMERIA POPOVI MACHUL'SKII, 1949

This species occurs in the Mongolian 5-toed jerboa *Allactaga saltator mongolica* in the USSR.

The oocysts are ovoid, 21—25 × 17—21 (mean 23 × 19) μm, with a two-layered wall 2 μm thick, the outer layer light brown, the inner layer colorless, with a micropyle, apparently without a residuum or polar granule. The sporocysts are ellipsoidal (illustrated as ovoid), 8—10 × 4—5 μm, with a Stieda body and residuum. The sporozoites are comma-shaped, with a clear globule.

References. Machul'skii (1949); Musaev & Veisov (1965).

EIMERIA WILLIAMSI MUSAEV & VEISOV, 1965

This species occurs in Williams' jerboa *Allactaga williamsi* in the USSR.

The oocysts are ellipsoidal, 16—20 × 12—16 (mean 18 × 14) μm, with a smooth, two-layered wall 2 μm thick, the outer layer colorless, the inner layer yellow-brown, without a micropyle or residuum, with a polar granule. The sporocysts are ellipsoidal, 6—8 × 4—6 (mean 6.5 × 4) μm, without a Stieda body, with a residuum. The sporozoites are bean-shaped, apparently without a clear globule. The sporulation time is 3 d.

Reference. Musaev & Veisov, 1965.

SKRJABINELLA MONGOLICA MACHUL'SKII, 1949

This species occurs in the Mongolian jerboa *Allactaga saltator* in the USSR.

The oocysts are spherical, 25—29 (mean 26) μm in diameter, with a one-layered, colorless wall 1.5 μm thick, without a micropyle, residuum, or polar granule. The sporocysts are ellipsoidal with tapering ends, 8 × 4—6 μm, without a Stieda body or residuum. The sporozoites have a clear globule at each end. Machul'skii (1949) also found many free sporocysts.

References. Machul'skii (1949); Musaev & Veisov (1965).

HOST FAMILY HYSTRICIDAE

HOST GENUS *TRICHYS*

EIMERIA LANDERSI COLLEY, 1971

This species occurs in the long-tailed porcupine *Trichys lipura* in Asia.

The oocysts are subspherical, 22—28 × 22—27 (mean 25 × 24) μm, with a rough, three-layered wall, the outer layer yellowish brown, about 0.5 μm thick, the middle layer yellow, about 1 μm thick, the inner layer dark brown, about 1 μm thick, without a micropyle or residuum, with a polar granule. The sporocysts are ovoid, 10—14 × 7—9 (mean 12 × 8) μm, without a Stieda body, with a residuum. The sporozoites are comma-shaped, apparently without a clear globule, and lie head to tail at opposite ends of the sporocysts.

Reference. Colley, 1971.

EIMERIA LIPURA COLLEY, 1971

This species occurs in the long-tailed porcupine *Trichys lipura* in Asia.

The oocysts are ovoid, 29—37 × 23—25 (mean 33 × 24) µm, with a two-layered wall, the outer layer smooth, greenish yellow, about 1.5 µm thick, the inner layer light brown, about 0.5 µm thick, with a micropyle and several polar granules, without a residuum. The sporocysts are ovoid to subspherical, 9—13 × 8—10 (mean 11 × 9) µm, without a Stieda body, with a residuum. The sporozoites are comma-shaped, apparently without clear globules, and lie at opposite ends of the sporocysts.

Reference. Colley (1971).

HOST GENUS *HYSTRIX*

SARCOCYSTIS SP. VILJOEN, 1921

This form was found in the muscles of the Old World porcupine *Hystrix africaeaustralis* in South Africa.

Reference. Viljoen (1921).

HOST FAMILY CAVIIDAE

HOST GENUS *CAVIA*

EIMERIA CAVIAE SHEATHER, 1924

Synonym. *E. acuminatus* Bacigalupo & Roueda, 1953.

This species occurs in the epithelial cells of the colon, occasionally ceca, and rarely small intestine of the domestic guinea pig *Cavia porcellus* (type host) and wild guinea pig *C. aperea* throughout the world. The oocysts are ovoid, ellipsoidal, or subspherical, 13—26 × 12—23 µm, with a smooth, often brownish, two-layered wall, without a micropyle or polar granule, with a residuum. The sporocysts are 11—13 × 6—7 µm, with a residuum. The sporulation time is 2—11 d.

The number of merozoite generations is unknown. The meronts produce less than 12—32 merozoites 6—16 µm long. There may or may not be a residuum in the meront.

The microgamonts produce a large number of biflagellate, curved, spindle-shaped microgametes 3 µm long. The prepatent period is 11—12 d.

This species is generally nonpathogenic, but it may sometimes cause diarrhea and even death.

Rattus norvegicus cannot be infected.

References. Alvez de Souza (1951); Bacigalupo & Roveda (1953); Becker (1933); Bugge & Heinke (1921); Ellis & Wright (1961); Henry (1932); Lapage (1940); Nie (1950); Sheather (1924); Muto, Sugisaki, Yusa, & Noguchi (1985); Muto, Yusa, Sugisaki, Tanaka, Noguchi, & Taguchi (1985).

EIMERIA SP. KLEEBERG & STEENKEN, 1963

This form was found in the domestic guinea pig *Cavia porcellus* in South Africa.

The oocysts were not described. The sporulation time is apparently 6 d at room temperature.

This form causes necrotic yellowish areas filled with parasites in the liver and occasionally lungs.

Reference. Kleeberg & Steenken (1963).

CRYPTOSPORIDIUM MURIS TYZZER, 1907

The type host of this species is the house mouse *Mus musculus*; it also occurs in the guinea pig *Cavia porcellus*. See under *Mus musculus* for a discussion.

References. Jervis, Merrill, & Sprinz (1966); Vetterling, Jervis, Merrill, & Sprinz (1971); Vetterling, Takeuchi, & Madden (1971).

SARCOCYSTIS CAVIAE DE ALMEIDA, 1928

Synonym. *S. acuminatus* Bacigalupo & Roveda, 1953.

This species occurs in the striated muscles of the guinea pig *Cavia porcellus* in South America. Pre-muscle meronts and merozoites, if any, are unknown. The sarcocysts have curved bradyzoites rounded at one end and more pointed at the other, 4.5 × 1 μm.

References. De Almeida (1928); Bacigalupo & Roveda (1958).

BESNOITIA BESNOITI (MAROTEL, 1913) HENRY, 1913

While the type intermediate host of this species is the ox *Bos taurus*, it has been found that the guinea pig *Cavia porcellus* can be infected experimentally.

References. See under *Mus.*

BESNOITIA JELLISONI FRENKEL, 1955

The guinea pig *Cavia porcellus* is an experimental intermediate host of this species.

References. See under *Peromyscus.*

TOXOPLASMA BAHIENSIS (DE MOURA COSTA, 1956 EMEND. LEVINE, 1978) LEVINE, 1983 IN LEVINE, 1985

Synonyms. *Isospora heydorni* Tadros & Laarman, 1976 in part; *I. bigemina* Stiles, 1891 var. *bahiensis* de Moura Costa, 1956; [non] *I. bigemina* Stiles, 1891; *I. bahiensis* de Moura Costa, 1956 emend. Levine, 1978; small form of *I. bigemina* (Stiles, 1891) Lühe, 1906 in part; *I. wallacei* Dubey, 1976; *Hammondia heydorni* (Tadros & Laarman, 1976) Dubey, 1977; *Toxoplasma heydorni* (Tadros & Laarman, 1976) Levine, 1977.

The definitive host of this species is the dog *Canis familiaris*. However, Matsui et al. (1981) found that the guinea pig *Cavia porcellus* can act as the obligatory intermediate host of this heteroxenous parasite. They could not transmit it to the dog via the laboratory mouse, rat, rabbit, or hamster.

Reference. Matsui et al. (1981).

KLOSSIELLA COBAYAE SEIDELIN, 1914

Synonym. *Klossia caviae* Sangiorgi, 1916.

This species occurs in the kidneys and other organs of the domestic guinea pig *Cavia porcellus* (type host) and wild guinea pig *C. aperea* throughout the world.

It is uncertain whether there is a true oocyst. Gamogony and sporogony with the formation of sporocysts containing 8—30 sporozoites each occur in the endothelial cells of the straight kidney tubules and loops of Henle.

The first merogony takes place in the endothelial cells of the capillaries in the kidneys and other organs, especially of the kidney glomeruli. The meronts attain a diameter of about 2.5 µm, their nuclei divide repeatedly, and each forms a number of merozoites 2×1 µm. This cycle is presumably repeated when the merozoites enter other endothelial cells. Eventually some of them enter the lumen of the kidney tubules, pass to the convoluted portion, and penetrate the tubule cells. They become meronts which produce about 100 merozoites each. The last generation merozoites pass down to the straight kidney tubules or loops of Henle, enter the cells, and become microgamonts on macrogametes. The microgamonts form two microgametes each, one of which fertilizes a macrogamete. The resultant zygote grows and becomes a sporont, causing its host cell to swell, and reaching a diameter of 30—40 µm. The sporont becomes a sporocyst containing 8—30 sporozoites. The sporocysts pass out in the urine.

K. cobayae apparently causes no clinical signs, although it may cause slight degenerative lesions in the kidneys.

References. Alvez de Souza (1931); Bonciu, Dincolesco, & Petrovici (1958), Hofmann & Hanichen (1970); Pearce (1916); Reichenow (1953); Sangiorgi (1916); Seidelin (1914); Stojanov & Cvetanov (1965); Wenyon (1926).

HOST GENUS *DOLICHOTIS*

EIMERIA DOLICHOTIS MORINI, BOERO & RODRIGUEZ, 1955

This species occurs in the small intestine and possibly the large intestine of the marra *Dolichotis p. patagonum* in South America.

The oocysts are subspherical to ellipsoidal, 16—27 × 14—22 µm, with a smooth, colorless wall illustrated as one-layered, without a micropyle, residuum, or polar granule. The sporocysts are apparently 11 ×7 µm with a Stieda body and residuum. The sporozoites are 8—13 × 5—7 (mean 11 × 6) µm, with a clear globule, and lie lengthwise head to tail in the sporocysts. The sporulation time is 72—80 h.

This species apparently causes enteritis, which may be accompanied by catarrh, ulceration, necrosis, edema, and hemorrhage.

The guinea pig cannot be infected with this species.

References. Morini, Boero, & Rodriguez (1955); Zwart & Strik (1961).

HOST FAMILY HYDROCHAERIDAE

HOST GENUS *HYDROCHAERUS*

EIMERIA CAPIBARAE CARINI, 1937
This species occurs in the capybara *Hydrochaerus hydrochaerus* (syn., *H. capibara*) in South America.

The oocysts were said to be oval but were illustrated as ellipsoidal, 25—33 × 20—28 (mean 30 × 26) μm, with a yellowish wall described as double but illustrated as single, 2 μm thick, with fine radial striations, without a micropyle, residuum, or polar granule. The sporocysts are ovoid, 14—15 × 8 μm, with a Stieda body and residuum. The sporulation time is 12 d in 1% chromic acid solution at room temperature.

Reference. Carini (1937).

EIMERIA HYDROCHAERI CARINI, 1937 EMEND. LEVINE & IVENS, 1987
Synonym. *Eimeria hydrochoeri* Carini, 1937.

This species occurs in the capybara *Hydrochaerus hydrochaerus* (syn., *H. capibara*) in South America.

The oocysts are ovoid, 20—22 × 16—18 μm, with a smooth, colorless, "double-contoured" wall 0.8 μm thick, without a micropyle, residuum, or polar granule. The sporocysts are ovoid, 10—11 × 6—7 μm, with a Stieda body and residuum. The sporulation time is 12 d in 1% chromic acid solution at room temperature.

Reference. Carini (1937); Levine & Ivens (1987).

HOST FAMILY DASYPROCTIDAE

HOST GENUS *DASYPROCTA*

EIMERIA AGUTI CARINI, 1935
This species occurs in the cotia or agouti *Dasyprocta leporina* (syns., *D. aguti*, "*Cotia vermelha* [*Aguti aguti*]" of Carini (1935) ("*Cotia vermelha*" is simply the Portuguese name for red cotia, and was probably italicized because Carini's paper was written in French) in South America.

The oocysts are spherical, 16—17 μm in diameter, with a smooth, colorless, thin wall, apparently without a micropyle, residuum, or polar granule. The sporocysts are 10×6 μm, with a residuum, illustrated with what might be a Stieda body.

Reference. Carini (1935).

EIMERIA COTIAE CARINI, 1935
This species occurs in the cotia or agouti *Dasyprocta leporina* (syns., *D. aguti*,

"*Cotia vermelha* [*Aguti aguti*]" of Carini (1935) (see under *E. aguti* for a discussion of the host name) in South America.

The oocysts were described as oval but illustrated as ellipsoidal, 28 or 29 × 18 μm, presumably with a two-layered wall, the outer layer sometimes very slightly rough, pale brownish yellow, quite thick, slightly striated, sometimes with a smooth, colorless, thinner inner layer, without a micropyle, apparently without a residuum or polar granule. The sporocysts are ovoid, 13 × 8—9 μm, with a Stieda body and residuum.

Reference. Carini (1935).

EIMERIA PARAENSIS CARINI, 1935

This species occurs in the cotia or agouti *Dasyprocta leporina* (syns., *D. aguti*, "*Cotia vermelha* [*Aguti aguti*]" of Carini (1935) (see under *E. aguti* for a discussion of the host name).

The oocysts are spherical or slightly ovoid, 33—40 × 30—35 μm, with a two-layered wall, the outer layer rough, punctate, brownish yellow, the inner layer radially striated, apparently without a micropyle, without a residuum or polar granule. The sporocysts are ovoid, 20 × 11 μm, with a Stieda body and residuum.

Reference. Carini (1935).

HOST GENUS *AGOUTI*

EIMERIA (?) NOELLERI (RASTEGAIEFF, 1930) BECKER, 1956

Synonym. (Gen?) *nolleri* Rastegaieff, 1930.

This species occurs in the paca *Agouti* (syns., *Cuniculus*, *Coelogenus*) *paca* in South America.

The oocysts are spherical, 19 μm in diameter. No sporulated oocysts were seen.

Hill (1952) reported that coccidiosis caused the death of a paca in the London Zoo, but he did not describe the organism.

References. Becker (1956); Hill (1952); Rastegaieff (1930).

HOST FAMILY CHINCHILLIDAE

HOST GENUS *CHINCHILLA*

EIMERIA CHINCHILLAE DE VOS & VAN DER WESTHUIZEN, 1968

This species occurs in the cecum of the chinchilla *Chinchilla laniger* (type host), cecum, colon, and sometimes small intestine of *Rhabdomys pumilio*, cecum and sometimes small intestine of *Praomys* (syn., *Rattus*) (*Mastomys*) *natalensis*, *Mus musculus*, *Otomys irroratus*, *Rattus norvegicus*, *Mystromys albicaudatus*, and *Arvicanthis niloticus* (all except *C. laniger* and *R. pumilio* experimental) in Africa.

The oocysts are ovoid, subspherical, or spherical, 13—24 × 11—21 µm, with a light brown, one-layered wall 0.7—1 µm thick, without a micropyle or residuum, with or without a polar granule. The sporocysts are ellipsoidal to ovoid, 10—13 × 4—8 µm, with a Stieda body and residuum. The sporozoites are elongate, about 8 µm long, with a clear globule at the large end, and lie lengthwise head to tail in the sporocysts. The sporulation time is 3 d at 28°C in 2% potassium bichromate solution. The prepatent period is 7—9 d.

This species causes marked inflammation and thickening of the cecum and colon mucosae together with white spots in the intestinal mucosa, but no diarrhea.

References. De Vos (1970); De Vos & Dobson (1970); De Vos & van der Westhuizen (1968).

FRENKELIA SP. BURTSCHER & MEINGASSNER, 1976

This form occurs in the brain of the chinchilla *Chinchilla laniger* in Europe.

The brain meronts are lobed, up to 0.6 mm in diameter, with a wall 0.5-0.8 µm thick, compartmented, with metrocytes, intermediate cells, and bradyzoites. The metrocytes are spherical, 4—6 µm in diameter, and the bradyzoites are banana-shaped, with one end pointed and the other rounded, 5—8 × 1—2 µm.

References. Burtscher & Meingassner (1976); Fankhauser (1978); Meingassner & Burtscher (1977).

HOST FAMILY CAPROMYIDAE

HOST GENUS *CAPROMYS*

EIMERIA CAPROMYDIS RYŠAVÝ, 1967

This species occurs commonly in the conga hutia *Capromys pilorides* in Cuba.

The oocysts are broadly ellipsoidal, 21—27 × 19—21 µm, with a smooth, one-layered wall about 0.8 µm thick, without a micropyle or polar granule, with a residuum. The sporocysts are ellipsoidal, 10—12 × 6—8 µm, without a Stieda body or residuum. The sporozoites are comma-shaped, 9 × 3 µm, with a clear globule at the large end, and lie longitudinally head to tail in the sporocysts. The sporulation time is 5 d at 28—30°C in 2% potassium bichromate solution.

Reference. Ryšavý (1967).

EIMERIA GARRIDOI RYŠAVÝ, 1967

This species occurs commonly in the conga hutia *Capromys pilorides* in Cuba.

The oocysts are broadly ellipsoidal to subspherical, 16—18 × 14—16 (mean 17 × 15) µm, with a smooth, colorless, apparently one-layered thin wall, without a micropyle or polar granule, ordinarily with a residuum. The sporocysts are subspherical or broadly ellipsoidal, 7—8 × 6 µm, without a Stieda body or residuum. The sporozoites are bean-shaped, 6 × 3 µm, with a clear globule at one end. The sporulation time is 5—7 d at 28—30°C in 2% potassium bichromate solution.

Reference. Ryšavý 1967).

EIMERIA JIROVECI RYŠAVÝ, 1967

This species occurs commonly in the conga hutia *Capromys pilorides* in Cuba.

The oocysts are ellipsoidal, 25—27 × 17—21 µm, with a clear wall 1.5 µm thick, without a micropyle or polar granule, with a residuum. The sporocysts are ovoid, 12—13 × 3—9 µm, with and occasionally without a Stieda body and residuum. The sporozoites are piriform, with a clear globule at the broad end, and lie lengthwise head to tail in the sporocysts. The sporulation time is 7 d at 28—30°C in 2% potassium bichromate solution.

Reference. Ryšavý (1967).

EIMERIA NORMANLEVINEI RYŠAVÝ, 1967

This species occurs commonly in the middle part of the small intestine of the conga hutia *Capromys pilorides* in Cuba.

The oocysts are ellipsoidal, 31—37 × 27—29 (mean 35 × 28) µm, with a brownish yellow wall 1.5 µm thick, without a micropyle or polar granule, with a residuum. The sporocysts are ellipsoidal, 12—13 × 8—9 µm, without a Stieda body or residuum. The sporozoites are piriform, with a clear globule at the large end, and lie lengthwise head to tail in the sporocysts. The sporulation time is 5—6 d at 28—30°C in 2% potassium bichromate solution.

Reference. Ryšavý (1967).

HOST GENUS *MYOCASTOR*

EIMERIA COYPI OBITZ & WADOWSKI, 1937

This species occurs commonly in the nutria *Myocastor coypus* in Europe and the USSR.

The oocysts are ovoid, 21—26 × 12—16 (mean 23 × 15) µm, with a smooth wall, without a micropyle or residuum, with a polar granule. The sporocysts are ellipsoidal, 9—12 × 5—7 (mean 11 × 6) µm, without a Stieda body, with a residuum.

This species seems to be pathogenic in some degree for young animals.

Others have given this name to coccidia with more or less different oocysts, and it is difficult to be sure what species they were dealing with. Prasad (1960a) thought that *E. coypi* is a synonym of *E. myopotami*.

References. Nukerbaeva & Svanbaev (1973); Obitz & Wadowski (1937); Pellérdy (1960); Prasad (1960a); Seidel (1954, 1956); Ball & Lewis (1984); Lewis & Ball (1984).

EIMERIA FLUVIATILIS LEWIS & BALL, 1984

This species occurs in the coypu *Myocastor coypus* in England.

The oocysts are subspherical, unpigmented, generally smooth, 17—22 × 16— 20 (mean 19 × 16.5) μm, with a two-layered wall 1 μm thick, without a micropyle or residuum, with one or two polar granules. The sporocysts are 13—18 × 6— 8 (mean 15 × 7) μm, with a Stieda body and residuum.

Reference. Lewis & Ball (1984).

EIMERIA MYOCASTORI PRASAD, 1960

This species occurs commonly in the nutria *Myocastor coypus* in Europe.

The oocysts are broad ovoid, 13—15 × 11—13 (mean 14 × 12) μm, with two-layered wall, the outer layer smooth, colorless, thinner than the colorless inner layer, with a micropyle, without a residuum or polar granule. The sporocysts are cigar-shaped, 9—11 × 3—5 (mean 10 × 4) μm, without a Stieda body, with a residuum. The sporozoites are spindle-shaped, 9—11 × 2.5—3 μm, with a clear globule.

Reference. Prasad (1960).

EIMERIA MYOCASTORIS RINGUELET & COSCARON, 1961

This species occurs in the nutria *Myocastor coypus bonariensis* in South America.

The oocysts are ellipsoidal, 22—32 × 10—18 (mean 25 × 15) μm, with a smooth, thin wall, without a micropyle, residuum, or polar granule. The sporocysts are ellipsoidal, 8—15 × 5—8 μm, without a Stieda body or residuum. The sporozoites are presumably at the ends of the sporocysts.

Reference. Ringuelet & Coscaron (1961).

EIMERIA MYOPOTAMI YAKIMOFF, 1933

This species occurs fairly commonly in the nutria *Myocastor coypus* in Europe and the USSR.

The oocysts are ovoid, sometimes ellipsoidal, very seldom spherical or subspherical, 22—27 × 12—23 μm, with a two- or three-layered wall, the outer layer smooth, yellowish brown and relatively thin, without a micropyle, residuum, or polar granule. The sporocysts are pointed at one end, presumably with a Stieda body, with a residuum. The sporozoites lie longitudinally in the sporocysts.

Different authors' descriptions of this species' oocysts differ, and a restudy of the oocysts of this and other nutria coccidia is needed.

References. Hohner (1966); Pellérdy (1960); Prasad (1960a); Seidel (1954, 1956); Yakimoff (1933); Zajíček (1955); Ball & Lewis (1984); Lewis & Ball (1984).

EIMERIA NUTRIAE PRASAD, 1960

This species occurs moderately commonly to commonly in the nutria *Myocastor coypus* in Europe and the USSR.

The oocysts are broadly ovoid or subspherical, 19—23 × 15—18 μm, with a yellow, one-layered wall pitted like a thimble, without a micropyle or residuum, with a polar granule. The sporocysts are ovoid, 10—12×4—6 (mean 11×5) μm, without a Stieda body, with a residuum. The sporozoites are spindle-shaped, 9— 11 × 2—3 μm, and have a clear globule at one end.

References. Nukerbaeva & Svanbaev (1973); Prasad (1960a); Ball & Lewis (1984); Lewis & Ball (1984).

EIMERIA PELLUCIDA YAKIMOFF, 1933

This species occurs quite commonly in the small intestine of the nutria *Myocastor coypus* in Europe and the USSR.

The oocysts are ovoid or almost cylindrical, 29—40×16—25 μm, with a two-layered wall, the outer layer smooth, with a micropyle, without a residuum. The sporocysts are piriform, 9 × 6 μm, with a residuum.

Sexual stages are in the small intestine. Mature macrogametes are 22×15 μm. The prepatent period is perhaps 11—12 d.

Different authors' descriptions of this species' oocysts differ, and a restudy of the oocysts of this and other nutria coccidia is needed.

References. Hohner (1966); Nukerbaeva & Svanbaev (1973); Pellérdy (1960); Prasad (1960a); Seidel (1954); Yakimoff (1933); Zajiček (1955).

EIMERIA QUIYARUM RINGUELET & COSCARON, 1961

This species occurs in the nutria *Myocastor coypus bonariensis* in South America.

The oocysts are subspherical, 14—22 × 12—20 (mean 18 × 15) μm, with a rough, thick wall, without a micropyle, residuum, or polar granule. The sporocysts are ellipsoidal, 8—14 × 5—7 μm, without a Stieda body or residuum.

Reference. Ringuelet & Coscaron (1961).

EIMERIA SEIDELI PELLÉRDY, 1957

Synonyms. *Eimeria (Globidium) fulva* Seidel, 1954; *E. fulva* Seidel, 1954; *E. (Globidium) perniciosa* Sprehn in Seidel, 1954; *non E. fulva* Farr, 1953; *non E. (Tyzzeria) perniciosa* (Allen, 1936) Seidel, 1954.

This species occurs rarely to commonly in the small intestine of the nutria *Myocastor coypus* in Europe.

The oocysts are spherical, 38—45 μm in diameter, or subspherical, 45—48 × 38—42 μm, with a three-layered wall 2—5 μm thick, the outer layer (present only in young oocysts) thin and whitish, the middle layer light to dark brown, almost opaque, thick, and very rough (*"gekornt"*), the inner layer colorless and thin, without a micropyle, residuum, or polar granule. The sporocysts are piriform to ovoid, 26—29 × 13 μm, with a Stieda body and residuum. The sporozoites are bean-shaped, 10 μm long (free sporozoites 14—16×4 μm). The sporulation time is 12—37 d at room temperature.

The meronts are in the epithelial cells of the small intestine. Sizes of 4.5—20 × 3—14 μm have been reported, as have 8—25 merozoites per meront.

Gamonts are beneath the epithelium, often in the lacteals and connective tissue. Mature macrogametes are 32—38 × 29—35 μm and mature microgamonts are 50—100 μm and contain a very great number of minute nuclei. The nuclei combine in groups to form islets up to 15 μm in diameter which produce many comma-shaped microgametes 3—5 μm long. The prepatent period is 14 d.

This is the most pathogenic species of *Eimeria* in the nutria. It causes diarrhea and inflammatory changes in the small intestine.

References. Farr (1953); Hohner (1966); Pellérdy (1957, 1960); Scheuring (1973); Seidel (1954, 1956); Ball & Lewis (1984).

ISOSPORA SP. NUKERBAEVA & SVANBAEV, 1973

This form occurs uncommonly in the nutria *Myocastor coypus* in the USSR.

The oocysts are short-oval, 28 × 25 μm or spherical, 25—28 μm in diameter, with a two-layered wall 1.5 μm thick, without a micropyle or polar granule, with a residuum. The sporocysts are 14 × 11 μm, with a residuum. Sporulation occurs outside the host.

Reference. Nukerbaeva & Svanbaev (1973).

HOST FAMILY ECHIMYIDAE

HOST GENUS *PROECHIMYS*

EIMERIA CARIPENSIS ARCAY-DE-PERAZA, 1964

This species occurs in the epithelial cells of the small intestine of the spiny rat *Proechimys guyanensis* in South America.

The oocysts are spherical, 20 μm in mean diameter, with a "double-contoured," striated wall, without a micropyle or residuum. The sporocysts are ovoid, 9 × 5.5 μm, with a Stieda body and residuum. The sporozoites are falciform. The sporulation time is 5 d at room temperature in 2% potassium bichromate solution.

Meronts are 1—13 μm in diameter and contain merozoites 2 × 2 μm arranged in a corona.

The microgamonts are more or less spherical, average 16.5 × 15 μm, and contain many microgametes 3 × 0.8 μm. The prepatent period is 11 d.

Reference. Arcay-de-Peraza (1964).

EIMERIA PROECHIMYI ARCAY-DE-PERAZA, 1964

This species occurs in the epithelial cells of the small intestine of the spiny rat *Proechimys guyanensis* in South America.

The oocysts are ellipsoidal, 23 × 17 μm, with a two-layered wall, without a micropyle or residuum. The sporocysts are spherical, 7 μm in diameter, without a Stieda body, with a residuum. The sporozoites are reniform, with a clear globule at one end. The sporulation time is 3 d at room temperature in 2% potassium bichromate solution. The prepatent period is 11 d.

This species may cause dense focal infiltration in the ileum with total sloughing of the epithelium, discontinuity of the longitudinal musculature, and substitution of granulation tissue for the submucosa.

The guinea pig cannot be infected with this species.

Reference. Arcay-de-Peraza (1964).

HOST FAMILY BATHYERGIDAE

HOST GENUS *HETEROCEPHALUS*

EIMERIA HETEROCEPHALI LEVINE & IVENS, 1965

Synonym. *Eimeria muris* Galli-Valerio, 1932 of Porter (1957), *non E. muris* Galli-Valerio, 1932.

This species occurs in the epithelial cells of the cecum of the naked mole rat *Heterocephalus glaber*; it was found in the London Zoo.

The oocysts are ovoid, without a micropyle. The sporocysts have a residuum.

Reference. Levine & Ivens (1965); Porter (1957).

HOST FAMILY CTENODACTYLIDAE

HOST GENUS *CTENODACTYLUS*

EIMERIA GUNDII MISHRA & GONZALEZ, 1978

This species occurs commonly in the mucosa of the ileum of the gondi *Ctenodactylus gundi* in Africa.

The oocysts are broadly ellipsoidal to subspherical, 20—27 × 18—23 (mean 24 × 20) μm, with a two-layered wall 1—1.5 μm thick, the outer layer transparent, the inner layer dark and thicker than the outer one, without a micropyle or polar granule, with a residuum. The sporocysts are ovoid, 9—12 × 5—7 (mean 11 × 7) μm, with a Stieda body and residuum. The sporozoites are 6—8 × 2.5—3 μm. The sporulation time is 4 d at room temperature in saturated sucrose solution plus a few drops of 2.5% sulfuric acid.

Reference. Mishra & Gonzalez (1978).

TOXOPLASMA GONDII (NICOLLE & MANCEAUX, 1908) NICOLLE & MANCEAUX, 1909

Synonyms. *Leishmania gondii* Nicolle & Manceaux, 1908; *Toxoplasma cuniculi* Splendore, 1908; *Toxoplasma canis* Mello, 1910; *Toxoplasma talpae* von Prowazek, 1910; *Toxoplasma columbae* Yakimoff & Kohl-Yakimoff, 1912; *Toxoplasma pyrogenes* Castellani, 1913; *Toxoplasma musculi* Sangiorgi, 1913;

Toxoplasma sciuri Coles, 1914; *Toxoplasma ratti* Sangiorgi, 1915; *Toxoplasma francae* (de Mello, 1915) Wenyon, 1926; *Toxoplasma caviae* Carini & Migliano, 1916; *Toxoplasma nicanorovi* Zasukhin & Gaisky, 1930; *Toxoplasma laidlawi* Coutelen, 1932; *Toxoplasma wenyoni* Coutelen, 1932; *Toxoplasma crociduri* Galli-Valerio, *Toxoplasma fulicae* de Mello, 1935; *Toxoplasma hominis* Wolf, Cowan, & Paige, 1939; *Toxoplasma gallinarum* Hepding, 1939.

The definitive hosts of this species are the domestic cat *Felis catus* (type definitive host) and the jaguarundi *Felis yagouaroundi*, ocelot *F. pardalis*, mountain lion *F. concolor*, Asian leopard cat *F. bengalensis*, bobcat *Lynx rufus*, and probably the cheetah *Acinonyx jubatus*. The intermediate hosts are the gondi *Ctenodactylus gundi* (type intermediate host) and about 200 other species of mammals and birds. Gamonts and one type of meront are found in the villar epithelial cells of the definitive hosts. Meronts and merozoites are found in many types of cell (including neurons and leukocytes) of the intermediate hosts. This species occurs throughout the world, and is common in both intermediate and presumably definitive hosts.

The oocysts are spherical to subspherical, $11—14 \times 9—11$ (mean 12.5×11) μm, without a micropyle, residuum, or polar granule. The sporocysts are ellipsoidal, about 8.5×6 μm, without a Stieda body, with a residuum. The sporozoites are about 8×2 μm. The sporulation time is 2—3 d at 24°C, 5—8 d at 15°C, or 14—21 d at 11°C.

The intermediate hosts can be infected by ingesting sporulated oocysts or infected meat or animals, or congenitally via the placenta. Congenital toxoplasmosis of the newborn resulting from infection of the mother while she is pregnant is probably the most common form in man and perhaps sheep. Mice can be infected congenitally for generation after generation. Experimental infections can be established by any type of parenteral inoculation or by feeding. Following experimental inoculation, the protozoa proliferate for a time at the site of injection and then invade the bloodstream and cause a generalized infection. Susceptible tissues all over the body are invaded, and the parasites multiply in them, causing local necrosis. The parasitemia continues for some time, until antibodies appear in the plasma, after which the parasites disappear from the blood and more slowly from the other tissues. They finally remain only in meronts and in the most receptive tissues. In general, the spleen, lungs, and liver are cleared of parasites relatively rapidly, the heart somewhat more slowly, and the brain much more slowly. Residual infections may persist for a number of years.

When the sporulated oocysts are ingested by a susceptible animal, the sporozoites emerge and pass to the parenteral tissues via the blood and lymph; any type of cell may be invaded. Here they multiply by endodyogeny or endopolygeny. The stage in which this occurs has been called many names, but "group stage" is perhaps best. The merozoites within it are tachyzoites ("fast," i.e., rapidly developing zoites). This is the stage found in the leukocytes in peritoneal exudate, but it also occurs in other parenteral locations such as the liver, lungs, and submucosa; this is the stage occurring in acute toxoplasmosis.

There is an indefinite number of tachyzoite generations. Eventually they enter other cells and induce the host cell to form a wall around them, forming a pseudocyst or meront. Within it a large number of bradyzoites ("slow," i.e., slowly developing zoites) is formed by endodyogeny. These meronts and bradyzoites may remain viable in the tissues for years. This is the stage found commonly in the brain, and it also occurs in other tissues such as muscle; it is the stage found in chronic infections.

The above meront is the end of the life cycle in all animals except felids so far as is known. In the cat and other felids, the bradyzoites enter the intestinal tissues and multiply. Five types of multiplicative stages have been reported in the intestinal epithelial cells.

The merozoites are 5—8 × 1—2 μm and have an apical complex consisting of 2 polar rings, a similar ring at the posterior end, a conoid, a variable number of rhoptries, about 50 micronemes, and 22 subpellicular microtubules. They also have a nucleus containing a nucleolus, Golgi apparatus, one or more micropores, mitochondria, ribosomes, and rough endoplasmic reticulum.

Gamonts are formed in the epithelial cells of the small intestine of felids. The microgamonts produce 12—32 slender, crescentic microgametes about 3 μm long which have two flagella plus the rudiments of a third. The macrogametes simply grow. The prepatent period in cats is 3—5 d after ingesting brain meronts, 7—10 d after ingesting merozoites, or 7—24 d after ingesting oocysts. The patent period in cats is 1—2 weeks.

The sexual stages are apparently not pathogenic for felids, but the parenteral stages may or may not cause symptoms. Most infections are inapparent, but they may vary between this and an acutely fatal disease.

References. This has been a much abbreviated review of *T. gondii* infections. Over 8000 papers have been written on this species. Jira & Kozojed (1970, 1983) published a 4-volume bibliography of papers from 1908—1975. Various aspects of *Toxoplasma* and toxoplasmosis have been reviewed since 1975 by Beverley (1976), Beyer et al. (1979), Dubey (1977), Overdulve (1978), Scott (1978), and Zasukhin (1980).

SUMMARY

This monograph summarizes the known information on taxonomy, synonymy, structure, life cycle, hosts, location in the host, reported geographic distribution, sporulation, merogony, gamogony, prepatent period, patent period, and pathogenicity of the 473 named species of coccidia of rodents. These include 372 species of *Eimeria*, 39 of *Isospora*; 28 of *Sarcocystis*; 5 each of *Besnoitia*, *Toxoplasma*, and *Wenyonella*; 4 each of *Caryospora* and *Klossiella*; 2 each of *Dorisa*, *Frenkelia*, *Klossia*, and *Cryptosporidium*; and 1 each of *Mantonella*, *Pythonella*, *Skrjabinella*, and *Tyzzeria*. In addition, similar data are given for the 99 forms for which insufficient information is available to justify assigning them species names.

Coccidia have been named from only 52% of the 29 families, 25% of the 380 genera, and 15% of the 1687 extant species of rodents. All the coccidian species in most host species have yet to be described, and most of the present descriptions of oocysts are far from complete. Furthermore, the life cycles of only a few rodent *Eimeria* species are quite fully known. Even the location in the host is known for only about a quarter of the species of *Eimeria*.

ADDENDUM

The following additional references refer to new species of coccidia which are not included in the present book:

Bandyopadhyay, S. 1986. *Eimeria biswapartii* sp. n., a new coccidium from the Indian bandicoot rat, *Bandicota indica* (Beckstein). Acta Protozool. 25:109-114.

Bandyopadhyay, S., R. Ray & A. Bhattarcharjee. 1986. On the occurrence of a new coccidium *Dorisa indica* sp. n. in a common house rat, *Rattus rattus arboreus* (Horsefield) from India. Acta Protozool. 35:115-118.

Bandyopadhyay, S., R. Ray & B. Das Gupta. 1986. A new coccidium, *Wenyonella levinei* n. sp. from a common house rat *Rattus rattus arboreus*. Arch. Protistenkd. 131:303-307.

Dubey, J. P. & H. G. Sheffield. 1988. *Sarcocystis sigmodontis* from the cotton rat (*Sigmodon hispidus*). J. Parasitol. 74:889-891.

Hafner, U. & F.-R. Matuschka. 1984. Life cycle studies on *Sarcocystis dirumpens* sp. n. with regard to host specificity. Z. Parasitenkd. 70:715-720.

Levine, N. D. & V. Ivens. 1987. Corrections in the names of rodent coccidia (Apicomplexa, Coccidiasina). J. Protozool. 34:371.

Lindsay, D. S., C. M. Hendrix, & B. L. Blagburn. 1988. Experimental *Cryptosporidium parvum* infections in opossums (*Didelphis virginiana*). J. Wildl. Dis. 24:157-159.

McAllister, G. T. & S. J. Upton. 1988. *Eimeria taylori* n. sp. (Apicomplexa: Eimeriidae) and *E. baiomysis* from the northern pigmy mouse, *Baiomys taylori* (Rodentia: Cricetidae) from Texas, U.S.A. Trans. Am. Micr. Soc. 107:296-300.

Rose, M. E. & B. J. Millard. 1980. Host specificity in eimerian coccidia: development of *Eimeria vermiformis* of the mouse, *Mus musculus*, in *Rattus norvegicus*. Parasitology. 90:557-563.

Sundermann, C. A. & D. S. Lindsay. 1989. Ultrastructure of in vivo-produced caryocysts containing the coccidian *Caryospora bigenetica* (Apicomplexa: Eimeriidae). J. Protozool. 36:81-86.

REFERENCES

Abbas, M. & E. C. Powell. 1983. Immunology of *Sarcocystis muris*. Prog. Abstr. Am. Soc. Parasitol. 58:46.

Abenov, D. B. & S. K. Svanbaev. 1979. Koktsidii zhellrogo suslike v Kazakhstane. [The coccidia of *Spermophilus fulvus* in Kazakhstan.] Izv. Akad. Nauk Kaz. SSR Ser. Biol. Nauk. 1979(4):22-30.

Abenov, D. B. & S. K. Svanbaev. 1982. Materialy po izucheniyu koktsidii reliktovogo suslika (*Citellus reliktus*) v Kazakhstane. In Zayanchkauskas et al., Eds. *Kishechnye Prosteishie*. Inst. Zool. Parazit. Litovsk SSR Akad. Nauk SSSR Vil'nyus. 3-5.

Adams, J. H. & K. S. Todd, Jr. 1983. Transmission electron microscopy of meront development of *Eimeria vermiformis* Ernst, Chobotar and Hammond, 1971 (Apicomplexa, Eucoccidiorida) in the mouse, *Mus musculus*. J. Protozool. 30:114-118.

Akinchina, G. T. & D. N. Zasukhin. 1971. Intranuclear localization of *Toxoplasma gondii* and *Besnoitia jellisoni* in conditions of tissue culture and some remarks on the intranuclear parasitism. Acta Protozool. 8:341-348.

Allen, E. A. 1936. *Tyzzeria perniciosa* gen. et sp. nov., a coccidium from the small intestine of the Pekin duck, *Anas domesticus* L. Arch. Protistenkd. 87:262-267.

de Almeida, F. P. 1928. Sobre um protozoario encontrado no coracao de cobaya. Ann. Fac. Med. Sao Paulo. 3:65-67.

Alves de Souza, M. 1931. Coccidiose em cobaya. Rev. Zootech. Vet. 17:11-14.

Anderson, L. C. & J. A. Hess. 1980. Development of first generation schizonts of *Eimeria bilamellata* in the Uinta ground squirrel, *Spermophilus armatus*. Prog. Abstr. Am. Soc. Parasitol. 55:63.

Anderson, L. C. & J. A. Hess. 1982. Ultrastructure of the macrogamonts of *Eimeria bilamellata* from the thirteen-lined ground squirrel, *Spermophilus tridecemlineatus*. In Müller, M., W. Gutteridge & P. Köhler, Eds. *Molecular and Biochemical Parasitology*. Elsevier, Amsterdam, pp. 614-615.

Anderson, L. C. & J. Hess. 1985. The development of first generation schizonts of *Eimeria bilamellata* in the Uinta ground squirrel *Spermophilus armatus*. J. Parasitol. 71:374-375.

Andreassen, J. & 0. Behnke. 1968. Fine structure of merozoites of a rat coccidian *Eimeria miyairii* with a comparison of the fine structure of other Sporozoa. J. Parasitol. 54:150-163.

Andrews, J. M. 1927. Host-parasite specificity in the coccidia of mammals. J. Parasitol. 13:183-194.

Andrews, J. M. 1928. New species of coccidia from the skunk and prairie dog. J. Parasitol. 14:193-194.

Ankrom, S. L., B. Chobotar, & J. Ernst. 1975. Life cycle of *Eimeria ferrisi* Levine & Ivens, 1965 in the mouse, *Mus musculus*. J. Protozool. 22:317-323.

Araujo, F. G., J. P. Dubey, & J. S. Remington. 1984. Antigenic similarity between the coccidian parasites *Toxoplasma gondii* and *Hammondia hammondi*. J. Protozool. 31:145-147.

Arcay, L. 1981. Nuevo coccidia de gato: *Cystoisospora frenkeli* sp. nova (Toxoplasmatinae) su desarrolo en la membrana corioalantoidea de embrion de pollo. Acta Cient. Venez. 32:401-410.

Arcay, L. 1982. Genital coccidiosis in the golden hamster (*Cricetus cricetus*). In Muller, M., W. Gutteridge & P. Kohler, Eds. *Molecular and Biochemical Parasitology*. Elsevier, Amsterdam. p. 404.

Arcay-de-Peraza, L. 1964. Tres nuevas especies de *Eimeria* (Protozoa, Coccidia, Eimeriidae) de roedores silvestres de Venezuela. Acta Biol. Venez. 4:185-203.

Arcay-de-Peraza, L. 1970. Coccidia de roedores silvestres de Venezuela: Ciclo evolutivo de *Eimeria guerlingueti* sp. nov. de *Sciurus* (*Guerlinguetus*) *granatensis* (Rodentia, Sciuromorpha, Sciuroidea, Sciuridae), y descripcion de *Eimeria akodoni* sp. nov. parasito de *Akodon urichi venezuelensis* (Rodentia, Cricetomorpha, Cricetidae). Rev. II Cong. Latinam. Parasit.:3.

Arnastauskene, T. V. 1977. K voprosy o zarazhennosti zhivotnykh koktsidiyami Taimyre. In Kornulova, G. V., V. K. Yastrebov & V. A. Klebanevskii, Eds. *Problemy Epidemiologii i Profilaktiki Prirodnoochagovykh Boleznei v Zapolyar'e.* Ministerstvo Zdravookhraneniya RSFSR, Omskii Ordena Trudovogo Krasnogo Znameni Meditsinskii Institut. Omsk, USSR, pp. 135-143.

Arnastauskene, T. V. 1980. Kharakteristika zarazhennosti melkikh mlekopitayushchikh koktsidiyami na Timyre v 1974-1975. [Characteristics of the infection of small mammals with coccidia in the Taimir Peninsula in 1974-1975.] T. Akad. Nauk Litovskoi SSR Ser. B. 1980(2):53-60.

Arnastauskene, T., Y. Kazlauskas & S. Mal'dzhyunaite. 1978. Estestvennye gruppirovki kishechnykh parazitov y myshevidnykh gryzunov zakaznika kamsha i vliyanie na nikh biotopa vida i struktury populyatsii khozyaina. [The natural groupings of the intestinal parasites of mouse rodents of the Kamsa Preserve and their dependence on host biotope, species and its population structure.] Acta Parasitol. Litu. 16:15-32.

Ashford, R. W. 1978. *Sarcocystis cymruensis* n. sp., a parasite of rats *Rattus norvegicus* and cats *Felis catus.* Ann. Trop. Med. Parasitol. 72:37-43.

Babudieri, B. 1932. I sarcosporidi e le sarcosporidiose. (Studio monografico). Arch. Protistenkd. 76:421-580.

Bacigalupo, L. & R. J. Roveda. 1953. Sarcosporidiosis del cobayo en la Argentina. Rev. Med. Vet. Buenos Aires. 35:111-113.

Ball, S. J. & D. C. Lewis. 1984. *Eimeria* (Protozoa: Coccidia) in wild populations of some British rodents. J. Zool. London. 202:373-381.

Ball, S. J. & K. R. Snow. 1984. *Eimeria confusa* Joseph, 1969 and *E. ontarioensis* Lee and Dorney, 1971, from grey squirrels in England. J. Parasitol. 70:390.

Ballard, N. B. 1970. *Eimeria ochrogasteri* n. sp. from *Microtus ochrogaster.* J. Protozool. 17:271-273.

Bandyopadhyay, S. & B. Dasgupta. 1982. A new coccidium *Eimeria bandicota* n. sp. from *Bandicota bengalensis* (Gray) in Darjeeling. J. Beng. Nat. Hist. Soc. 1(2) n.s.:28-36.

Bandyopadhyay, S. & R. Ray. 1982. On a new coccidium, *Dorisiella bengalensis*, from an Indian palm squirrel, *Funambulus pennanti* Wroughton. Indian J. Parasitol. 6:265-266.

Barker, I. K. & P. L. Carbonell. 1974. *Cryptosporidium agni* sp. n. from lambs, and *Cryptosporidium bovis* sp. n. from a calf, with observations on the oocyst. Z. Parasitenkd. 44:289-298.

Barnard, W. P., J. V. Ernst, & C. F. Dixon. 1974. Coccidia of the cotton rat, *Sigmodon hispidus*, from Alabama. J. Parasitol. 60:406-414.

Barnard, W. P., J. V. Ernst, & R. A. Roper. 1971. *Eimeria kinsellai* sp. n. (Protozoa: Eimeriidae) in a marsh rice rat *Oryzomys palustris* from Florida. J. Protozool. 18:546-547.

Barnard, W. P., J. V. Ernst, & R. 0. Stevens. 1971. *Eimeria palustris* sp. n. and *Isospora hammondi* sp. n. (Coccidia: Eimeriidae) from marsh rice rat *Oryzomys palustris* (Harlan). J. Parasitol. 57:1293-1295.

Beaver, P. C. & J. R. Maleckar. 1981. *Sarcocystis singaporensis* Zaman and Colley, (1975) 1976, *Sarcocystis villivillosi* sp. n., and *Sarcocystis zamani* sp. n.: Development, morphology, and persistence in the laboratory rat, *Rattus norvegicus.* J. Parasitol. 67:241-250.

Becker, E. R. 1933. Cross-infection experiments with coccidia of rodents and domesticated animals. J. Parasitol. 19:230-234.

Becker, E. R. 1934. *Coccidia and Coccidiosis of Domesticated, Game, and Laboratory Animals and of Man.* Collegiate Press, Ames, IA. 146 pp.

Becker, E. R. 1939. Effect of thiamin chloride on *Eimeria nieschulzi* infection of the rat. Proc. Soc. Exp. Biol. Med. 42:597-598.

Becker, E. R. 1941. Effect of parenteral administration of vitamin B_1 and vitamin B_6 on a coccidium infection. Proc. Soc. Exp. Biol. Med. 46:494-495.

Becker, E. R. 1942. Nature of *Eimeria nieschulzi*-growth-promoting potency of feeding stuffs. 4. Riboflavin and nicotinic acid. Proc. Iowa Acad. Sci. 49:503-506.

Becker, E. R. 1956. Catalog of Eimeriidae in genera occurring in vertebrates and not requiring intermediate hosts. Iowa State Coll. J. Sci. 31:85-139.

Becker, E. R. & R. D. Burroughs. 1933. Rediscovery of *Eimeria carinii* Pinto, 1928. J. Parasitol. 20:123.

Becker, E. R. & R. I. Dilworth. 1941. Nature of *Eimeria nieschulzi* growth-promoting potency of feeding stuffs. II. Vitamins B$_1$ and B$_6$. J. Infect. Dis. 68:285-290.

Becker, E. R. & P. R. Hall. 1931. *Eimeria separata*, a new species of coccidium from the Norway rat (*Epimys norvegicus*). Iowa State Coll. J. Sci. 6:131.

Becker, E. R. & P. R. Hall. 1933. Cross-immunity and correlation of oocyst production during immunization between *Eimeria miyairii* and *Eimeria separata* in the rat. Am. J. Hyg. 18:220-223.

Becker, E. R., P. R. Hall, & A. Hager. 1932. Quantitative biometric and host-parasite studies on *E. miyairii* and *E. separata* in rats. Iowa State Coll. J. Sci. 6:299-316.

Becker, E. R., M. Manresa, Jr. & L. Smith. 1943. Nature of *Eimeria nieschulzi*-growth-promoting potency of feeding stuffs. 5. Dry-heating ingredients of the ration. Iowa State Coll. J. Sci. 17:257-262.

Becker, E. R. & L. Smith. 1942. Nature of *Eimeria nieschulzi* growth-promoting potency of feeding stuffs. III. Pantothenic acid. Iowa State Coll. J ·i. 16:443-449.

Beltran, E. & R. Perez. 1950. Protozoarios parasit. ratas en la Ciudad de Mexico. Rev. Inst. Salubr. Enferm. Trop. Mexico City. 11:71-78.

van den Berghe, L. 1938. Two new coccidia, *Wenyonella uelensis* n. sp. and *Wenyonella parva* n. sp., from two Congolese rodents. Parasitology. 30:275-277.

van den Berghe, L. & M. Chardome. 1956. Une nouvelle coccidie, *Eimeria arvicanthis* n. sp., chez *Arvicanthis abyssinicus rubescens* (Wroughton). Rev. Zool. Bot. Afr. 53:65-66.

van den Berghe, L. & M. Chardome. 1956a. Une coccidie nouvelle du fuku, rat taupe du Congo Belge (*Tachyoryctes ruandae* Lomb. et Gyld.) *Eimeria tachyoryctis* n. sp. Rev. Zool. Bot. Afr. 53:67-69.

van den Berghe, L. & M. Chardome. 1957. *Eimeria schoutedeni*, n. sp., une coccidie du rat de Gambie, *Cricetomy* ····im·lis (Rochebrune). Rev. Zool. Bot. Afr. 56:1-3.

Bestetti, G. & R. Frank. ·978. Doppelinfektion des Gehirns mit Frenkelia und Toxoplasma bei einem Chinchi... Lichtund el-ektronenmikroskopische Untersuchung. Schweiz. Arch. Tierheilkd. 120:591-601.

Beverley, J. K. A. 1976. Toxoplasmosis in animals. Vet. Rec. 99:123-127.

Beyer, T. V., N. A. Bezukladnikova, I. G. Galuzo, S. I. Konovalova, & S. M. Pak, Eds. 1979. *Toksoplazmidy.* [*The Toxoplasmids.*] Leningrad "Nauka." 118 pp.

Biocca, E. 1968. Class Toxoplasmatea: critical review and proposal of the new name *Frenkelia* gen. n. for M-organism. Parassitologia (Rome). 10:89-98.

Bird, R. G. 1981. Protozoa and viruses. Human cryptosporidiosis and concomitant viral enteritis. In Canning. E. U., Ed. *Parasitological Topics.* Soc. Protozool. Spec. Publ. No. 1:39-47.

Blagburn, B. L., J. H. Adams & K. S. Todd, Jr. 1982. First asexual generation of *Eimeria vermiformis* Ernst, Chobotar, and Hammond, 1971 in *Mus musculus*. J. Parasitol. 68:1178-1180.

Blagburn, B. L., B. Chobotar & R. T. Smith. 1979. Clinical and histologic observations of actively induced resistance to *Eimeria ferrisi* (Protozoa: Eimeriidae) in the mouse (*Mus musculus*). Z. Parasitenkd. 59:1-14.

Blagburn, B. L. & K. S. Todd, Jr. 1984. Pathological changes and immunity associated with experimental *Eimeria vermiformis* infections in *Mus musculus*. J. Protozool. 31:556-561.

Blanchard, R. A. E. 1885. Sur un nouveau type de sarcosporidies. C. R. Acad. Sci. 100:1599-1601.

Bledsoe, B. 1977. Life cycle of *Sarcocystis* sp. from deer mice and gopher snakes. Prog. Abstr. Am. Soc. Parasitol. 52:49.

Bledsoe, B. 1979. Life Cycle of *Sarcocystis* sp. in Deer Mice and Gopher Snakes. Ph.D. thesis, University of Illinois, Urbana, viii + 137 pp.

Bledsoe, B. 1980. *Sarcocystis idahoensis* sp. n. in deer mice *Peromyscus maniculatus* (Wagner) and gopher snakes *Pituophis melanoleucus* (Daudin). J. Protozool. 27:93-102.

Bledsoe, B. 1980a. Transmission studies with *Sarcocystis idahoensis* of deer mice (*Peromyscus maniculatus*) and gopher snakes (*Pituophis melanoleucus*). J. Wildl. Dis. 16:195-200.

Bogovsky, P. A. 1956. [Histological changes in the kidneys of albino mice infected by *Klossiella muris.*] Byull. Eksper. Biol. Med. 41:365-368.

Bonciu, C., M. Dincolesco, & M. Petrovici. 1958. Contribution a l'étude de la maladie d'Armstrong chez la souris blanche. Arch. Roum. Pathol. Exp. Microbiol. 17:455-470.

Bond, B. B. & E. C. Bovee. 1958. A redescription of an eimerian coccidian from the flying squirrel, *Glaucomys volans*, designating it *Eimeria parasciurorum* nov. sp. J. Protozool. 4:225-229.

Bonfante, G. R., E. C. Faust, & L. E. Giraldo. 1961. Parasitologic surveys in Cali, Departamento del Valle, Colombia. IX. Endoparasites of rodents and cockroaches in Ward Siloe, Cali, Colombia. J. Parasitol. 47:843-846.

Bonne, C. 1925. *Klossiella muris*, parasite général des souris badigeonnées au Goudron. C. R. Soc. Biol. 92:1190-1192.

Bornand, M. 1937. Sur quelques affections parasitaires du gibier observées en 1936. Bull. Soc. Vaudoise Sci. Nat. 59:509-514.

Boulard, Y. 1973. Cycle biologique du sporozoaire *Klossiella mabokensis* (Adeleidea) parasite de muridés africains. Bull. Mus. Natl. Hist. Nat., Ser. 3, Zool. 99:721-746.

Boulard, Y. & I. Landau. 1971. Note preliminaire sur la description et la cycle biologique de *Klossiella mabokensis* n. sp., Adeleidea, parasite de muridés africains. C. R. Acad. Sci. 273(D):2271-2274.

Boyer, C. D. & J. V. Scorza B. 1957. Encuesta sobre parasitos microscopicos de algunos Cricetinae silvestres de Venezuela (ratas silvestres). Bol. Venez. Lab. Clin. 2:59-67.

Brändler, U. 1982. Inaug. Diss., München. 56 pp.

Bray, R. S. 1958. On the parasitic protozoa of Liberia. I. Coccidia of some small animals. J. Protozool. 5:81-83.

Brehm, H. & W. Frank. 1980. Der Entwicklungskreislauf von *Sarcocystis singaporensis* Zaman und Colley 1976 in End und Zwischenwirt. Z. Parasitenkd. 62:15-30.

Bristol, J. R., A. J. Piñon, & L. F. Mayberry. 1983. Interspecific interactions between *Nippostrongylus brasiliensis* and *Eimeria nieschulzi* in the rat. J. Parasitol. 69:372-374.

Brösigke, S., J. Heine, & J. Boch. 1981. Der Nachweis extraintestinaler Entwicklungsstadien (Dormozoiten) in experimentell mit Cystoisosporarivolta-Oozysten infizierten Mäusen. Kleintier Prax. 17:25-34.

Brown, R. J., W. P. Carney, P. F. D. Van Peenen, & M. Sudomo. 1974. Southeast Asian J. Trop. Med. Public Health. 5:451-452.

Brumpt, E. 1913. *Précis de Parasitologie*. 2nd ed. 1011 pp.

Brunelli, G. 1935. Su di alcune varieta di coccidi. Arch. Ital. Sci. Med. Colon. 16:354-366.

Bugge, G. & P. Heinke. 1921. Ueber das Vorkommen von Kokzidien beim Meerschweinchen. Dtsch. Tieraerztl. Wochenschr. 29:41-42.

Bullock, W. L. 1959. The occurrence of *Eimeria tamiasciuri* in New Hampshire. J. Parasitol. 45(Suppl.):39-40.

Bump, G. 1942. 31st annual report New York State Conservation Department. (*non vidi*; from Doran, 1954).

Burtscher, E. & J. G. Meingassner. 1976. *Frenkelia*-infektion bei Chinchillas. Z. Parasitenkd. 50:220.

Cannarella, O. 1931. L'azione patogena della *Klossiella muris* nel rene del topolino e transmissione del coccidio dal toppo alla cavia. Boll. Ist. Sieroter. Milan. 10:620-684.

Carini, A. 1932. Eimeriose intestinal de um serelepe por *Eimeria botelhoi* n. sp. Rev. Biol. Hyg. 3:80-82.

Carini, A. 1935. Sur trois nouvelles *Eimeria* de *cotia vermelha*. Ann. Parasitol. 13:342-344.

Carini, A. 1937. Sobre uma nova Eimeria (*E. oryzomysi* n. sp.) do intestino de um ratinho do campo. IX Reun. Soc. Argent. Patol. Region. Buenos Aires 2:624-627.

Carini, A. 1937a. Sur deux nouvelles *Eimeria* d'*Hydrochoerus capibara*. Ann. Parasitol. 15:367-369.

Carini, A. 1937b. Sur une nouvelle *Eimeria*, parasite de l'intestin du *Caluromys philander*. Ann. Parasitol. 15:453-455.

Castro, G. A. & D. W. Duszynski. 1984. Local and systemic effects on inflammation during *Eimeria nieschulzi* infection. J. Protozool. 31:283-287.

Čatar, G., M. Zachar, M. Valent, J. Vrablik, R. Hynie-Holkova, & M. Pavlina. 1967. Nalezy tkanivovych protozoi u micromamalii. Bratisl. Lek. Listy. 47:226-234 (from Kalyakin, Kovalevsky, & Nikitina, 1973).

Cawthorn, R. J. & R. N. Brooks. 1985. Endogenous development of *Sarcocystis rauschorum* (Protozoa: Sarcocystidae) in varying lemmings *Dicrostonyx richardsoni*: Light microscopy. Prog. Abstr. Am. Soc. Parasitol. 60:36-37.

Cawthorn, R. J., A. A. Gajadhar, & R. J. Brooks. 1984. Description of *Sarcocystis rauschorum* sp. n. (Protozoa: Sarcocystidae) with experimental cyclic transmission between varying lemmings (*Dicrostonyx richardsoni*) and snowy owls (*Nyctea scandiaca*). Can. J. Zool. 62:217-225.

Cawthorn, R. J. & P. H. G. Stockdale. 1981. Description of *Eimeria bubonis* sp. n. (Protozoa: Eimeriidae) and *Caryospora bubonis* sp. n. (Protozoa: Eimeriidae) is the great horned owl, *Bubo virginianus* (Gmelin), of Saskatchewan. Can. J. Zool. 59:170-173.

Cawthorn, R. J. & P. H. G. Stockdale. 1982. The developmental cycle of *Caryospora bubonis* Cawthorn and Stockdale 1981 (Protozoa: Eimeriidae) in the great horned owl, *Bubo virginianus* (Gmelin). Can. J. Zool. 60:152-157.

Cawthorn, R. J., G. A. Wobeser, & A. A. Gajadhar. 1983. Description of *Sarcocystis campestris* sp. n. (Protozoa: Sarcocystidae): A parasite of the badger *Taxidea taxus* with experimental transmission to the Richardson ground squirrel, *Spermophilus richardsonii*. Can J. Zool. 61:370-377.

Černá, Z. 1962. Contribution to the knowledge of coccidia in Muridae. Vestn. Česk. Spol. Zool. 26:1-13.

Černá, Z. 1975. On the problem of antigenic identity between the coccidian *Eimeria contorta* Haberborn, 1971 and *E. falciformis* Eimer, 1870. J. Protozool. 22:60A.

Černá, Z. 1976. Two new coccidians form passeriform birds. Folia Parasitol. (Prague). 23:277-279.

Černá, Z. 1977. Cycle de développement sarcosporidien d'une coccidie, chez la souris, après infestation des animaux par des oocystes-sporocystes isolés de l'intestin de la chouette effraie (*Tyto alba*). Protistologica. 13:401-405.

Černá, Z. 1983. Multiplication of merozoites of Sarcocystis dispersa Černá, Kolářová et Šulc, 1978 and Sarcocystis cernae Levine, 1977 in the blood stream of the intermediate host. Folia Parasitol. (Prague). 30:5-8.

Černá, Z. & A. B. Ally. 1979. Two sarcosporidian species from *Microtus arvalis*. J. Protozool. 26:40A.

Černá, Z. & M. Daniel. 1956. K poznání kokcidií rolné žijících drobných ssavcn. Česk. Parazitol. 3:10-23.

Černá, Z. & I. Kolářová. 1978. Contribution to the serological diagnosis of sarcocystosis. Folia Parsitol. (Prague). 25:289-292.

Černá, Z., L. Kolářová, & P. Šulc. 1978. Contribution to the problem of cyst-producing coccidians. Folia Parasitol. (Prague). 25:9-16.

Černá, Z. & M. Loučková. 1976. Microtus arvalis as the intermediate host of a coccidian from the kestrel (Falco tinnunculus). Folia Parasitol. (Prague). 23:110.

Černá, Z. & J. Sénaud. 1969. *Eimeria pragensis* sp. n., a new coccidian parasite from the intestine of mice (*Mus musculus*). Folia Parasitol. (Prague). 16:171-175.

Černá, Z. & J. Sénaud. 1977. Sur un type nouveau de multiplication asexuée d'une sarcosporidie, dans le foie de la souris. C. R. Acad. Sci. 285(D): 347-349.

Černá, Z., J. Sénaud, H. Mehlhorn, & E. Scholtyseck. 1974. Étude comparée des rélations morphologiques et immunologiques chez les coccidies de la souris: *Eimeria falciformis* et *Eimeria pragensis* (Coccidia, Eimeriidae). Folia Parasitol. (Prague). 21:301-309.

Černá, Z. & L. Vánová. 1979. Detection of *Sarcocystis* antibody in dried blood samples on filter paper. J. Protozool. 26:40A.

Chbouki, N, & J.-F. Dubremetz. 1982. Redistribution des recepteurs superficiels chez le sporozoite d'*Eimeria nieschulzi* (Coccidia) induite par différents ligands. J. Protozool. 29:299-300.

Chobotar, B., E. Scholtyseck, J. Sénaud, & J. V. Ernst. 1975. A fine structural study of asexual stages of the murine coccidium *Eimeria ferrisi* Levine and Ivens. Z. Parasitenkd. 45:291-306.

Chobotar, B., J. Sénaud, J. V. Ernst, & E. Scholtyseck. 1980. Ultrastructure of macrogametogenesis and formation of the oocyst wall of *Eimeria papillata* in *Mus musculus*. Protistologica. 16:115-124.

Chobotar, B., J. Sénaud, & E. Scholtyseck. 1978. The ultrastructure of macrogamete development and oocyst wall formation. Short Com. Fourth Int. Congr. Parasit. B:70.

Chowattukunnel, J. T. 1979. *Eimeria malabaricas* sp. n. and *Eimeria bandipurensis* from the South Indian tree squirrel, *Funambulus tristriatus*. J. Prtozool. 26:36-38.

Clarke, J. J. 1895. A study of coccidia met with in mice. Q. J. Microsc. Sci. 37:277-283.

Colley, F. C. 1967. Fine structure of sporozoites of *Eimeria nieschulzi*. J. Protozool. 14:217-220.

Colley, F. C. 1967a. Fine structure of microgametocytes and macrogametes of *Eimeria nieschulzi*. J. Protozool. 14:663-674.

Colley, F. C. 1968. Fine structure of schizonts and merozoites of *Eimeria nieschulzi*. J. Protozool. 15:374-382.

Colley, F. C. 1971. *Eimeria callosciuri* n. sp. (Protozoa: Eimeriidae) from Prevost's squirrel *Callosciurus prevostii* Demarest, 1822 in Malaysia. J. Protozool. 18:199-200.

Colley, F. C. 1971a. *Eimeria lipura* and *E. landersi* n. spp. (Protozoa, Eimeriidae) from the long tailed porcupine *Trichys lipura* Gunther, 1876 in Malaysia. J. Protozool. 18:473-474.

Colley, F. C. & S. W. Mullin. 1971. New species of *Eimeria* (Protozoa, Eimeriidae) from Malaysian squirrels. J. Protozool. 18:400-402.

Colley. F. C. & S. W. Mullin. 1971a. New species of *Eimeria* (Protozoa: Eimeriidae) from Malaysian rats. J. Protozool. 18:601-604.

Collins, G. H. 1981. *Sarcocystis* in rats from Stewart Island. N. Z. J. Zool. 8:129.

Cordero del Campillo, M. 1959. Estudios sobre *Eimeria falciformis* (Eimer, 1870) parasito del raton. I. Observaciones sobre el periodo prepatente, esporulacion, morfologia de los ooquistes y estudio biometrico de los mismos, produccion de ooquistes y patogenicidad. Rev. Iber. Parasitol. 19:351-368.

Cross, J. H., J. W. Fresh, G. S. Jones, & S. Gunawan. 1973. *Sarcocystis* from rats of central Java. Southeast Asian J. Trop. Med. Public Health. 4:435.

Crouch, H. B. & E. R. Becker. 1931. Three species of coccidia from the woodchuck, *Marmota monax*. Iowa State Coll. J. Sci. 5:127-131.

Current, W. L. 1985. Cryptosporidiosis. J. Am. Vet. Med. Assoc. 187:1334-1338.

Current, W. L., J. V. Ernst, & G. W. Benz. 1981. Endogenous stages of *Eimeria tuskegeenensis* (Protozoa: Eimeriidae) in the cotton rat, *Sigmodon hispidus*. J. Parasitol. 67:204-213.

Danforth, H. D., P. C. Augustine, & B. Chobotar. 1984. Production and utilization of hybridoma antibodies in the study of avian and mammalian coccidian. Prog. Abstr. Annu. Meet. Am. Soc. Parasitol. 59:47-48.

Danforth, H. D., B. Chobotar, & R. Entzeroth. 1984. Cellular pathology in mouse embryonic brain cells following in vitro penetration by sporozoites of *Eimeria papillata*. Z. Parasitenkd. 70:165-171.

Davidson, W. R. 1976. Endoparasites of selected populations of gray squirrels (*Sciurus carolinensis*) in the southeastern United States. Proc. Helminthol. Soc. Wash. 43:211-217.

Davis, B. S. 1967. *Isospora peromysci* n. sp., *I. californica* n. sp., and *I. hastingsi* n. sp. (Protozoa; Eimeriidae) from four sympatric species of white-footed mice (*Peromyscus*) in central California. J. Protozool. 14:575-585.

Davronov, O. 1973. Koktsidii gryzunov Uzbekistana. Parasitologiya. 7:79-82.

De Vos, A. J. 1970. Studies on the host range of *Eimeria chinchillae* De Vos & Van der Westhuizen, 1968. Onderstepoort J. Vet. Res. 37:29-36.

De Vos, A. J. & L. D. Dobson. 1970. *Eimeria chinchillae* De Vos & Van der Westhuizen, 1968 and other *Eimeria* spp. from three South African rodent species. Onderstepoort J. Vet. Res. 37:185-190.

De Vos, A. J. & E. B. Van der Westhuizen. 1968. The occurrence of *Eimeria chinchillae* n. sp. (Eimeriidae) in *Chinchilla laniger* (Molina, 1782) in South Africa. J. S. Afr. Vet. Med. Assoc. 39:81-82.

Dieben, C. P. A. 1924. Over der morphologie en biologie van het rattencoccidium *Eimeria nieschulzi* n. sp., en zijne verspreiding in Nederland. Proefschr. Veeartsenijk. Hoogesch. Utrecht. 119 pp.

Dobell, C. C. 1919. A revision of the coccidia parasitic in man. Parasitology. 11:147-149.

Dobell, C. C. & P. W. O'Connor. 1921. *The Intestinal Protozoa of Man*. xii + 211 pp. London.

Doby, J. M., A. Jeannes, & B. Rault. 1965. Systematical research of toxoplasmosis in the brain of small mammals by a histological method. Czech. Parasitol. 12:133-144.

Doflein, F. J. T. 1916. *Lehrbuch der Protozoenkunde. Ein Darstellung der Naturgeschichte der Protozoen mit besonderer Berucksichtigung der parasitischen und pathogenen Formen*. 4th ed. xv + 1190 pp.

Doran, D. J. 1951. Studies on coccidiosis in the kangaroo rats of southern California. Proc. Am. Soc. Protozool. 2:18.

Doran, D. J. 1953. Coccidiosis in the kangaroo rats of California. Univ. Calif. Berkeley Publ. Zool. 59:31-60.

Doran, D. J. & T. L. Jahn. 1949. Observations on *Eimeria mohavensis* sp. nov. from the kangaroo rat *Dipodomys mohavensis*. Anat. Rec. 105:631-632.

Doran, D. J. & T. L. Jahn. 1952. Preliminary observations on *Eimeria mohavensis* n. sp. from the kangaroo rat *Dipodomys panamintinus mohavensis* (Grinnell). Trans. Am. Microsc. Soc. 71:93-101.

Dorney, R. S. 1962. A survey of the coccidia of some Wisconsin Sciuridae with descriptions of three new species. J. Protozool. 9:258-261.

Dorney, R. S. 1963. Coccidiosis — incidence, epizootiology in two Wisconsin Sciuridae. Trans. North Am. Wildl. Nat. Resour. Conf. 28:207-215.

Dorney, R. S. 1965. *Eimeria tuscarorensis* n. sp. (Protozoa: Eimeriidae) and redescriptions of other coccidia of the woodchuck, *Marmota monax*. J. Protozool. 12:423-426.

Dorney, R. S. 1966. Quantitative data on four species of *Eimeria* in eastern chipmunks and red squirrels. J. Protozool. 13:549-550.

Dubey, J. P. 1975. Experimental *Isospora canis* and *Isospora felis* infection in mice, cats, and dogs. J. Protozool. 22:416-417.

Dubey, J. P. 1975a. *Isospora ohioensis* sp. n. proposed for *I. rivolta* of the dog. J. Parasitol. 61:462-465.

Dubey, J. P. 1977. *Toxoplasma, Hammondia, Besnoitia, Sarcocystis*, and other tissue cyst-forming coccidia of man and animals. In Kreier, J. P., Ed. *Parasitic Protozoa* 3:101-237. Academic Press, New York.

Dubey, J. P. 1977a. Taxonomy of Sarcocystis and other coccidia of cats and dogs. J. Am. Vet. Med. Assoc. 170:778, 782.

Dubey, J. P. 1983. *Sarcocystis montanaensis* and *S. microti* n. spp. from the meadow vole (*Microtus pennsylvanicus*). Proc. Helminthol. Soc. Wash. 50:318-324.

Dubey, J. P. 1983. *Sarcocystis bozemanensis* sp. nov. (Protozoa: Sarcocystidae) and *S. campestris* from the Richardson's ground squirrel (*Spermophilus richardsonii*), in Montana, U.S.A. Can. J. Zool. 61:942-946.

Dubey, J. P. 1983. *Sarcocystis peromysci* n. sp. and *S. idahoensis* in deer mouse (*Peromyscus maniculatus*) in Montana. Can. J. Zool. 61:1180-1182.

Dubey, J. P. & J. K. Frenkel. 1972. Cyst-induced toxoplasmosis in cats. J. Protozool. 19:155-177.

Dubey, J. P., N. L. Miller, & J. K. Frenkel. 1970. Characterization of the new fecal form of *Toxoplasma gondii*. J. Parasitol. 56:447-456.

Dubremetz, J.-F., D. Colwell, & J. Mahrt. 1975. Etude de l'oocyste de la coccidie *Eimeria nieschulzi* par cryodecapage. C. R. Acad. Sci. 280(D):2117-2119.

Dubremetz, J.-F. & G. Torpier. 1978. Freeze fracture study of the pellicle of an eimerian sporozoite (Protozoa, Coccidia). J. Ultrastruct. Res. 62:94-109.

Duncan, S. 1968. *Eimeria ovata* n. sp. and other coccidia of the eastern chipmunk *Tamias striatus* in Massachusetts. J. Protozool. 15:319-320.

Duncan, S. 1973. *Eimeria ontarioensis* Lee and Dorney, 1971, from the gray squirrel, *Sciurus carolinensis*, in Massachusetts. J. Parasitol. 59:330.

Duszynski, D. W. 1971. Increase in size of *Eimeria separata* oocysts during patency. J. Parasitol. 57:948-952.

Duszynski, D. W. 1972. Host and parasite interactions during single and concurrent infections with *Eimeria nieschulzi* and *E. separata* in the rat. J. Protozool. 19:82-88.

Duszynski, D. W. & G. A. Conder. 1977. External factors and self-regulating mechanisms which may influence the sporulation of oocysts of the rat coccidium, *Eimeria nieschulzi*. Int. J. Parasitol. 7:83-88.

Duszynski, D. W., G. Eastham, & T. L. Yates. 1982. *Eimeria* from jumping mice (*Zapus* spp.): A new species and genetic and geographic features of *Z. hudsonius luteus*. J. Parasitol. 68:1146-1148.

Duszynski, D. W., K. Ramaswamy, & G. A. Castro. 1982. Intestinal absorption of b-methyl-D-glucoside in rats infected with *Eimeria nieschulzi*. J. Parasitol. 68:727-729.

Duszynski, D. W., S. A. Roy, & G. A. Castro. 1978. Intestinal disaccharidase and peroxidase deficiencies during *Eimeria nieschulzi* infections in rats. J. Protozool. 25:226-231.

Duszynski, D. W., S. A. Roy, J. Stewert, & G. A. Castro. 1978. Intestinal transit time during infection with *Eimeria nieschulzi* in rats. J. Protozool. 25:370-374.

Duszynski, D. W., D. Russell, S. A. Roy, & G. A. Castro. 1978. Suppressed rejection of *Trichinella spiralis* in immunized rats concurrently infected with *Eimeria nieschulzi*. J. Parasitol. 64:83-88.

Dykova, I. & J. Lom. 1983. Fish coccidia: An annotated list of described species. Folia Parasitol. (Prague). 30:193-208.

Dyl'ko, M. I. 1962. Dokl. Akad. Nauk BSSR. 6:399-400 (from Kalyakin & Zasukhin, 1975).

Dzerzhinskii, V. A. 1981. Fauna koktsidii tyan'-shan'skoi myshouki v Kazakhstane. [The coccidia of *Sicista tianschanica* in Kazakhstan.] Parazitologiya. 15:290-292.

Dzerzhinskii, V. A. 1982. Koktsidii lesnoi soni v Kazakhstane (Coccidiida). [The coccidia of the forest dormouse in Kazakhstan.] Parazitologiya. 16:333-335.

Dzerzhinskii, V. A. & S. K. Svanbaev. 1980. [Coccidia of rodents in the foothills of the Zailiiskii Altai mountains, Kazakh SSR]. IX Konf. Ukrain. Parazit. Obshchestva. Tezisy Dokladov. Chast' 2, Kiev, USSR "Naukova Dumka." 1980:20-21.

Eimer, G. H. T. 1880. *Ueber die Ei oder kugelformigen sogenannten Psorospermien der Wirbelthiere. Würzburg.*

Elaut, L. 1932. Infection rénale chez la souris blanche par Coccidium Klossiella. C. R. Soc. Biol. 110:1012-1014.

Ellis, P. A. & A. E. Wright. 1961. Coccidiosis in guinea-pigs. J. Clin. Pathol. 14:394-396.

Elton, C., E. B. Ford, J. R. Baker, & A. D. Gardner. 1931. The health and parasites of a wild mouse population. Proc. Zool. Soc. London. 1931:657-721.

Enemar, A. 1965. M-organisms in the brain of the Norway lemming, *Lemmus lemmus*. Ark. Zool. 18(Ser. 2):9-16.

Entzeroth, R. 1984. Electron microscope study of host-parasite interactions of *Sarcocystis muris* (Protozoa, Coccidia) in tissue culture and in vitro. Z. Parasitenkd. 70:131-134.

Entzeroth, R. 1985. Invasion and early development of *Sarcocystis muris* (Apicomplexa, Sarcocystidae) in tissue cultures. J. Protozool. 32:446-453.

Entzeroth, R., B. Chobotar, & E. Scholtyseck. 1985. *Sarcocystis crotali* sp. n. with the Mojave rattlesnake (*Crotalus scutulatus scutulatus*) mouse (*Mus musculus*) cycle. Arch. Protistenkd. 129:19-23.

Entzeroth, R., E. Scholtyseck, & B. Chobotar. 1983. Ultrastructure of *Sarcocystis* sp. n. from the eastern chipmunk (*Tamias striatus*). Z. Parasitenkd. 69:823-826.

Erhardova, B. 1955. Hepatozoon microti Coles, 1914 bei unseren kleinen Säugetieren. Folia Biol. (Prague). 1:282-287.

Ernst, J. V. & B. Chobotar. 1978. The endogenous stages of *Eimeria utahensis* (Protozoa: Eimeriidae) in the kangaroo rat, *Dipodomys ordii*. J. Parasitol. 64:27-34.

Ernst, J. V., B. Chobotar, & L. C. Anderson. 1967. *Eimeria balphae* n. sp. from the Ord kangaroo rat *Dipodomys ordii*. J. Protozool. 14:547-548.

Ernst, J. V., B. Chobotar, & D. M. Hammond. 1971. The oocysts of *Eimeria vermiformis* sp. n. and *E. papillata* sp. n. (Protozoa; Eimeriidae) from the mouse *Mus musculus*. J. Protozool. 18:221-223.

Ernst, J. V., B. Chobotar, E. C. Oaks, & D. M. Hammond. 1968. *Besnoitia jellisoni* (Sporozoa: Toxoplasmea) in rodents from Utah and California. J. Parasitol. 54:545-549.

Ernst, J. V., W. L. Cooper, & M. J. Frydendall. 1970. *Eimeria sprehni* Yakimoff, 1934, and *E. causeyi* sp. n. (Protozoa: Eimeriidae) from the Canadian beaver, *Castor canadensis*. J. Parasitol. 56:30-31.

Ernst, J. V., W. L. Current, & J. A. Moore. 1980. *Eimeria tuskegeensis* (Protozoa: Eimeriidae) in the cotton rat, *Sigmodon hispidus*. Prog. Abstr. Am. Soc. Parasitol. 55:63.

Ernst, J. V., M. J. Frydendall, & D. M. Hammond. 1967. *Eimeria scholtysecki* n. sp. from Ord's kangaroo rat (*Dipodomys ordii*). J. Protozool. 14:181-182.

Ernst, J. V., D. M. Hammond, & B. Chobotar. 1968. *Eimeria utahensis* sp. n. from kangaroo rats (*Dipodomys ordii* and *D. microps*) in northwestern Utah. J. Protozool. 15:430-432.

Ernst, J. V., E. C. Oaks, and J. R. Sampson. 1970. *Eimeria reedi* sp. n. and *E. chobotari* sp. n. (Protozoa: Eimeriidae) from heteromyid rodents. J. Protozool. 17:453-455.

Ernst, J. V., K. S. Todd, Jr., & W. P. Barnard. 1977. Endogenous stages of *Eimeria sigmodontis* (Protozoa: Eimeriidae) in the cotton rat, *Sigmodon hispidus*. Int. J. Parasitol. 7:373-381.

Fantham, H. B. 1909. The sporozoan *Eimeria tenella* (*Coccidium tenellum*) parasitic in the alimentary canal of the grouse. Proc. Zool. Soc. London. Abstr. 43-44.

Fantham, H. B., 1926. Some parasitic protozoa found in South Africa, IX. S. Afr. J. Sci. 23:560-570.

Farr, M. M. 1953. Three new species of coccidia from the Canada goose, *Branta canadensis* (Linné, 1758). J. Wash. Acad. Sci. 43:336-340.

Fayer, R. & J. K. Frenkel. 1979. Comparative infectivity for calves of oocysts of feline coccidia: *Besnoitia, Hammondia, Cystoisospora, Sarcocystis*, and *Toxoplasma*. J. Parasitol 65:756-762

Fayer, R., D. M. Hammond, B. Chobotar, & Y. Y. Elsner. 1969. Cultivation of *Besnoitia jellisoni* in bovine cell cultures. J. Parasitol. 55:645-653.

Findlay, G. M. & A. D. Middleton. 1934. Epidemic disease among voles (Microtus) with special reference to Toxoplasma. J. Anim. Ecol. 3:150-160.

Firlotte, W. R. 1948. A survey of the parasites of the brown Norway rat. Can. J. Comp. Med. 12:187-191.

Fish, F. 1930. Coccidia of rodents: *Eimeria monacis*, n. sp. from the woodchuck. J. Parasitol. 17:98-100.

Fleming, W. J., J. R. Georgi, & J. W. Caslick. 1979. Parasites of the woodchuck (*Marmota monax*) in central New York state. Proc. Helminthol. Soc. Wash. 46:115-127.

Frandsen, J. C. 1983. Effects of low-level infections by coccidia and roundworms on the nutritional status of rats fed an adequate diet. J. Anim. Sci. 57:1487-1497.

Frank, C. 1978. Klein-säuger-protozoon in Neusiedler see — gebiet. Ang. Parasitol. 19:23:87.

Frenkel, J. K. 1955 (1953). Infections with organisms resembling *Toxoplasma*. Rias. Communicaz. VI Cong. Intern. Microbiol. 2:556-557.

Frenkel, J. K. 1973. Toxoplasmosis: Parasite life cycle, pathology, and immunology. In Hammond, D. M. with P. L. Long, Eds. *The Coccidia*. University Park Press, Baltimore. 343-410.

Frenkel J. K. 1977. *Besnoitia wallacei* of cats and rodents: with a reclassification of other cyst-forming isosporoid coccidia. J. Parasitol. 63:611-628.

Frenkel, J. K. 1977a. Animal model: chronic besnoitiosis of golden hamsters (*Mesocricetus auratus*). Am. J. Pathol. 86:749-752.

Frenkel, J. K. & J. P. Dubey, 1975. *Hammondia hammondi* gen. nov., sp. nov., from domestic cats, a new coccidian related to *Toxoplasma* and *Sarcocystis*. Z. Parasitenkd. 46:3-12.

Frenkel, J. K., J. P. Dubey, & N. L. Miller. 1970. *Toxoplasma gondii* in cats: fecal stages identified as coccidian oocysts. Science. 167:893-896.

Frenkel, J. K. & J. K. Reddy. 1977. Induction of liver tumors by 3'-methyl-dimethylaminoazobenzene (3'-Me-DAB) in rats chronically infected with Toxoplasma or Besnoitia. J. Reticulo. Soc. 21:61-68.

Galli-Valerio, B. 1905. Einige Parasiten von *Arvicola nivalis*. Zool. Anz. 28:519-522.

Galli-Valerio, B. 1922. Parasitologische Untersuchungen und Beiträge zur parasitologischen Technik. Zentralbl. Bakteriol. II Abt. 56:344-347.

Galli-Valerio, B. 1923. Parasitologische Untersuchungen und Beiträge zur parasitologischen Technik. Zentralbl. Bakteriol. I. Orig. 91:120-125.

Galli-Valerio, B. 1931. Notes de parasitologie. Zentralbl. Bakteriol. I. Orig. 120:90-106.

Galli-Valerio, B. 1932. Notes de parasitologie et de technique parasitologique. Zentralbl. Bakteriol. I. Orig. 125:129-142.

Galli-Valerio, B. 1935. Notes parasitologiques. Schweiz. Arch. Tierheilkd. 77:643-647.

Galli-Valerio, B. 1940. Notes de parasitologie et de technique parasitologique. Schweiz. Arch. Tierheilkd. 82:352-358.

Galli-Valerio, B. 1940a. Notes de parasitologie et de technique parasitologique. Schweiz. Arch. Tierheilkd. 82:387-392.

Gard, S. 1945. A note on the coccidium *Klossiella muris*. Acta Path. Microbiol. Scand. 22:424-434.

Garnham, P. C. C. 1966. *Besnoitia* (Protozoa: Toxoplasmea) in lizards. Parasitology. 56:329-334.

Geisel, O., E. Kaiser, H. E. Krampitz, & M. Rommel. 1978. Beitrage zum Lebenszyklus der Frenkelien. IV. Pathomorphologische Befunde an den Organen experimentell infizierter Rötelmäuse. Vet. Pathol. 15:621-630.

Geisel, O., E. Kaiser, O. Vogel, H. E. Krampitz, & M. Rommel. 1979. Pathomorphologic findings in short-tailed voles (*Microtus agrestis*) experimentally-infected with *Frenkelia microti*. J. Wildl. Dis. 15:267-270.

Gerard, G., B. Chobotar, & J. V. Ernst. 1977. *Eimeria zapi* sp. n. from the meadow jumping mouse, *Zapus hudsonius* Zimmerman in southwestern Michigan. J. Protozool. 24:362-363.

Gjerde, B. 1983. Shedding of *Hammondia heydorni*-like oocysts by foxes fed muscular tissue of reindeer (*Rangifer tarandus*). Acta Vet. Scand. 24:241-243.

Glebezdin, V. S. 1969. Novye vidy koktsidii iz bol'shoi peschanki *Rhombomys opimus* Licht. v Turkmenii. Izv. Akad. Nauk Turkm. SSR Ser. Biol. Nauk. 1:86-88.

Glebezdin, V. S. 1970. Koktsidii iz malago tushkanchika (*Allactaga elater* Licht.) yuzhnoi Turkmenii. Izv. Akad. Nauk Turkm. SSR Ser. Biol. Nauk. 5:85-89.

Glebezdin, V. S. 1971. K faune koktsidii myshevidnogo khomyachka (*Calomyscus bailwardi* Thom.) yuzhnoi Turkmenii. [On the fauna of *Calomyscus bailwardi* Thom. in south Turkestan.] Izv. Akad. Nauk Turkm. SSR (Biol.). 1971(4):74-76.

Glebezdin, V. S. 1971a. Koktsidii dikikh mlekopitayushchikh Turkmenii. Mat-lyu 1 s'ezda, vsesoynz o-va protozool, Baku, USSR (from Glebezdin & Babich, 1974).

Glebezdin, V. S. 1973. Zarazhennost' koktsidiyami obyknovennoi lesnoi myshi (*Apodemus sylvaticus* L.) v Kara-Kalinskom Raione TCCP. [Coccidia of *Apodemus sylvaticus* in the Kara-Kalinskom Rayon, Turkestan.] Izv. Akad. Nauk Turkm. SSR Ser. Biol. Nauk. 1973(4):57-62.

Glebezdin, V. S. 1974. K faune koktsidii lesnoi soni (*Dryomys nitedula* Pall.) yuzhnoi Turkmenii. [On the coccidial fauna of *Dryomys nitedula* from southern Turkmenistan.] Izv. Akad. Nauk Turkm. SSR (Biol.). 1974(5):76-78.

Glebezdin, V. S. 1974a. [On the coccidial fauna of *Mus musculus* L. and *Meriones persicus* Blanf. in Turkmenistan.] Izv. Akad. Nauk Turkm. SSR Ser. Biol. Nauk. 1974(6):40-43.

Glebezdin, V. S. 1978. K faune koktsidii dikikh mlekopitayushchikh yugo-zapadnoi Turkmenii. [The coccidia of wild mammals in southwest Turkmenia.] Izv. Akad. Nauk Turkm. SSR Biol. Nauk. 1978(3):71-78.

Glebezdin, V. S. & V. V. Babich. 1974. K izucheniyu fauny koktsidii krasnokhvostoi peschanki (*Meriones erythrourus* Gray) v Turkmenistane. [Investigation of the coccidial fauna of *Meriones erythrourus* in Turkmenistan.] Izv. Akad. Nauk Turkm. SSR Ser. Biol. Nauk. 1974(2):84-87.

Goebel, E. & U. Braendler. 1982. Gamogonic and sporogonic stages of the development of Cryptosporidium and their relationship to the epithelial cell of the intestine. *In* Müller, M., W. Gutteridge and P. Köhler, Eds. *Molecular and Biochemical Parasitology*. Elsevier, Amsterdam. pp. 402-403.

Golemanski, V. 1978. Description de neuf nouvelles espèces de coccidies (*Coccidia*: Eimeriidae), parasites de micromammiferes en Bulgarie. Acta Protozool. 17:261-270.

Golemanski, V. G. 1979. V'rkhu koktsidiite (Coccidia, Eimeriidae) na drebnite bozainitsi ot rezervatite "Parangalitsa", "Ropotamo" i "Sreb'rna" v B'lgariya. B'lgar. Akad. Nauk, Acta Zool. Bulgarica. 12:(June):12-26.

Golemanski, V. & D. Duhlinska. 1973. On the coccidian parasites (Sporozoa, Coccidia) of the squirrels (*Sciurus vulgaris*) in Bulgaria. Bull. Inst. Zool. Musee. 38:61-66.

Golemanski, V. & P. Yankova. 1973. Izsledvaniya v'rkhu vidoviya s'stav i razprostranenieto na koktsidiite (Sporozoa, Coccidia) po nyako drebni bozainitsi v B'lgariya. Izv. Zool. Inst. Muzei B'lgar. Akad. Nauk. 37:5-31.

Gonzales-Mugaburu, L. 1942. Estudio sobre *Eimeria phyllotis* n. sp. Rev. Med. Exp. 1:137-151.

Gonzales-Mugaburu, L. 1946. *Emeria weissi* nov. sp. (Coccidiida, Eimeriidae). Rev. Cienc. Lima. 48:91-100.

Grassi, B. 1879. Dei protozoi parassiti e specialmente di quelli che sono nell'uomo. Gazz. Med. Ital. Lomb. 39:445-448.

Gunders, A. E. 1985. Cutaneous nodules of *Besnoitia* sp. in a feral rodent. S. Afr. J. Sci. 81:48.

Gut, J. 1982. Infection of mice immunized with formolized cystozoites of Sarcocystis dispersa ('Cerna', Kolářová et sulc, 1978.) Folia Parasitol. (Prague). 29:285-288.

Haberkorn, A. 1970. Zur Empfänglichkeit nicht spezifischer Wirte für Schizogonie-Stadien verschiedener *Eimeria*-Arten. Z. Parasitenkd. 35:156-161.

Haberkorn, A. 1971. Zur Wirtsspezifität von *Eimeria contorta* n. sp. (Sporozoa: Eimeriidae). Z. Parasitenkd. 37:303-314.

Hall, P. R. 1934. The relation of the size of the infective dose to the number of oocysts eliminated, duration of infection, and immunity to *Eimeria miyairii* Ohira infections in the white rat. Iowa State Coll. J. Sci. 9:115-124.

Hall, P. R. & F. Knipling. 1935. *Eimeria franklinii* and *Eimeria eubeckeri*, two new species of coccidia from the Franklin ground squirrel, *Citellus franklinii* Sabine. J. Parasitol. 21:128-129.

Hammond, D. M., C. A. Speer, & W. Roberts. 1970. Occurrence of refractile bodies in merozoites of *Eimeria* species. J. Parasitol. 56:189-191.

Hampton, J. C. & B. Rosario. 1966. The attachment of protozoan parasites to intestinal epithelial cells of the mouse. J. Parasitol. 52:939-949.

Hartig, F. & G. Hebold. 1970. Das Vorkommen von Klossiellen in der Niere der weissen Ratte. Exp. Pathol. (Jena). 4:367-377.

Hartmann, M. & C. Schilling. 1917. *Die pathogenen Protozoen und die durch sie verursachten Krankheiten*. Berlin. ¥ + 462 pp.

Hayden, D. W., N. W. King, & A. S. K. Murthy. 1976. Spontaneous Frenkelia infection in a laboratory-reared rat . Vet . Pathol. 13:337-342.

Heine, J. 1981. Die tryptische Organverdauung als Methode zum Nachweis extraintestinaler Stadien der Cystoisospora spp. — Infektionen (Kurzmitteilung). Berl. Muench. Tieraertzl. Wochenschr. 94:103-104.

Hendricks, L. D., J. V. Ernst, C. H. Courtney, & C. A. Speer. 1979. *Hammondia pardalis* sp. n. (Sarcocystidae) from the ocelot, *Felis pardalis*, and experimental infection of other felines. J. Protozool. 26:39-43.

Henry, A. C. L. 1913. [*Besnoitia besnoiti* (Marotel, 1913)]. Rec. Med. Vet. 90:328.

Henry, A. & C. Leblois. 1926. Essai de classification des coccidies de la famille des Diplosporidae Léger, 1911. Ann. Parasitol. 4:22-28.

Henry, D. P. 1932. Observations on the coccidia of small mammals in California, with descriptions of seven new species. Univ. Calif. Berkeley Publ. Zool. 37:279-290.

Hiepe, T., S. Nickel, R. Jungmann, U. Hansel, & C. Unger. 1980. Untersuchungen zur Ausscheidung von Sporozoen-Fäkalformen bei Jagdhunde, Rotfüchsen und streunenden Hauskatzen sowie zum Vorkommen von Muskelsarkosporidien bei Wildtieren. Monath. Vet. Med. 35:335-338.

Higgs, S. & F. Nowell. 1983. Infection patterns of *Eimeria hungaryensis* in the woodmouse *Apodemus sylvaticus*. Parasitology. 87:vi.

Hill, T. P. & T. L. Best. 1985. Coccidia from California kangaroo rats (*Dipodomys* spp.). J. Parasitol. 71:682-683.

Hill, T. P. & D. W. Duszynski. 1986. Coccidia (Apicomplexa: Eimeriidae) from sciurid rodents (*Eutamias, Sciurus, Tamiasciurus* spp.) from the western United States and northern Mexico with descriptions of two new species. J. Protozool. 33:282-288.

Hill, W. C. O. 1952. Report of the Society's prosector for the year 1951. Proc. Zool. Soc. London. 122:515-532.

Hilton, D. F. J. & J. L. Mahrt. 1971. *Eimeria spermophili* n. sp. and other *Eimeria* spp. (Protozoa, Eimeriidae) from three species of Alberta *Spermophilus* (Rodentia, Sciuridae). Can. J. Zool. 49:699-701.

Hohner, L. 1966. Untersuchungen über die Agamogonie und Frühstadien der Gamogonie von *Eimeria perniciosa* aus der Globidium-Gruppe. Arch. Protistenkd. 109:289-296.

Hoogenboom, I., S. Daan, & J. J. Laarman. 1984. A field study on the transmission dynamics of the parasite *Sarcocystis cernae* in the common vole (*Microtus arvalis*) and its predator the kestrel (*Falco tinnunculus*). Trop. Geogr. Med. 36: 390.

Inman, L. U. & A. Takeuchi. 1979. Spontaneous cryptosporidiosis in an adult female rabbit. Vet. Pathol. 16:89-95.

Iseki, M. 1979. *Cryptosporidium felis* sp. n. (Protozoa: Eimeriorina) from the domestic cat. Jpn. J. Parasitol. 28:285-307.

Ismailov, S. G. & G. D. Gaibova. 1983. *Eimeria akeriana* n. sp. (Eimeriidae, Coccidiida) from *Meriones blackleri*. Protozool. Issled. Azerbaid., Baku, USSR. 1983:33-35.

Ito, S. & K. Shimura. 1986. The comparison of *Isospora bigemina* large type of the cat and *Besnoitia wallacei*. Jpn. J. Vet. Sci. 48: 433-435.

Ito, J., K. Tsunoda, & K. Shimura. 1978. Life cycle of the large type of *Isospora bigemina* of the cat. Nat. Inst. Anim. Health Q. 18:69-82.

Ivens, V., F. J. Kruidenier, & N. D. Levine. 1959(1958). Further studies of Eimeria (Protozoa: Eimeriidae) from Mexican rodents. Trans. Ill. Acad. Sci. 51:53-57.

Iwanoff-Gobzem, P. S. 1934. Zum Vorkommen von Coccidien bei kleinen wilden Saügetiere. Dtsch. Tieraerztl. Wochenschr. 42:149-151.

Jawdat, S. Z. & A. R. Al-Jafary. 1979. The endogenous phase of the life cycle of *Eimeria nesokiai* Mirza (1975) (Protozoa, Eimeriidae) in the bandicoot rats, *Nesokia indica*. Bull. Biol. Res. Cen. Iraq. 11:81-99.

Jellison, W. L., L. Glesne, & R. S. Peterson. 1960. *Emmonsia*, a fungus, and *Besnoitia*, a protozoan, reported for South America. Bol. Chil. Parasitol. 15:46-47.

Jervis, H. R., T. G. Merrill, & H. Sprinz. 1966. Coccidiosis in the guinea pig small intestine due to a Cryptosporidium. Am. J. Vet. Res. 27:408-414.

Jewell, M. I., J. K. Frenkel, K. M. Johnson, V. Reed., & A. Ruiz. 1972. Development of *Toxoplasma* oocysts in neotropical Felidae. Am. J. Trop. Med. Hyg. 21:512-517.

Jira, J. & V. Kozojed. 1970. *Toxoplasmose — 1908-1967*. Gustav Fischer Verlag, Stuttgart, Germany. 2 vols., 396 + 464 pp.

Jira, J. & V. Kozojed. 1983. *Toxoplasmose — Toxoplasmosis 1968-1975*. Vols. 1 & 2. Gustav Fischer Verlag, Stuttgart, Germany. 207 + 395 pp.

Joseph, T. 1969. The Coccidia of the Grey Squirrel *Sciurus carolinensis* with Descriptions of Two New Species. Ph.D. thesis, Boston University, Boston.

Joseph, T. 1971. Coccidial immunity studies in the grey squirrel. Proc. Indiana Acad. Sci. 81:341.

Joseph, T. 1972. *Eimeria lancasterensis* Joseph, 1969 and *Eimeria confusa* Joseph, 1969 from the grey squirrel *Sciurus carolinensis*. J. Protozool. 19:143-150.

Joseph, T. 1972a. Observations on the endogenous stages of *Eimeria confusa* Joseph, 1969 from the grey squirrel *Sciurus carolinensis*. J. Protozool. 19:408-413.

Joseph, T. 1973. Eimerians occurring in or infective to both the fox squirrel *Sciurus niger rufiventer* and the gray squirrel *S. carolinensis*. J. Protozool. 20:509.

Joseph, T. 1975. Experimental transmission of *Eimeria confusa* Joseph 1959 to the fox squirrel. J. Wildl. Dis. 11:402-403.

Joyeux, C. E. 1927. Deux parasites nouveaux pour la marmotte des alpes, *Marmota marmota* L. Ann. Parasitol. 5:381-382.

Kalyakin, V. N., Yu. V. Kovalevsky, & N. A. Nikitina. 1973. Some epizootiological characteristics of *Toxoplasma glareoli* Erhardova, 1955 infection in redbacked voles (Clethrionomys). Folia Parasitol. (Prague). 20:119-129.

Kalyakin, V. N. & D. N. Zasukhin. 1975. Distribution of Sarcocystis (Protozoa: Sporozoa) in vertebrates. Folia Parasitol. 22:289-307.

Kan, S. P. 1979. Ultrastructure of the cyst wall of *Sarcocystis* spp. from some rodents in Malaysia. Int. J. Parasitol. 9:475-480.

Kan, S. P. & A. S. Dissanaike. 1977. Ultrastructure of *Sarcocystis* sp. from the Malaysian house rat, *Rattus rattus diardii*. Z. Parasitenkd. 52:219-227.

Karstad, L. 1963. *Toxoplasma microti* (the M-organism) in the muskrat (*Ondatra zibethica*). Can. Vet. J. 4:249-251.

Kartchner, J. A. & E. R. Becker. 1930. Observations on *Eimeria citelli*, a new species of coccidium from the striped ground squirrel. J. Parasitol. 17:90-94.

Kasim, A. A. & Y. R. Al Shawa. 1985. *Isospora dawadimiensis* n. sp. (Protozoa: Eimeriidae) from the jerboa (*Jaculus jaculus*) in Saudi Arabia. J. Protozool. 32:575-576.

Kelley, G. L. & N. N. Youssef. 1977. Development in cell cultures of *Eimeria vermiformis* Ernst, Chobotar and Hammond, 1971. Z. Parasitenkd. 53: 23-29.

Kepka, 0. & E. Scholtyseck. 1970. Weitere Untersuchungen der Feinstruktur von *Frenkelia* spec. (= M-Organismus, Sporozoa). Protistologica. 6:249-266.

Kheisin, E. M. 1959. [Observations of the residual bodies of oocysts and spores of several species of *Eimeria* from the rabbit and *Isospora* from the fox, skunk and hedgehog.] Zool. Zh. 38:1776-1784. (In Russian).

Kisskalt, K. J. & M. Hartmann. 1907. *Praktikum der Bakteriologie und Protozoologie*. Jena. vi + 174 pp.

Kleeberg, H. H. & W. Steenken, Jr. 1963. Severe coccidiosis in guinea-pigs. J. S. Afr. Vet. Med. Assoc. 34:49-52.

Klesius, P. H., A. L. Elston, W. H. Chambers, & H. H. Fudenberg. 1979. Resistance to coccidiosis (*Eimeria ferrisi*) in C57BL6 mice: effects of immunization and transfer factor. Clin. Immunol. Immunopathol. 12:143-149.

Klesius, P. H. & S. E. Hinds. 1979. Strain dependent differences in murine susceptibility to coccidia. Infect. Immun. 26:1111-1115.

Klesius, P. H., D. F. Qualls, A. L. Elston, & H. H. Fundenberg. 1979. Effects of bovine transfer factor (TFd) in mouse coccidiosis (*Eimeria ferrisi*). Clin. Immunol. Immunopathol. 10:214-221.

Knipling, E. F. & E. R. Becker. 1935. A coccidium from the fox squirrel *Sciurus niger rufiventer*, Geoffrey. J. Parasitol. 21:417-418.

Koffman, M. K. 1946. Om parasiter hos möss. Skand. Vet. Tidskr. Uppsala. 36:424-442.

Koldřová, L. 1986. Mouse (Mus musculus) as intermediate host of Sarcocystis sp. from the goshawk (Accipiter gentilis). Folia Parasitol. (Prague). 33:15-19.

Kotlan, S. & L. Pospesch. 1934. A hazinyul coccidiosisanak ismeretehez. Egy uj Eimeria-faj (*Eimeria piriformis* sp. n.) hazinyulbol. Allatorv. Lapok. 57:215-217.

Krampitz, H. E. & M. Rommel. 1977. Experimentelle Untersuchungen über das Wirtsspektrum der Frenkelien der Erdmaus. Berl. Muench. Tieraerztl. Wochenschr. 90:17-19.

Krampitz, H. E., M. Rommel, 0. Geisel, & E. Kaiser. 1976. Beiträge zum Lebenszyklus der Frenkelien. II. Die ungeschlechtliche Entwicklung von *Frenkelia clethrionomyobuteonis* in der Rötelmaus. Z. Parasitenkd. 51:7-14.

Krasnova, A. M. 1971. Tr. Sarat. Zootekh. Vet. Inst. 21:210-212 (from Kalyakin & Zasukhin, 1975).

Krishnamurthy, R. & H. S. Kshirsagar. 1980. Two new species of *Eimeria* Schneider, 1875 from rats in India. Arch. Protistenkd. 123:215-220.

Kruidenier, F. J., N. D. Levine, & V. Ivens. 1960 (1959). *Eimeria* (Protozoa: Eimeriidae) from the rice rat and pygmy mouse in Mexico. Trans. Ill. State Acad. Sci. 52:100-101.

Kshirsagar, H. S. 1980. Two new species of *Isospora* Schneider, 1881 from rats in India. Arch. Protistenkd. 123:267-271.

Kutzer, E., H. Frey, & J. Kotremba. 1980. Zur Parasitenfauna österreichischer Greifvögel (Falconiformes). Angew. Parasitol. 21:183-204.

Kyle, J. E. & C. A. Speer. 1984. Proteins of oocysts, sporozoites and merozoites of *Eimeria falciformis*. Prog. Abstr. Annu. Meet. Am. Soc. Parasitol. 59:40.

Labbé, A. 1896. Recherches zoologiques, cytologiques et biologiques sur les coccidies. Arch. Zool. Exp. Gen., 3rd ser. 4:517-654.

Labbé, A. 1899. Sporozoa. *Das Tierreich*. 5. Lief., xx + 180 pp.

Lai, P. F. 1977. *Sarcocystis* in Malaysian field rats. Southeast Asian J. Trop. Med. Public Health. 8:417-419.

Lainson, R. 1968. Parasitological studies in British Honduras. III. Some coccidial parasites of mammals. Ann. Trop. Med. Parasitol. 62:252-259.

Landers, E. J., Jr. 1960. Studies on excystation of coccidial oocysts. J. Parasitol. 46:195-200.

LaPage, G. 1940. The study of coccidians (*Eimeria caviae* [Sheather, 1924]) in the guinea pig. Vet. J. 96:144-154, 190-202, 242-254, 280-295.

Law, R. G. & A. H. Kennedy. 1932. Parasites of Fur-Bearing Animals. Bull. No. 4, Ontario Department of Game & Fish.

LeCompte, E. L. 1933. Scientific investigation of the muskrat industry in Maryland. Md. Conserv. 10:1-3.

Lee, B. L. & R. S. Dorney. 1971. *Eimeria ontarioensis* n. sp., *E. confusa* Joseph, 1969 and *E.* sp. (Protozoa: Eimeriidae) from the Ontario gray squirrel *Sciurus carolinensis*. J. Protozool. 18:587-592.

Léger, L. 1904. Protozoaires parasites des vipères. B. Mens. Assoc. Franc. Avanc. Sci. 9:208.

Levine, N. D. 1952 (1951). *Eimeria dicrostonicis* n. sp., a protozoan parasite of the lemming, and other parasites from Arctic rodents. Trans. Ill. State Acad. Sci. 44:205-208.

Levine, N. D. 1973. *Protozoan Parasites of Domestic Animals and of Man*. 2nd ed. Burgess Publishing, Minneapolis. ix + 406 pp.

Levine, N. D. 1977. Taxonomy of *Toxoplasma*. J. Protozool. 24:36-41.

Levine, N. D. 1977a. Sarcocystis cernae sp. n., replacement name for Sarcocystis sp. Černá and Loučková, 1976. Folia Parasitol. (Prague). 24:316.

Levine, N. D. 1977b. Nomenclature of *Sarcocystis* in the ox and sheep and of fecal coccidia of the dog and cat. J. Parasitol. 63:36-51.

Levine, N. D. 1978. Recent advances in classification of Protozoa. In Romberger, J. A., Ed. *Beltsville Symposia in Agricultural Research. (2) Biosystematics in Agriculture*. Allanheld, Osmun & Co., Montclair, N.J. pp. 71-87.

Levine, N. D. 1980. *Dorisa* n. gen. (Protozoa, Apicomplexa, Eimeriidae). J. Parasitol. 66:11.

Levine, N. D. 1981. Some corrections of coccidian (Apicomplexa: Protozoa) nomenclature. J. Parasitol. 66:830-834.

Levine, N. D. 1984. Taxonomy and review of the coccidian genus *Cryptosporidium* (Protozoa, Apicomplexa). J. Protozool. 31:94-98.

Levine, N. D. 1984. Nomenclatural corrections and new taxa in the apicomplexan protozoa. Trans. Am. Microsc. Soc. 103:195-204.

Levine, N. D. 1985. *Erhardovina* n.g., *Ascogregarina polynesiensis* n. sp., *Eimeria golemanski* n. sp., *Isospora tamariscini* n. sp., *Gregarina kazumii* n. nom., new combinations and emendations in the names of apicomplexan protozoa. J. Protozool. 32:359-363.

Levine, N. D. 1985. *Veterinary Protozoology*. Iowa State University Press. 414 pp.

Levine, N. D. 1988. *The Protozoan Phylum Apicomplexa*. CRC Press, Boca Raton, FL. 2 volumes.

Levine, N. D., R. S. Bray, V. Ivens, & A. E. Gunders. 1959. On the parasitic protozoa of Liberia. V. Coccidia of Liberian rodents. J. Protozool. 6:215-222.

Levine, N. D. & T. J. Husar. 1979. Rediscovery and redescription of *Eimeria miyairii* Ohira, 1912 from the Norway rat. Proc. Helminthol. Soc. Wash. 46:135-137.

Levine, N. D. & V. Ivens. 1960. *Eimeria* and *Tyzzeria* (Protozoa; Eimeriidae) from deermice (*Peromyscus* spp.) in Illinois. J. Parasitol. 46:207-212.

Levine, N. D. & V. Ivens. 1963. *Eimeria siniffi* sp. n. and *E. arizonensis* (Protozoa: Eimeriidae) from deermice in British Columbia. J. Parasitol. 49:660-661.

Levine, N. D. & V. Ivens. 1965. *The Coccidian Parasites* (*Protozoa, Sporozoa*) *of Rodents*. Illinois Biol. Monogr. #33. University of Illinois Press, Urbana. 365 pp.

Levine, N. D. & V. Ivens. 1981. *The Coccidian Parasites* (*Protozoa, Apicomplexa*) *of Carnivores*. Illinois Biol. Monogr. #51. University of Illinois Press, Urbana. pp. 248.

Levine, N. D., V. Ivens, & F. J. Kruidenier. 1955. Two new species of *Klossia* (Sporozoa: Adeleidae) from a deer mouse and a bat. J. Parasitol. 41:623-629.

Levine, N. D., V. Ivens, & F. J. Kruidenier. 1955a. *Dorisiella arizonensis* n. sp., a coccidium from the desert woodrat, *Neotoma lepida*. J. Protozool. 2:52-53.

Levine, N. D., V. Ivens, & F. J. Kruidenier. 1957. New species of *Eimeria* from Arizona rodents. J. Protozool. 4:80-88.

Levine, N. D., V. Ivens, & F. J. Kruidenier. 1957a. *Isospora citelli* n. sp. from the rock squirrel, *Citellus variegatus utah*. J. Protozool. 4:143-144.

Levine, N. D., V. Ivens, & F. J. Kruidenier. 1958 (1957). New species of *Eimeria* from Mexican rodents. Trans. Ill. Acad. Sci. 50:291-298.

Levine, N. D. & W. Tadros. 1980. Named species and hosts of *Sarcocystis* (Protozoa: Apicomplexa: Sarcocystidae). Syst. Parasitol. 2:41-59.

Lewis, D. C. & S. J. Ball. 1982. The life-cycle of *Eimeria cernae* Levine and Ivens, 1965 in the bank vole, *Clethrionomys glareolus*. Parasitology. 85:443-449.

Lewis, D. C. & S. J. Ball. 1984. *Eimeria fluviatilis* n. sp. and other species of *Eimeria* in wild coypus in England. Syst. Parasitol. 6:191-198.

Lewis, D. C. & S. J. Ball. 1983. Species of Eimeria of small wild rodents from the British Isles, with descriptions of two new species. Syst. Parasitol. 5:259-270.

Liburd, E. M., H. F. Pabst, & W. D. Armstrong. 1972. Transfer factor in rat coccidiosis. Cell. Immunol. 5:487-489.

Lindberg, R. E. & J. K. Frenkel. 1977. Cellular immunity to Toxoplasma and Besnoitia in hamsters: specificity and the effects of cortisol. Infect. Immun. 15:855-862.

Lindemann, K. 1865. Weiteres über Gregarinen. Bull. Soc. Imp. Nat. Moscow 38:381-387.

Litvenkova, E. A. 1969. Coccidia of wild mammals in Byelorussia. Prog. Protozool. 3:340.

Long, P. L. & C. A. Speer. 1977. Invasion of host cells by coccidia. In Taylor, A. E. R. & R. Moller, Eds. *Parasite Invasion*. Blackwell Scientific, Oxford, 1-26.

Lubimov, M. P. 1934. [*Biology of the Hares and Squirrels and Their Diseases*.] Manteufel, P., Ed. Moscow and Leningrad. p. 108 (from Kotlan, 1950).

Lunde, M. N. & A. H. Gelderman. 1971. Resistance of AKR mice to lymphoid leukemia associated with a chronic protozoan infection, *Besnoitia jellisoni*. J. Nat. Cancer Inst. 47:485-488.

Machul'skii, S. N. 1941. Koktsidiozy promyslovykh zhivotnykh v Buryat-Mongol'skoi ASSR. Tr. Buryat Mong. Zoovet. In-ta, 2, Ulan- Ude. 134-142 (from Musaev & Veisov, 1965).

Machul'skii, S. N. 1947. K voprosu o koktsidiyakh pushnykh zverei v Buryat-Mongol'skoi ASSR. Soobshchenie vtoroe. Tr. Buryat Mongol. Zoovet. In-ta, 3, Burmongiz. 40-56 (from Musaev & Veisov, 1965).

Machul'skii, S. N. 1949. K voprosu o koktsidioze gryzunov yuzhkykh raionov Buryat-Mongol'skoi ASSR. Zoovet. Inst., Ulan-Ude. 5 (from Svanbaev, 1956).

Machul'skii, S. N. & N. D. Miskaryan. 1958. Tr. Buryat Mong. Zoovet. Inst. 3:87-92 (from Kalyakin & Zasukhin, 1975).

Mackinnon, D. L. & M. J. Dibb. 1938. Report on intestinal protozoa of some mammals in the zoological gardens at Regent's Park. Proc. Zool. Soc. London. 108:323-345.

Mahrt, J. L. & S.-J. Chai. 1972. Parasites of red squirrels in Alberta, Canada. J. Parasitol. 58:639-640.

Mandal, A. K. 1976. Coccidia of Indian vertebrates. Rec. Zool. Surv. India. 70: 39-120.

Mandour, A. M. 1965. *Sarcocystis garnhami* n. sp. in the skeletal muscle of an opossum, *Didelphis marsupialis*. J. Protozool. 12:606-609.

Mandour, A. M., R. G. Bird, & M. Morris. 1965. Electron microscopic studies on *Sarcocystis* (sp. inq.) in *Lophuromys flavopunctatus* from Kenya. Trans. R. Soc. Trop. Med. Hyg. 59:360.

Mandour, A. M. & I. F. Keymer. 1972. Proc. 1st Sci. Symp. Rod. Control Egypt. 129-133 (from Markus, Killick-Kendrick, & Garnham, 1974).

Mantovani, A. 1965. Osservazioni sulla coccidiosi delle volpi in Abruzzo. Parassitologia. 7:9-17.

Marchiondo, A. A., D. W. Duszynski, & G. O. Maupin. 1976. Prevalence of antibodies to *Toxoplasma gondii* in wild and domestic animals of New Mexico, Arizona, and Colorado. J. Wildl. Dis. 12:226-232.

Marchiondo, A. A., D. W. Duszynski, & C. A. Speer. 1978. Fine structure of the oocyst wall of *Eimeria nieschulzi*. J. Protozool. 25:434-437.

Markus, M. B. 1982. *Sarcocystis* of African birds and mammals: Specificity for the intermediate host. In Müller, M., W. Gutteridge, & P. Köhler, Eds. *Molecular and Bio-chemical Parasitology*. Elsevier, Amsterdam. p. 345.

Markus, M. B. 1983. The hypnozoite of *Isospora canis*. S. Afr. J. Sci. 79:117.

Markus, M. B., R. Killick-Kendrick, & P. C. C. Garnham. 1974. The coccidial nature and life-cycle of *Sarcocystis*. J. Trop. Med. Hyg. 77:248-259.

Marotel, G. 1913 (1912). [*Sarcocystis besnoiti*]. Rec. Med. Vet. 90:328.

Marotel, G. 1921. Sur une nouvelle coccidie du chat. Bull. Soc. Sci. Vet. Lyon. 1921:86.

Marquardt, W. C. 1966. The living, endogenous stages of the rat coccidium, *Eimeria nieschulzi*. J. Protozool. 13:509-514.

Marquardt, W. C. 1966a. Attempted transmission of the rat coccidium *Eimeria nieschulzi* to mice. J. Parasitol. 52:691-694.

Marquardt, W. C., A. Y. Osman, & T. A. Muller. 1983. Arrested development in a coccidium: single and multiple infections with *Eimeria nieschulzi*. J. Protozool. 30:8A.

Martin, H. M. 1930. A species of *Eimeria* from the muskrat *Ondatra zibethica* (Linnaeus). Arch. Protistenkd. 70:273-278.

Mason, R. W. 1980. The discovery of *Besnoitia wallacei* in Australia and the identification of a free-living intermediate host. Z. Parasitenkd. 61:173-178.

Matsui, T., T. Morii, T. Iijima, S. Ito, K. Tsunoda, W. M. Correa, & T. Fujino. 1981. Cyclic transmission of the small type of *Isospora bigemina* of the dog. Jpn. J. Parasitol. 30:179-186.

Matuschka, F. R. & U. Häfner. 1984. Cyclic transmission of an African *Besnoitia* species by snakes of the genus *Bitis* to several rodents. Z. Parasitenkd. 70:471-476.

Matuschka, F.-R. 1986. *Sarcocystis clethrionomyelaphis* n. sp. from snakes of the genus *Elaphe* and different voles of the family Arvicolidae. J. Parasitol. 72:226-231.

Matubayasi, H. 1983. Studies on parasitic protozoa in Japan. IV. Coccidia parasitic in wild rats (*Epimys rattus alexandrinus* and *E. norvegicus*) Annot. Zool. Jpn. 17:144-163.

Mayberry, L. F. 1973. Incorporation of nucleic acid precursors by rat intestine infected with *Eimeria nieschulzi*. J. Protozool. 20:504-505.

Mayberry, L. F. & W. C. Marquardt. 1973. Transmission of *Eimeria separata* from the normal host, *Rattus*, to the mouse, *Mus musculus*. J. Parasitol. 59:198-199.

Mayberry, L. F. & W. C. Marquardt. 1974. Nucleic acid precursor incorporation by *Eimeria nieschulzi* (Protozoa: Apicomplexa) and jejunal villus epithelium. J. Protozool. 21:599-603.

Mayberry, L. F., W. C. Marquardt, D. J. Nash, & B. Plan. 1982. Genetic dependent transmission of *Eimeria separata* from *Rattus* to three strains of *Mus musculus*, an abnormal host. J. Parasitol. 68:1124-1126.

Mayberry, L. F., B. Plan, D. J. Nash, & W. C. Marquardt. 1975. Genetic dependence of coccidial transmission from rat to mouse. J. Protozool. 22:28A.

McKenna, P. B. & W. A. G. Charleston. 1980. Coccidia (Protozoa: Sporozoasida) of cats and dogs. II. Experimental induction of *Sarcocystis* infection in mice. N. Z. Vet. J. 28:117-119.

McKenna, P. B. & W. A. G. Charleston. 1980a. Coccidia (Protozoa: Sporozoasida) of cats and dogs. III. The occurrence of a species of *Besnoitia* in cats. N. Z. Vet. J. 28:120-122.

McMillan, B. 1958. A new species of *Eimeria* (Coccidiida, Eimeriidae) from the ground squirrel *Xerus (Euxerus) erythropus* (Geoffroy, 1803). Ann. Trop. Med. Parasitol. 52:20-23.

McQuistion, T. E. & K. M. Schurr. 1978. The effect of *Eimeria nieschulzi* infection on leukocyte levels in the rat. J. Protozool. 25:374-377.

McQuistion, T. E. & J. M. Wright. 1985 (1984). The prevalence and seasonal distribution of coccidial parasites of woodchucks (*Marmota monax*). J. Parasitol. 70:994-996.

Mehlhorn, H. & J. K. Frenkel. 1980. Ultrastructural comparison of cysts and zoites of *Toxoplasma gondii, Sarcocystis muris*, and *Hammondia hammondi* in skeletal muscle of mice. J. Parasitol. 66:59-67.

Mehlhorn, H., W. J. Hartley, & A.-O. Heydorn. 1976. A comparative ultrastructural study of the cyst wall of 13 Sarcocystis species. Protistologica. 12:451-467.

Mehlhorn, H. & E. Scholtyseck. 1974. Cytochemistry of the Toxoplasmatea *Sarcocystis, Frenkelia,* and *Besnoitia* at the ultrastructural level. Prog. Protozool. 4:275.

Mehlhorn, H., J. Sénaud, & E. Scholtyseck. 1973. La schizogonie chez *Eimeria falciformis* (Eimer, 1870) Coccidie, Eimeriidae, parasite de l'epithelium intestinal de la souris (*Mus musculus*). Étude au microscopie electronique des mérozoites et de leur développement au cours d'infections expérimentales. Protistologica. 9:269-291.

Meingassner, J. G. & H. Burtscher. 1977. Doppelinfektion des Gehirns mit *Frenkelia* species und *Toxoplasma gondii* bei *Chinchilla laniger*. Vet. Pathol. 14:146-153.

Mesfin, G. M. & J. E. C. Bellamy. 1978. The life cycle of *Eimeria falciformis* var. *pragensis* (Sporozoa: Coccidia) in the mouse, *Mus musculus*. J. Parasitol. 64:696-705.

Mesfin, G. M. & J. E. C. Bellamy. 1979. Migration of sporozoites of *Eimeria falciformis* var. *pragensis* from the absorptive to the crypt epithelium of the colon. J. Parasitol. 65:469-471.

Mesfin, G. M. & J. E. C. Bellamy. 1979a. The life cycle of *Eimeria falciformis* var. *pragensis* (Sporozoa: Coccidia) in the mouse, *Mus musculus*. J. Parasitol. 64:696-705.

Mesfin, G. M. & J. E. C. Bellamy. 1979b. Effects of acquired resistance on infection with *Eimeria falciformis* var. *pragensis* in mice. Infect. Immun. 23:109-114.

Mesfin, G. M. & J. E. C. Bellamy. 1979c. Thymic dependence of immunity to *Eimeria falciformis* var. *pragensis* in mice. Infect. Immun. 23:460-464.

Mesfin, G. M., J. E. C. Bellamy, & P. H. G. Stockdale. 1978. The pathological changes caused by *Eimeria falciformis* var. *pragensis* in mice. Can. J. Comp. Med. 42:496-510.

Meshorer, A. 1970. Interstitial nephritis in the spiny mouse (*Acomys cahirinus*) associated with *Klossiella* sp. infection. Lab. Anim. 4:227-232.

Mikeladze, L. G. 1971. [Coccidia of the mouse, *Microtus arvalis* Pall. in Georgia. Summary.] Soobshch. Akad. Nauk Gruz. SSR. 71:209-211.

Mikeladze, L. G. 1973. Novyi vid koktsidii *Isospora arvalis* sp. n. iz obyknovennoi polevki *Microtus arvalis* Pall. Soobshch. Akad. Nauk Gruz. SSR. 72:477-479.

Miller, N. L., J. K. Frenkel, & J. P. Dubey. 1972. Oral infections with toxoplasma cysts and oocysts in felines, other mammals, and in birds. J. Parasitol. 58:928-937.

Mir, N. A., M. B. Chhabra, R. M. Bhardwaj, & O. P. Gautam. 1982. *Toxoplasma* infection and some other protozoan parasites of the wild rat in India. Indian Vet. J. 59:60-63.

Mirza, M. Y. 1970. Incidence and Distribution of Coccidia (Sporozoa: Eimeriidae) in Mammals from Baghdad Area. M. S. thesis, Baghdad, Iraq. 195 pp.

Mirza, M. Y. 1975. Three new species of coccidia (Sporozoa: Eimeriidae). Bull. Nat. Res. Cent. 6:39-51.

Mirza, M. Y. & A. Y. Al-Rawas. 1975. *Eimeria taterae* sp. n. and other intestinal parasites from the antelope rat *Tatera indica* in Baghdad District. J. Protozool. 22:23-24.

Mirza, M. Y. & A. Y. Al-Rawas. 1978. *Wenyonella baghdadensis* sp. n. from the bandicoot rat *Nesokia indica* in the Baghdad area. J. Protozool. 25:285-286.

Mishra, G. S. & J .P. Gonzalez. 1978. *Eimeria gundii* n. sp. (Protozoa: Eimeriidae) from Tunisian gundi (*Ctenodactylus gundi*). Ann. Parasitol. 53:241-243.

Möller, J. 1923. Coccidien bei den Säugetieren des zoologischen Gartens zu Berlin. Inaug. -Diss., Berlin. 23 pp.

Morini, E. G., J. J. Boero, & A. Rodriquez. 1955. Parasitos intestinales en el "marra" (Dolichotis patagonum patagonum). (Coccidia). Publ. Mision Estud. Patol. Reg. Argent. 26 (85-86):83-89.

de Moura Costa, M. D. 1956. Isosporose do cão—com a descrição de uma nova variedade (*Isospora bigemina* Stiles, 1891 *bahiensis* n. var. Bol. Inst. Biol. Bahia. 3(1):107-112.

Mueller, B. E. G. 1975. *In vitro* development from sporozoites to first-generation merozoites in *Eimeria contorta* Haberkorn, 1971: A fine structural study. Z. Parasitenkd. 47:23-34.

Mueller, B., D. M. Hammond, & E. Scholtyseck. 1973. *In vitro* development of first- and second-generation schizonts of *Eimeria contorta* Haberkorn, 1971 (Coccidia, Sporozoa). Z. Parasitenkd. 41:173-185.

Mueller, B. E. G. & E. Scholtyseck. 1974. Electron microscope studies on the asexual development of *Eimeria contorta* (Sporozoa, Coccidia). Proc. III Int. Congr. Parasit. 1:102.

Mullin, S. W. & F. C. Colley. 1972. *Eimeria* and *Klossia* spp. (Protozoa: Sporozoa) from wild mammals in Borneo. J. Protozool. 19:406-408.

Mullin, S. W., F. C. Colley, & G. S. Stevens. 1972. Coccidia of Malaysian mammals: new host records and description of three new species of *Eimeria*. J. Protozool. 19:260-263.

Mullin, S. W., F. C. Colley, & Q. B. Welch. 1975. Ecological studies on coccidia of Malaysian forest mammals. Southeast Asian J. Trop. Med. Public Health. 6:93-98.

Munday, B. L. 1977. A species of *Sarcocystis* using owls as definitive hosts. J. Wildl. Dis. 13:205-207.

Munday, B. L. 1983. An isosporan parasite of masked owls producing sarcocysts in rats. J. Wildl. Dis. 19:146-147.

Munday, B. L. & R. W. Mason. 1980. *Sarcocystis* and related organisms in Australian wildlife: III. *Sarcocystis murinotechis* sp. n. life cycle in rats (*Rattus*, *Pseudomys* and *Mastocomys* spp.) and tiger snakes (*Notechis ater*). J. Wildl. Dis. 16:83-87.

Munday, B. L., R. W. Mason, W. J. Hartley, P. J. A. Presidente, & D. Obendorf, 1978. *Sarcocystis* and related organisms in Australian wildlife. I. Survey findings in mammals. J. Wildl. Dis. 14:417-433.

Musaev, M. A. 1967. Novyi vid koktsidii roda *Eimeria* ot polevki shelkovnikova. Izv. Akad. Nauk Az. SSR, Ser. Biol. Nauk. 1:44-46.

Musaev, M. A. & F. K. Alieva. 1961. Novye vidy koktsidii roda *Eimeria* iz krasnokhvostoi peschanki *Meriones erythrourus* Gray. Izv. Akad. Nauk Az. SSR. 1961(5):53-59.

Musaev, M. A. & F. K. Alieva. 1963. Some questions of the ecology of coccidia of Meriones erythrourus. Prog. Protozool. 1:331-333.

Musaev, M. A. & S. G. Ismailov. 1969. Life cycles of *Eimeria erythrourica* and *E. schamchorica* (Sporozoa, Coccidia), parasites of the red-tailed Libyan jird (Meriones erythrourus). Prog. Protozool. 3:345-346.

Musaev, M. A. & S. G. Ismailov. 1969a. Ismenchivost' ootsist *Eimeria scnamchorica* [sic] Musajev et Alijeva, 1961 — parazita krasnokhvostoi peschanki. Parazitologiya. 3:176-184.

Musaev, M. A. & S. G. Ismailov. 1971. Endogenous stages of the life cycle of *Eimeria schamchorica* Musajev et Alijeva, 1961 (Sporozoa, Coccidia) the parasite of *Meriones erythrourus* Gray. Acta Protozool. 9:223-233.

Musaev, M. A. & S. G. Ismailov. 1973. The life cycle of *Eimeria martunica* Musajev et Alijeva, 1961. Prog. Protozool. 4:288.

Musaev, M. A., S. G. Ismailov, & G. D. Gaibova. 1978. Life cycle and cytochemistry of endogenic development stages of *Eimeria dzhahoriana* Vejsov, 1961. Short Com. Fourth Internat. Congr. Parasit. B:70-71.

Musaev, M. A., S. G. Ismailov, & G. D. Gaibova. 1982. Zhiznennyi tsikl i tsitckhimkcheskoe issledovaie studii endogennogo razvitiya *Eimeria akeriana* (Coccidiida, Eimeriidae) iz maloaziiskoi peschanki. [Life cycle and cytochemical study of the endogenous stages of *Eimeria akeriana* from *Meriones blackleri*.] Parazitologiya. 16(1):18-23.

Musaev, M. A. & A. M. Veisov. 1959. Novye vidy koktsidii roda *Eimeria* iz obshchestvennoi polevki *Microtus socialis* Pall. v Azerbaidzhanskoi SSR. Izv. Akad. Nauk Az. SSR Ser. Biol. Sel'khoz. Nauk. 3:45-50.

Musaev, M. A. & A. M. Veisov. 1959a. Koktsidii lesnoi soni *Dyromys nitedula* Pall. v Azerbaidzhane. Dokl. Akad. Nauk Az. SSR. 15(6):535-539.

Musaev, M. A. & A. M. Veisov. 1960. Novye vidy koktsidii iz maloaziiskikh peschanok *Meriones tristrami* Thomas. Izv. Akad. Nauk Az. SSR Ser. Biol. Med. Nauk. 6:80-85.

Musaev, M. A. & A. M. Veisov. 1960a. Novye vidy koktsidii iz vodyanykh polevok *Arvicola terrestris* L. Izv. Akad. Nauk Az. SSR Ser. Biol. Med. Nauk. 1:51-61.

Musaev, M. A. & A. M. Veisov. 1961. Novye vidy koktsidii roda *Eimeria* iz serogo khomyachka (*Cricetulus migratorius* Pallas, 1770). Zool. Zh. 40:971-975.

Musaev, M. A. & A. M. Veisov. 1961a. Novyi vid koktsidii iz soni polchka — *Glis glis* (1766). Dokl. Akad. Nauk Az. SSR. 17:1085-1088.

Musaev, M. A. & A. M. Veisov. 1962. Novye vidy koktsidii roda *Eimeria* iz zakavkazskogo khomyachka *Cricetus auratus* Water (1939). Dokl. Akad. Nauk Az. SSR. 18(10):65-68.

Musaev, M. A. & A. M. Veisov. 1963. Novyi vid koktsidii roda *Eimeria* iz gornoi (zakavkazskoi) slepushonki *Ellobius lutescens* Thomas (1897). Dokl. Akad. Nauk Az. SSR. 19(7):75-77.

Musaev, M. A. & A. M. Veisov. 1963a. Novye vidy koktsidii roda *Eimeria* iz malogo tushkanchika (*Allactaga dater* [*sic*] Licht., 1825). Zool. Zh. 42:126-128.

Musaev, M. A & A. M. Veisov. 1963b. Koktsidii lesnoi myshi *Apodemus sylvaticus* L. v Azerbaidzhane. Izv. Akad. Nauk Az. SSR Ser. Biol. Nauk. 5:3-14.

Musaev, M. A. & A. M. Veisov. 1965. *Koktsidii Gryzunov SSR.* Izdat. Akad. Nauk Azerbaid. SSR, Baku. 154 pp.

Musaev, M. A., A. M. Veisov, & F. K. Alieva. 1963. Pyat' novykh vidov koktsidii roda *Eimeria* iz obshchestvennol polevki *Microtus socialis* Pall. Zool. Zh. 42:809-813.

Muto, T., M. Sugisaki, T. Yusa, & Y. Noguchi. 1985. Studies on coccidiosis in guinea pigs. I. Clinicopathological observation. Exp. Anim. 34:23-30.

Muto, T., T. Yusa, M. Sugisaki, K. Tanaka, Y. Noguchi, & K. Taguchi. 1985. Studies on coccidiosis in guinea pigs. 2. Epizootiological survey. Exp. Anim. 34:31-39.

Naciri, M. & P. Yvoré. 1982. Développement d'*Eimeria tenella* agent d'une coccidiose caecale de poulet, chez un hote non spécifique: existence d'une forme exointestinale infectante. C. R. Acad. Sci. III, 294:219-221.

Nash, P. V., G. M. Callis, & C. A. Speer. 1985. B cell responses in the gut and mesenteric lymph nodes to infection with *Eimeria falciformis*. Prog. Abstr. Am. Soc. Parasitol. 60:56.

Neméseri, L. 1959. Adatok a kutyz coccidiosisakoz. I. *Isospora canis*. Magy. Allatorv. Lapja. 14:91-92.

Neuman, M. & T. A. Nobel. 1981. Observations on the pathology of besnoitiosis in experimental animals. Zbl. Vet. Med. 828:345-354.

Neuman, M., T. A. Nobel, & B. Z. Perelman. 1979. The neuropathogenicity of *Besnoitia besnoiti* (Marotel, 1912) in experimental animals. J. Protozool. 26:51A.

Nicolle, C. & L. Manceaux. 1908. Sur une infection a corps de Leishman (ou organismes voisins) de gondi. C. R. Acad. Sci. 147:763-766.

Nicolle, C. & L. Manceaux. 1909. Sur un protozoaire nouveau du gondi. C. R. Acad. Sci. 148:369-372.

Nie, D. 1950. Morphology and taxonomy of the intestinal protozoa of the guinea-pig, *Cavia porcella*. J. Morphol. 86:381-494.

Nieschulz, O. & A. Bos. 1931. Ueber den Infektionsverlauf der Mausekokzidiose. Z. Infektnskr. Hyg. Haust. 39:160-168.

Nieschulz, O. & A. Bos. 1933. Ueber die Coccidien der Silberfüchse. Detsch. Tieraerztl. Wochennschr. 41:819-820.

Nöller, W. 1920. Kleine Beobachtungen an parasitischen Protozoen. Zur Kenntnis der Darmprotozoen des Hamsters (*Eimeria falciformis* var. *criceti* nov. var). Arch. Protistenkd. 41:169-189.

Nowak, R. M. & J. L. Paradiso. 1983. *Walker's Mammals of the World.* 4th ed. Johns Hopkins University Press, Baltimore. 2 vols., 1362 pp.

Nukerbaeva, K. K. & S. K. Svanbaev. 1973. Koktsidii pushnykh zverei v Kazakhstane. [Coccidia of fur-bearing mammals in Kazakhstan.] Vestn. Skh. Nauki Kaz. 1973(12):50-54.

Nukerbaeva, K. K. & S. K. Svanbaev. 1977. Rezul'taty izucheniya koktsidii pushnykh zverei. Tr. Inst. Zool. Akad. Nauk Kaz. SSR. 37:51-90.

Obitz, K. & S. Wadowski. 1937. *Eimeria coypi* n. sp., pasozyt nutrii (*Myocastor coypus* Mol.). Pamiet. Panstw. Inst. Nauk Gospodarst. Wiejsk. Pukawach. Wydz. Wet. 1:98-99.

O'Donoghue, P. J., C. H. S. Watts, & B. R. Dixon. 1987. Ultrastructure of *Sarcocystis* spp. (Protozoa, Apicomplexa) in rodents from North Solawesi and West Java, Indonesia. J. Wildl. Dis. 23:225-232.

O'Donoghue, P. J. & H. Weyreter. 1983. Detection of *Sarcocystis* antigens in the sera of experimentally-infected pigs and mice by an immunoenzymatic assay. Vet. Parasitol. 12:13-29.

O'Donoghue, P. J. & H. Weyreter. 1984. Examinations on the serodiagnosis of Sarcocystis infections. II. Class-specific immunoglobulin responses in mice, pigs and sheep. Zbl. Bakt. Hyg. A. 257:168-184.

Ohira, T. 1912. Ein Beitrag zur Coccidiumforschung. Ueber ein bei Ratten gefundenes Coccidium (*Eimeria miyairii*). Mitt. Med. Gesell. Tokio. 26:1045-1056.

Otto, H. 1957. Befunde an Mäsennieren bei Coccidiose. Frankf. Z. Pathol. 68:41-48.

Overdulve, J. P. 1978. *Prudish parasites. The discovery of sexuality, and studies on the life cycle, particularly the sexual stages in cats, of the sporozoan parasite* Isospora (Toxoplasma) gondii. Proefschrift, Rijksuniviersiteit te Utrecht, Netherlands. xxx + 83 pp.

Owen, D. 1974. A species of *Eimeria* in specific pathogen free and germfree mice. Proc. III Internat. Congr. Parasit. 1:94-95.

Owen, D. G. 1983. The effect of "Alcide" on 4 strains of rodent coccidial oocysts. Lab. Anim. 17:267-269.

Pak, S. M., V. V. Perminova, & N. V. Eshtokina. 1979. *Sarcocystis citellivulpes* sp. n. iz zheltykh suslikov (*Citellus fulvus*). [*Sarcocystis citellivulpes* n. sp. in the yellow suslik *Spermophilus fulvus*.] In Beyer, T. V. et al., Eds. *Toksoplazmidy.* USSR Acad. Sci., Soc. Protozool. USSR., Ser. *Protozoology* No. 4. Izdat. Nauka, Leningrad, USSR. pp. 111-114.

Parker, J. C., E. J. Rigg, & R. B. Holliman. 1974 (1972). Notes on parasites of gray squirrels from Florida. Q. J. Fla. Acad. Sci. 35:161-162.

Patton, W. S. & E. Hindle. 1926. Notes on three new parasites of the striped hamster (*Cricetulus griseus*). Proc. R. Soc. London Ser. B. 100:387-390.

Pearce, L. 1916. Klossiella infection of the guinea pig. J. Exp. Med. 23:431-442.

Pellérdy, L. 1954. Zur Kenntnis der Coccidien aus *Apodemus flavicollis.* Acta Vet. 4:187-191.

Pellérdy, L. 1954a. Contribution to the knowledge of coccidia of the common squirrel (*Sciurus vulgaris*). Acta Vet. 4:475-480.

Pellérdy, L. 1956. Catalogue of the genus *Eimeria* (Protozoa: Eimeriidae). Acta Vet. Acad. Sci. Hung. 6:75-102.

Pellérdy, L. 1957. On the homonymy of *Eimeria fulva* Farr, 1953 and *Eimeria fulva* Seidel, 1954. J. Parasitol. 43:591.

Pellérdy, L. 1960. Intestinal coccidiosis of the coypu. II. The endogenous development of *Eimeria seideli* and the present status of the group "Globidium". Acta Vet. Acad. Sci. Hung. 10:389-399.

Pellérdy, L. 1963. *Catalogue of Eimeriidea (Protozoa, Sporozoa).* Akad. Kiado, Budapest. 160 pp.

Pellérdy L. 1974. *Coccidia and Coccidiosis.* 2nd ed. Akad. Kiado, Budapest. 959 pp.

Pellérdy, L. 1974a. Studies on the coccidia of the domestic cat. *Isospora novocati* sp. n. Acta Vet. Acad. Sci. Hung. 24:127-131.

Pellérdy, L. & A. Babos. 1953. Zur Kenntnis der kokzidien aus *Citellus citellus.* Acta Vet. Acad. Sci. Hung. 3:167-172.

Péllerdy, L., A. Haberkorn, H. Mehlhorn, & E. Scholtyseck. 1971. Die Feinstruktur der Schizonten und Merozoiten des Mäusecoccids *Eimeria falciformis.* Acta Vet. Acad. Sci. Hung. 21:433-443.

Pelster, B. & G. Piekarski. 1971. Elektronenmikroskopische Analyse der Mikrogametenentwicklung bei *Toxoplasma gondii*. Z. Parasitenkd. 37:267-277.

Pérard, C. 1926. Sur la coccidiose du rat. Rec. Med. Vet. Bull. et Mem. Soc. Centr. Med. Vet. 102:120-124.

Peteshev, V. M., I. G. Galuso & A. P. Polotaoshnov. 1974. Koshki-definitivnye Khozyaeva besnoitii (*Besnoitia besnoiti*). Izv. Akad. Nauk Kaz. SSR. Ser Biol. 1:33-38.

Pop-Cenitch, S. & A. Bordjochki. 1957. Sur une nouvelle espèce de coccidie de l'écureuil. Arch. Inst. Pasteur Alger. 35:73-75.

Pope, J. H., V. A. Bicks, & I. Cook. 1957. Toxoplasma in Queensland. II. Natural infections in bandicoots and rats. Aust. J. Exp. Biol. Med. Sci. 35:481-490.

Porchet-Henneré, E., M. C. D'Hooghe, A. Sadek, & S. Frontier. 1985. Relations entre le coccidie *Besnotia jellisoni* et sa cellule-hôte, *in vitro*, étudiérs par microcinématographie. Protistologica. 21:39-45.

Porter, A. 1957. Morphology and affinities of entozoa and endophyta of the naked mole rat *Heterocephalus glaber*. Proc. Zool. Soc. London. 128:515-526.

Prasad, H. 1960. A new species of coccidia of the grey squirrel *Sciurus* (*Neosciurus*) *carolinensis*. J. Protozool. 7:135-139.

Prasad, H. 1960a. Two new species of coccidia of the coypu. J. Protozool. 7:207-210.

Prasad, H. 1960b. Studies on the coccidia of some rodents of the families *Muridae*, *Dipodidae*, and *Cricetidae*. Ann. Trop. Med. Parasitol. 54:321-330.

Railliet, A. & A. Lucet. 1891. Note sur quelques espèces de coccidies encore peu étudiées. Bull. Soc. Zool. France 16:246-250.

Rastegaieff, E. F. 1930. Zur Frage über Coccidien wilder Tiere. Arch. Protistenkd. 71:377-404.

Ray, H. N., D. C. Banik, & A. K. Mukherjea. 1965. A new coccidium, *Eimeria bandipurensis* n. sp., from the Indian palm squirrel, *Funambulus palmarum*. J. Protozool. 12:478-479.

Ray, H. N. & M. Das Gupta. 1937. A new coccidian, *Wenyonella hoarei* n. sp., from an Indian squirrel. Parasitology. 27:117-120.

Ray, H. N. & H. Singh. 1950. On a new coccidium, *E. petauristae* n. sp. from the intestine of a Himalayan flying squirrel, *Petaurista inornatus* (Geoffrey). Proc. Zool. Soc. Bengal. 3:65-70.

Reduker, D. W. & D. W. Duszynski. *Eimeria ladronensis* n. sp. and *E. albigulae* (Apicomplexa: Eimeriidae) from the woodrat *Neotoma albigula* (Rodentia: Cricetidae). J. Protozool. 32:548-550.

Reduker, D. W., L. Hertel, & D. W. Duszynski. 1985. *Eimeria* species (Apicomplexa: Eimeriidae) infecting *Peromyscus* rodents in the southwestern United States and northern Mexico with description of a new species. J. Parasitol. 71:604-613.

Reichenow, E. 1921. Die Coccidien. In Prowazek, S. *Handbuch der Pathogenen Protozoen*. Leipzig. 3:1136-1277.

Reichenow, E. 1953. *Doflein's Lehrbuch der Protozoenkunde*. 5th ed., G. Fischer Verlag, Jena.

Reimer, 0. 1923. *Zur Pathologie der Mausekokzidiose*. Diss. (Berlin). 20 pp.

Ringuelet, R. A. & S. Coscaron. 1961 (1960). Coccidios del quiya (*Myocastor coypus bonariensis*). Actas Trab. I Congr. Sudamer. Zool. 1959. 2:257-260.

Rivolta, S. 1878. Della gregarinosi dei polli e dell' ordinamento delle gregarine e dei psorospermi degli animali domestici. G. Anat. Fisiol. Patol. Anim. 10:220-235.

Roberts, W. L. & D. M. Hammond. 1973. Scanning electron microscopy study of invasion of host cells by *Eimeria larimerensis* sporozoites. Proc. Helminthol. Soc. Wash. 40:118-123.

Roberts, W. L., D. M. Hammond, L. C. Anderson, & C. A. Speer. 1970. Ultrastructural study of schizogony in *Eimeria callospermophilli*. J. Protozool. 17:584-592.

Roberts, W. L., D. M. Hammond, & C. A. Speer. 1970. Ultrastructural study of the intra- and extracellular sporozoites of *Eimeria callospermophili*. J. Parasitol. 56:907-917.

Roberts, W. L., C. A. Speer, & D. M. Hammond. 1970. Electron and light microscope studies of the oocyst walls, sporocysts, and excysting sporozoites of *Eimeria callospermophili* and *E. larimerensis*. J. Parasitol. 56:918-926.

Rodhain, J. 1954. *Eimeria vinckei* n. sp. parasite de l'intestin de *Thamnomys surdaster surdaster*. Ann. Parasitol. 29:327-329.

Romero Rodriguez, J. 1979. Contribucion al estudio de los protozoa-Eimeriidae parasitos de *Epimys norvegicus norvegicus* (B.) en Granada (España). Rev. Iber. Parasitol. 39:73-79.

Rommel, M. & A.-O. Heydorn. 1971. Versuche zur Uebertragung der Immunitaet gegen *Eimeria*-Infektionen durch Lymphozyten. Z. Parasitenkd. 36:242-250.

Rommel, M. & H. E. Krampitz. 1975. Beiträge zum Lebenszyklus der Frenkelien. I. Die identität von *Isospora buteonis* aus dem Mäusebussard mit einer Frenkelienart (*F. clethrionomyobuteonis* spec. n.) aus der Rötelmaus. Berl. Muench. Tieraerztl. Wochenschr. 88:338-340.

Rommel, M. & H. E. Krampitz. 1978. Weitere Untersuchungen über das Zwischenwirtsspektrum und den Entwicklungszyklus von Frenkelia microti aus der Erdmaus. Zbl. Vet. Med. B. 25:273-281.

Rommel, M., H. E. Krampitz, E. Göbel, O. Geisel, & E. Kaiser. 1976. Untersuchungen über den Lebenszyklus von *Frenkelia clethrionomyobuteonis*. Z. Parasitenkd. 50:204-205.

Rose, M. E. & P. Hesketh. 1987 (1986). Eimerian life cycles: the patency of *Eimeria vermiformis*, but not *Eimeria pragensis*, is subject to host (Mus musculus) influence. J. Parasitol. 72:949-954.

Rose, M. E., P. Hesketh, & B. M. Ogilvie. 1979. Peripheral blood leucocyte response to coccidial infections: a comparison of the response in rats and chickens and its correlation with resistance to reinfection. Immunology. 36:71-79.

Rose, M. E., B. M. Ogilvie, P. Hesketh, & M. F. W. Festing. 1979. Failure of nude (athymic) rats to become resistant to reinfection with the intestinal coccidian parasite *Eimeria nieschulzi* or the nematode *Nippostrongylus prasitiensis*. Parasit. Immunol. 1:125-132.

Rose, M. E., D. G. Owen, & P. Hesketh. 1984. Susceptibility to coccidiosis: Effect of strain of mouse on reproduction of *Eimeria vermiformis*. Parasitology. 88:45-54.

Rose, M. E., J. V. Peppard, & S. M. Hobbs. 1984. Coccidiosis: characterization of antibody responses with *Eimeria nieschulzi*. Parasit. Immunol. 6:1-12.

Rosenmann, M. & P. R. Morrison. 1975. Impairment of metabolic capability in feral house mice by *Klossiella muris* infection. Lab. Anim. Sci. 25:62-64.

Roudabush, R. L. 1937. The endogenous phases of the life cycle of *Eimeria nieschulzi, Eimeria separata,* and *Eimeria miyairii* coccidian parasites of the rat. Iowa State Coll. J. Sci. 11:135-163.

Roudabush, R. L. 1937a. Two *Eimeria* from the flying squirrel, *Glaucomys volens*. J. Parasitol. 23:107-108.

Ruiz, A. & J. K. Frenkel. 1976. Recognition of cyclic transmission of *Sarcocystis muris* by cats. J. Infect. Dis. 133:409-418.

Ruiz, A. & J. K. Frenkel. 1980. *Toxoplasma gondii* in Costa Rican cats. Am. J. Trop. Med. Hyg. 29:1150-1160.

Ryan, M. J., D. S. Wyand, & S. W. Nielsen. 1982. A *Hammondia*-like coccidian with a mink-muskrat life cycle. J. Wildl. Dis. 18:29-35.

Ryšavý, B. 1954. Příspévek k poznani kokcidii našich i dovezenych obratlovcǔ. Česk. Parazitol. 1:131-174.

Ryšavý, B. 1957. Poznamky k fauné kokcidii volné zijicich drobnych ssavcǔ v Československu. Cesk. Parasitol. 4:331-336.

Ryšavý, B. 1967. Coccidia from Conga hutia — Capromys pilorides Say, 1822. Folia Parasitol. (Prague). 14:93-96.

Ryšavý, B. & Z. Černá. 1979. Coccidia parasites (Coccidia, Eimeriidae) of the squirrel *Sciurus vulgaris* L. from Czechoslovakia. Folia Parasitol. (Prague). 26: 65-68.

Rzepczyk, C. M. 1974. Evidence of a rat-snake life cycle for *Sarcocystis*. Int. J. Parasitol. 4:447-449.

Rzepczyk, C. M. & E. Scholtyseck. 1976. Light and electron microscope studies on the *Sarcocystis* of *Rattus fuscipes*, an Australian rat. Z. Parasitenkd. 50:137-150.

Samish, M., U. Shkap, E. Pipano, & C. Bin. 1982. Cultivation of Besnoitia besnoiti in tick cell culture. J. Protozool. 29:313.

Sampson, J. R. 1969. *Eimeria yukonensis* n. sp. (Protozoa: Eimeriidae) from the arctic ground squirrel *Spermophilus undulatus*. J. Protozool. 16:45-46.

Sangiorgi, G. 1916. Di un coccidio parasita del rene della cavia. Pathologica. 8:49-53.

Saxe, L. H., N. D. Levine, & V. Ivens. 1960. New species of coccidia from the meadow mouse, *Microtus pennsylvanicus*. J. Protozool. 7:61-63.

Sayin, F. 1981 (1980). Eimeriidae of the herbivorous mole-rat, *Spalax enhrenbergi* Nehring. J. Protozool. 27:364-367.

Sayin, F., S. Dincer, & I. Meric. 1977. Coccidia (Protozoa: Eimeriidae) of the herbivorous mole-rat, *Spalax leucodon* Nordmann. J. Protozool. 24:210-212.

Schaudinn, F. 1900. Untersuchungen über den Generationswechsel bei Coccidien. Zool. Jahrb. Abt. Anat. Ontog. Tiere. 13:197-292.

Scheuring, W. 1973. Uwagi o chorobach inwazyjnych nutrii, z uwzglenienium badan w woj zielonogorskim. Med. Wet. 29:406-407.

Schneider, A.-C.-U. 1875. Note sur la psorospermie oviforme du poulpe. Arch. Zool. Exp. Gen. 4:xi-xiv.

Schneider, C. R. 1965. *Besnoitia panamensis* sp. n. (Protozoa: Toxoplasmatidae) from Panamanian lizards. J. Parasitol. 51:340-344.

Scholtyseck, E. 1954. Untersuchungen über die bei einheimischen Vogelarten vorkommenden Coccidien der Gattung *Isospora*. Arch. Protistenkd. 100:91-112.

Scholtyseck, E. 1973. Die Deutung von Endodyogenie und Schizogonie bei Coccidien und anderen Sporozoen. Z. Parasitenkd. 42:87-104.

Scholtyseck, E. & F. A. Ghaffar. 1981. *Eimeria falciformis* — merozoites with refractile bodies. Z. Parasitenkd. 65:117-120.

Scholtyseck, E. & H. Mehlhorn. 1970. Ultrastructural study of characteristic organelles (paired organelles, micronemes, micropores) of Sporozoa and related organisms. Z. Parasitenkd. 34:97-127.

Scholtyseck, E., H. Mehlhorn, & K. Friedhoff. 1970. The fine structure of the conoid of Sporozoa and related organisms. Z. Parasitenkd. 34:68-94.

Scholtyseck, E., H. Mehlhorn, & A. Haberkorn. 1971. Die Feinstruktur der Makrogameten der Maeusecoccids *Eimeria falciformis*. Z. Parasitenkd. 37:44-51.

Scholtyseck, E., H. Mehlhorn, & D. M. Hammond. 1971. Fine structure of macrogametes and oocysts of coccidia and related organisms. Z. Parasitenkd. 37:1-43.

Scholtyseck, E., H. Mehlhorn, & D. M. Hammond. 1972. Electron microscope studies of microgametogenesis in coccidia and related groups. Z. Parasitenkd. 38:95-131.

Scholtyseck, E., H. Mehlhorn, & B. E. G. Müller. 1973. Identifikation von Merozoiten der vier cystenbildenden Coccidien (*Sarcocystis, Toxoplasma, Besnoitia, Frenkelia*) auf Grund fein-struktureller Kriterien. Z. Parasitenkd. 42:185-206.

Scholtyseck, E., H. Mehlhorn, & B. E. G. Müller. 1974. Feinstruktur der Cyste und Cystenwand von *Sarcocystis tenella, Besnoitia jellisoni, Frenkelia* sp. und *Toxoplasma gondii*. J. Protozool. 21:284-294.

Scholtyseck, E., H. Mehlhorn, & J. Sénaud. 1972. Die subpellikulären Mikrotubuli in den merozoiten von *Eimeria falciformis* (Coccidia, Sporozoa). Z. Parasitenkd. 40:281-294.

Scholtyseck, E., L. Pellérdy, H. Mehlhorn, & A. Haberkorn. 1973. Elektronenmikroskopische Untersuchungen uber die Mikrogametenentwicklung des Mäusecoccids *Eimeria falciformis*. Acta Vet. Acad. Sci. Hung. 23:6173.

Schrecke, W. & U. Dürr. 1970. Excystations und Infektionsversuche mit Kokzidienoocysten bei neugeborenen Tieren. Zentralbl. Bakt. I. Orig. 215:252-258.

Schuberg, A. 1892. Uber Coccidien des Mäusedarmes. Sitzungsber. Phys. Med. Ges. Würzb. 65:65-72.

Scott, M. J. 1978. Toxoplasmosis. Trop. Dis. Bull. 75:809-827.

Šebek, Z. 1963. Sarcocystis und verwandte Organismen bei den in Insektenfressern und Nageti-eren. Proc. I. Int. Congr. Protozool. 473-477.

Šebek, Z. 1975. Parasitische Gewebeprotozoen der wildlebenden Kleinsauger in der Tschecho-slowakei. Folia Parasitol. (Prague). 22:111-124.

Seidel, E. 1954. Ein eigenartigen Parasit des Sumpfbibers: Eimeria (Globidium) fulva n. sp. Arch. Exp. Veterinaermed. 8:759-764.

Seidel, E. 1956. Parasitologische Befunde beim Sumpfbiber. In Borchert, A., Ed. *Probleme de Parasitologie. Vorträge der II Parasitologischen Arbeitstagung.* Akademie-Verlag, Berlin. 117-126.

Seidelin, H. 1914. *Klossiella* sp. in the kidney of a guinea pig. Ann. Trop. Med. Parasitol. 8:553-564.

Sela-Peréz, M. C., A. R. Martinez-Fernández, M. C. Arias-Fernández, & M. E. Ares-Mazas. 1982. Ultrastructura de *Sarcocystis muris* (Blanchard, 1885) Labbé, 1899. Rev. Iber. Parasitol. 42:9-19.

Sénaud, J. 1969. L'ultrastructure des kystes de *Besnoitia jellisoni* Frenkel 1953 — Sporozoa, (Toxoplasmea) chez la souris (*Mus musculus*). C. R. Acad. Sci. 268:816-819.

Sénaud, J. & Ž. Černá. 1968. Etude en microscopie électronique des mérozoites et de la mérogonie chez *Eimeria pragensis* (Černá et Sénaud, 1968), coccidie parasite de l'intestin de la souris (*Mus musculus*). Ann. Sta. Biol. Besse-en-Chandesse. 3:221-241.

Sénaud, J. & H. Mehlhorn. 1975. Etude ultrastructurale des coccidies formant des kystes: *Toxoplasma gondii, Sarcocystis tenella, Besnoitia jellisoni* et *Frenkelia* sp. (*Sporozoa*). II. Mise en evidence de l'ADN et de l'ARN au niveau des ultrastructures. Ann. Stat. Biol. Besse-en-Chandesse. 1975(9):111-156.

Sénaud, J. & H. Mehlhorn. 1977. Aspects ultrastructuraux du développement de *Besnoitia jellisoni* (Sporozoa, Apicomplexa) en culture de tissu. J. Protozool. 24:64A.

Sénaud, J. & H. Mehlhorn. 1978. *Besnoitia jellisoni* Frenkel 1953 (Sporozoa, Apicomplexa) en culture sur cellules de rein de chien (MDCK): étude au microscope électronique. Protistologica. 14:5-14.

Sénaud, J., H. Mehlhorn, & E. Scholtyseck. 1974. *Besnoitia jellisoni* in macrophages and cysts from experimentally infected laboratory mice. J. Protozool. 21:715-720.

Shaw, J. J. & R. Lainson. 1969. *Sarcocystis* of rodents and marsupials in Brazil. Parasitology. 59:233-244.

Sheather, A. L. 1924. Coccidiosis in the guinea pig. J. Comp. Pathol. 37:243-246.

Sheffield, H. G. 1966. Electron microscope study of the proliferative form of *Besnoitia jellisoni*. J. Parasitol. 52:583-594.

Sheffield, H. G., J. K. Frenkel, & A. Ruiz. 1977. Ultrastructure of the cyst of *Sarcocystis muris*. J. Parasitol. 63:629-641.

Sheppard, A. M. 1974. Ultrastructural pathology of coccidial infection. J. Parasitol. 60:369-371.

Shillinger, J. E. 1938. Coccidiosis in muskrats influenced by water levels. J. Wildl. Manage. 2:233-234.

Shkap, V., E. Pipano, & C. Greenblatt. 1983. Response of gerbils (*Meriones* tristrami) to inp. and S. C. inoculation with *Besnoitia benoiti*. J. Protozool. 30:57A.

Short, J. A., L. F. Mayberry, & J. R. Bristol. 1981 (1980). *Eimeria chihuahuaensis* sp. n. and other coccidia from *Dipodomys* spp. in El Paso County, Texas. J. Protozool. 27:361-364.

Sibert, G. J. & C. A. Speer. 1980. Fine structure of zygotes and oocysts of *Eimeria nieschulzi*. J. Protozool. 27:374-379.

Sinniah, B. 1979. Parasites of some rodents in Malaysia. Southeast Asian J. Trop. Med. Public Health. 10:115-121.

Skidmore, L. V. 1929. Note on a new species of coccidia from the pocket gopher (*Geomys bursarius*) (Shaw). J. Parasitol. 15:183-184.

Skofitsch, G. & O. Kepka. 1982. Evidence of circulating antibodies against *Frenkelia glareoli* (Apicomplexa). Z. Parasitenkd. 66:355-358.

Smith, D. D. & J. K. Frenkel. 1977. *Besnoitia darlingi* (Protozoa, Toxoplasmatidae): cyclic transmission by cats. J. Parasitol. 63:1066-1071.

Smith, T. & H. P. Johnson. 1902. On a coccidium (*Klossiella muris* gen. et sp. nov.) parasitic in the renal epithelium of the mouse. J. Exp. Med. 6:303-316.

Soon, B.-L. & R. S. Dorney. 1969. Occurrence of *Eimeria tamiasciuri* in Ontario red squirrels (*Tamiasciurus hudsonicus*). Can. J. Zool. 47:731-732.

Speer, C. A., L. R. Davis, & D. M. Hammond. 1972. Cinemicrographic observations on the development of *Eimeria larimerensis* in cultured bovine cells. J. Protozool. 18(Suppl.):11.

Speer, C. A. & D. M. Hammond. 1970. Development of *Eimeria larimerensis* from the Uinta ground-squirrel in cell cultures. Z. Parasitenkd. 25:105-118.

Speer, C. A. & D. M. Hammond. 1970a. Nuclear divisions and refractile-body changes in sporozoites and schizonts of *Eimeria callospermophili* in cultured cells. J. Parasitol. 56:461-467.

Speer, C. A., D. M. Hammond, & L. C. Anderson. 1970. Development of *Eimeria callospermophili* and *E. bilamellata* from the Uinta ground squirrel *Spermophilus armatus* in cultured cells. J. Protozool. 17:274-284.

Speer, C. A., D. M. Hammond, & Y. Y. Elsner. 1973. Development of second-generation schizonts and immature gamonts of *Eimeria larimerensis* in cultured cells. Proc. Helminthol. Soc. Wash. 40:147-153.

Speer, C. A. & F. L. Pollari. 1984. Surface changes in the intestinal mucosa of mice infected with *Eimeria falciformis*. Can. J. Zool. 62:1675-1679.

Splendore, A. 1918. Studi nell' interesse divna lotta biologica contro le arvicole. Boll. Min. Agric., Roma. S. B. an 17, n(1-6) 1-10 (Index-Cat. Med. Vet. Zool., p. 482).

Stabler, R. M. & K. Welch. 1961. *Besnoitia* from an oppossum. J. Parasitol. 47:576.

von Sternberg, C. 1929. *Klossiella muris*-einhaufiger, anscheinend wenig bekannter Parasit der weissen Maus. Wien. Klin. Wochenschr. 42:419-442.

Stevenson, A. C. 1915. *Klossiella muris*. Q. J. Microsc. Sci. 61:127-135.

Stockdale, P. H. G. & R. J. Cawthorn. 1981. The coccidian *Caryospora bubonis* in the great horned owl (*Bubo virginianus*). J. Protozool. 28:255-257.

Stockdale, P. H. G., G. B. Tiffin, G. Kozub, & B. Chobotar. 1979. *Eimeria contorta* Haberkorn, 1971: a valid species of rodent coccidium? Can. J. Zool. 57:264-270.

Stojanov, D. P. & J. L. Cvetanov. 1965. Uber die Klossiellose bei Meerschweinchen. Z. Parasitenkd. 25:350-358.

Stout, C. A. & D. W. Duszynski. 1983. Coccidia from kangaroo rats (*Dipodomys* spp.) in the western United States, Baja California, and northern Mexico with descriptions of *Eimeria merriami* sp. n. and *Isospora* sp. J. Parasitol. 69:209-214.

Straneva, J. E. & W. W. Gallati. 1980. *Eimeria strangfordensis* sp. n., *Eimeria barleyi* sp. n., and *Eimeria antonellii* sp. n. from the eastern woodrat (*Neotoma floridana*) from Pennsylvania. J. Parasitol. 66:329-332.

Straneva, J. E. & G. L. Kelley. 1979. *Eimeria clethrionomyis* sp. n., *Eimeria gallatii* sp. n., *Eimeria pileata* sp. n. and *Eimeria marconii* sp. n. from the red-backed vole *Clethrionomys gapperi* Vigors, from Pennsylvania. J. Protozool. 26:530-532.

Sureau, P. 1963. Infection spontanée des souris d'elevage a Tananarive par *Encephalitozoon cuniculi* et *Klosiella* [*sic*] *muris*. Arch. Inst. Pasteur Madagascar. 31:125-126.

Svanbaev, S. K. 1956. Materialy k faune koktsidii dikikh mlekopitayushchikh zapodnogo Kazakhstana. Tr. Inst. Zool. Akad. Nauk Kazakh. SSR 5:180-191.

Svanbaev, S. K. 1958. K poznaniyu fauny koktsidii gryzunov tsentral'nogo Kazakhstana. Trudy Inst. Zool. Akad. Nauk Kaz. SSR. 9:183-186.

Svanbaev, S. K. 1960. Koktsidii cerebristo-chernykh lisits Alma-Atinskoi oblasti. Tr. Inst. Zool. Akad. Nauk Kaz. SSR. 14:34-36.

Svanbaev, S. K. 1962. Novy vid koktsidii. Vestn. Akad. Nauk Kaz.. SSR. No. 5(206):86-87.

Svanbaev, S. K. 1962a. Obnaruzhenie koktsidii u ondatry, akklimatizirovannoi v Kazakhstane. Tr. Inst. Zool. Akad. Nauk Kaz. SSR. 16:206-207.

Svanbaev, S. K. 1963. Koktsidii surka menzbira (*Marmota menzbieri*, Kaschkarov, 1925). Parazity dikikh zhivotnykh Kazakhstana. Tr. Inst. Zool. Akad. Nauk Kaz. SSR. 19:51-54.

Svanbaev, S. K. 1969. *Koktsidii Dikikh Zhivotnykh Kazakhstana*. [*Coccidia of Wild Animals in Kazakhstan*.] Izdat. "Nauka", Kazakh. SSR, Alma-Ata, Kazakhstan, USSR. 212 pp.

Swellengrebel, N. H. 1914. Zur Kenntnis der Entwicklungsgeschichte von *Isospora bigemina* (Stiles). Arch. Protistenkd. 32:379-392.

Tadros, W. 1976. Contribution to the understanding of the life-cycle of *Sarcocystis* of the short-tailed vole *Microtus agrestis*. Folia Parasitol. (Prague). 23:193-199.

Tadros, W. 1981. Studies of Sarcosporidia of rodents with birds of prey as definitive hosts. In Canning, E. U., Ed. *Parasitological Topics*. Soc. Protozool. Spec. Publ. 1:248-259.

Tadros, W. A., R. G. Bird, & D. S. Ellis. 1972. The fine structure of cysts of Frenkelia (the M-organism). Folia Parasitol. (Prague). 19:203-209.

Tadros, W. & J. J. Laarman. 1976. *Sarcocystis* and related coccidian parasites: a brief general review, together with a discussion on some biological aspects of their life cycles and a new proposal for their classification. Acta Leiden. 44:1-107.

Tadros, W. & J. J. Laarman. 1978. Apparent congenital transmission of *Frenkelia* (Coccidia: Eimeriidae): first recorded incidence. Z. Parasitenkd. 58:41-46.

Tadros, W. & J. J. Laarman. 1979. Successful rodent to rodent transmission of *Sarcocystis sebeki* by inoculation of precystic schizogonic stages. Trans. R. Soc. Trop. Med. Hyg. 73:350.

Tadros, W. & J. J. Laarman. 1980. The tawny owl, *Strix aluco* as final host of a new species of *Sarcocystis* with *Mus musculus* as intermediate host. Trop. Geog. Med. 32:364.

Tanabe, M. & M. Okinami. 1940. On the parasitic protozoa of the ground squirrel, *Eutamias asiaticus* Utkensis, with special reference to *Sarcocystis eutamias* sp. nov. Keizyo J. Med. 10:126-134.

Tilahun, G. & P. H. G. Stockdale. 1981. Oocyst production of four species of murine coccidia. Can. J. Zool. 59:1796-1800.

Tilahun, G. & P. H. G. Stockdale. 1982. Sensitivity and specificity of the indirect fluorescent antibody test in the study of four murine coccidia. J. Protozool. 29:129-132.

Todd, K. S., Jr. & D. M. Hammond. 1968a. Life cycle and host specificity of *Eimeria larimerensis* Vetterling, 1964, from the Uinta ground squirrel *Spermophilus armatus*. J. Protozool. 15:268-275.

Todd, K. S., Jr. & D. M. Hammond. 1968b. Life cycle and host specificity of *Eimeria callospermophili* (Henry, 1932) from the Uinta ground squirrel *Spermophilus armatus*. J. Protozool. 15:1-8.

Todd, K. S., Jr., D. M. Hammond, & L. C. Anderson. 1968. Observations on the life cycle of *Eimeria bilamellata* Henry, 1932 in the Uinta ground squirrel *Spermophilus armatus*. J. Protozool. 15:732-740.

Todd, K. S., Jr., & D. L. Lepp. 1971. The life cycle of *Eimeria vermiformis* Ernst, Chobotar, and Hammond, 1971 in the mouse *Mus musculus*. J. Protozool. 18:332-337.

Todd, K. S., Jr. & D. L. Lepp. 1972. Completion of the life cycle of *Eimeria vermiformis* Ernst, Chobotar, and Hammond, 1971, from the mouse, *Mus musculus*, in dexamethasone-treated rats, *Rattus norvegicus*. J. Parasitol. 58:400-401.

Todd, K. S., Jr., D. L. Lepp, & C. V. Trayser. 1971. Development of the asexual cycle of *Eimeria vermiformis* Ernst, Chobotar, and Hammond, 1971, from the mouse, *Mus musculus*, in dexamethasone-treated rats, *Rattus norvegicus*. J. Parasitol. 57:1137-1138.

Todd, K. S., Jr., D. L. Lepp, & C. A. Tryon, Jr. 1971. Endoparasites of the northern pocket gopher from Wyoming. J. Wildl. Dis. 7:100-104.

Todd, K. S., Jr. & C. A. Tryon, Jr. 1970. *Eimeria fitzgeraldi* n. sp. from the northern pocket gopher, *Thomomys talpoides*. J. Wildl. Dis. 6:107-108.

Torbett, B. E., W. C. Marquardt, & A. C. Carey. 1982. A new species of *Eimeria* from the golden-mantled ground squirrel, *Spermophilus lateralis*, in northern Colorado. J. Protozool. 59:157-159.

Twort, J. M. & C. C. Twort. 1923. Disease in relation to carcinogenic agents among 60,000 experimental mice. J. Pathol. Bacteriol. 35:219-242.

Tyzzer, E. E. 1907. A sporozoön found in the peptic glands of the common mouse. Proc. Soc. Exp. Biol. Med. 5:12-13.

Tyzzer, E. E. 1910. An extracellular coccidium, *Cryptosporidium muris* (gen. et sp. nov.), on the gastric glands of the common mouse. J. Med. Res. 23:487-510.

Tyzzer, E. E. 1912. *Cryptosporidium parvum* (sp. nov.) a coccidium of the common mouse. Arch. Protistenkd. 26:394-412.

Tzipori, S., K. W. Angus, I. Campbell, & E. W. Gray. 1980. *Cryptosporidium:* evidence for a single-species genus. Infect. Immun. 30:884-886.

Tzipori, S. & I. Campbell. 1981. Prevalence of *Cryptosporidium* antibodies in 10 animal species. J. Clin. Microbiol. 14:455-456.

Upton, S. J. & W. L. Current. 1985. The species of *Cryptosporidium* (Apicomplexa: Cryptosporidiidae) infecting mammals. J. Parasitol. 71:625-629.

Upton S. J., W. L. Current, J. V. Ernst, & S. M. Barnard. 1984. Extraintestinal development of *Caryospora simplex* (Apicomplexa: Eimeriidae) in experimentally infected mice, *Mus musculus.* J. Protozool. 31:392-398.

Upton, S. J., D. S. Lindsay, W. L. Current, & J. V. Ernst. 1985. *Isospora masoni* sp. n. (Apicomplexa: Eimeriidae) from the cotton rat, *Sigmodon hispidus.* Proc. Helminthol. Soc. Wash. 52:60-63.

Vance, T. L. & D. W. Duszynski. 1985. Coccidian parasites (Apicomplexa: Eimeriidae) of *Microtus* spp. (Rodentia: Arvicolidae) from the United States, Mexico, and Japan, with descriptions of five new species. J. Parasitol. 71:302-311.

Vassiliades, M. G. 1966. Une nouvelle coccidie, *Eimeria moreli* n. sp. (Protozoa: Eimeriidae), du rat de Gambie (*Cricetomys gambianus*). C. R. Acad. Sci. 262(Ser. D):2473-2475.

Vassiliades, G. 1967. Une nouvelle coccidie *Eimeria xeri* n. sp. (Protozoa: Eimeriidae) de l'ecureuil fouisseur *Xerus (Euxerus) erythropus* Geoffroy, 1803. C. R. Acad. Sci. 265(Ser. D):882-884.

Veisov, A. M. 1961. Novye vidy koktsidii roda *Eimeria* iz peschanok Vinogradova (*Meriones vinogradovi* Hept.) v Nakhicheviskoi ASSR. Izv. Akad. Nauk Az. SSR Ser. Biol. Med. Nauk. 4:25-35.

Veisov, A. M. 1962. Novye vidy koktsidii roda *Eimeria* iz maloaziiskoi kustarnikovoi polevki *Microtus majori* Thomas (1906). Dokl. Akad. Nauk Az. SSR. 18(9):59-64.

Veisov, A. M. 1963. Koktsidii obyknoveinoi polevki *Microtus arvalis* Pall. (1778) v Azerbaidzhane. Izv. Akad. Nauk Az. SSR Ser. Biol. Med. Nauk. 4:61-75.

Veisov, A. M. 1964. Novy vidy koktsidii roda: *Eimeria* ot *Meriones meridianus* Pall., *Rattus turkestanicus* Satun., *Ellobius talpinus* Pall. i roda *Isospora*: ot *Meriones erythrourus* Gray ot gryzunov Tadzhikistana. Dokl. Akad. Nauk Az. SSR. 20(9):65-69.

Veisov, A. M. 1973. Materials on the life cycle of *Eimeria kriygsmanni* and examination of oocysts resembling those of *E. hindlei*, parasites of *Mus musculus.* Prog. Protozool. 4:422.

Veisov, A. M. 1975. Tri novykh vida koktsidii roda *Eimeria* iz gornogo ili belozobugo slepysha *Spalax leucodon* Nordman (1840). Izv. Akad. Nauk Az. SSR Ser. Biol. Nauk. 1975(4):82-85.

Veisov, A. M. 1977. A comparative study of the life cycles of three coccidian parasites of *Mus musculus --- Eimeria* sp. Vejsov, 1973, *E. kriygsmanni* and *E. keilini.* Abstr. V. Internat. Congr. Protozool., p. 17.

Vejsov (see Veisov).

Veluvolu, P. 1981. Coccidia (*Protozoa: Eimeriidae*) of the Ground Squirrel, *Spermophilus beldingi.* M. S. thesis, University of Illinois, Urbana. 46 pp.

Veluvolu, P. & N. D. Levine. 1984. *Eimeria beldingii* n. sp. and other coccidia (Apicomplexa) of the ground squirrel *Spermophilus beldingi.* J. Protozool. 31:357-358.

Vetterling, J. M. 1964. Coccidia (*Eimeria*) from the prairie dog, *Cynomys ludovicianus ludovicianus*, in northern Colorado. J. Protozool. 11:89-91.

Vetterling, J. M., H. R. Jervis, T. G. Merrill, & H. Sprinz. 1971. *Cryptosporidium wrairi* sp. n. from the guinea pig *Cavia porcellus*, with an emendation of the genus. J. Protozool. 18:243-247.

Vetterlius, J. M., A. Takeuchi, & P. A. Madden. 1971. Ultrastructure of *Cryptosporidium wrairi* from the guinea pig. J. Protozool. 18:248-260.

Viljoen, P. R. 1921. *Das Vorkommen von Sarkosporidien in süd-afrikanischen Tieren (Haustieren und Wild) mit besonderer Berücksichtigung ihrer Beziehung zu pathologischen Zuständen.* Diss., Bern (cited by Kalyakin & Zasukhin, 1975).

Von Zellen, B. W. 1959. *Eimeria carolinensis* n. sp., a coccidium from the white-footed mouse, *Peromyscus leucopus* (Rafinesque). J. Protozool. 6:104-105.

Von Zellen, B. W. 1961. *Eimeria leucopi* n. sp., a coccidium from the deer mouse *Peromyscus leucopus.* J. Protozool. 8:134-138.

Wacha, R. S. & H. L. Christiansen. 1982. Development of *Caryospora bigenetica* n. sp. (Apicomplexa, Eimeriidae) in rattle snakes laboratory mice. J. Protozool. 29:272-278.

Wallace, G. D. 1975. Observations on a feline coccidium with some characteristics of *Toxoplasma* and *Sarcocystis*. Z. Parasitenkd. 46:167-178.

Wallace, G. D. & J. K. Frenkel. 1975. Besnoitia species (Protozoa, Sporozoa, Toxoplasmatidae): recognition of cyclic transmission by cats. *Science.* 188:369-371.

Von Wasielewski, T. 1904. *Studien und Mikrophotogramme zur Kenntnis der pathogenen Protozoen.* Leipzig. 118 pp.

Webster, J. M. 1980. Investigations into the coccidia of the grey squirrel *Sciurus carolinensis* Gmelin. J. Protozool. 7:139-146.

Wenyon, C. M. 1923. Coccidiosis of cats and dogs and the status of *Isospora* of man. Ann. Trop. Med. Parasitol. 17:231-288.

Wenyon, C. M. 1926. *Protozoology: a Manual for Medical Men, Veterinarians and Zoologists.* 2 vols. Ballière, Tindall & Cox, London. xvi + 1563 pp.

Wheat, B. E. & J. V. Ernst. 1974. *Eimeria glauceae* sp. n. and *Eimeria dusii* sp. n. (Protozoa: Eimeriidae) from the eastern woodrat *Neotoma floridana,* from Alabama. J. Parasitol. 60:403-405.

Whitmire, W. M. & C. A. Speer. 1984. Ultrastructural localization and effects of parasite-specific IgG and IgA antibodies on various stages of *Eimeria falciformis.* Prog. Abstr. Ann. Meet. Am. Soc. Parasitol. 59:41.

Whitmire, W. M. & C. A. Speer. 1986. Ultrastructural localization of IgA and IgG receptors on oocysts, sporocysts, sporozoites, and merozoites of *Eimeria falciformis.* Can. J. Zool. 64:778-784.

Wiesner, J. 1980. A new sarcosporidian species of *Clethrionomys glareolus* inhabiting the owl *Aegolius funereus* as definitive host. J. Protozool. 27-72A.

Wilson, V. C. L. & G. H. Edrissian. 1974. An infection of *Klossiella muris,* Smith and Johnson, 1902, in a wild *Mus musculus* from Iran. Trans. R. Soc. Trop. Med. Hyg. 68:8.

Winchell, E. J. 1977. Parasites of the beach vole, *Microtus breweri* Baird 1858. J. Parasitol. 63:756-757.

Winter, H. & D. Watt. 1971. *Klossiella hydromyos* n. sp. from the kidneys of an Australian water rat (*Hydromys chrysogaster*). Vet. Pathol. 8:222-231.

Witte, H. M. & G. Piekarski. 1970. Die Oocysten-Ausscheidung bei experimentell infizierten Katzen in Abhängigkeit von *Toxoplasma*-Stamm. Z. Parasitenkd. 33:358-360.

Wobeser, G., R. J. Cawthorn, & A. A. Gajadhar. 1982. Pathology in Richardson's ground squirrels (*Spermophilus richardsoni*) infected with a *Sarcocystis* sp. from badgers (*Taxidea taxus*). In Muller, M., W. Gutteridge, & P. Köhler, Eds. *Molecular and Biochemical Parasitology.* Elsevier, Amsterdam. pp. 375-376.

Woodmansee, D. B. 1983. Cross-transmission and *in vitro* excystation experiments with *Sarcocystis muris.* 35th Annu. Midwest. Conf. Parasitol., p. 16.

Woodmansee, D. B. & E. C. Powell. 1984. Cross-transmission and *in vitro* excystation experiments with *Sarcocystis muris.* J. Parasitol. 70:182-183.

Yakimoff, W. L. 1933. Die coccidiose der Nutrien. Landw. Pelztierzucht. 4:189-190.

Yakimoff, W. L. 1934. Die Biberkokzidiose. Berl. Tieraerztl. Wochenschr. 50:294.

Yakimoff, W. L. 1935. A new coccidium, *Eimeria halli* n. sp., in rats. Proc. Helminthol. Soc. Wash. 2:81-83.

Yakimoff, W. L. & B. I. Buewitsch. 1932. Zur Frage der Coccidien wildlebender Vogeln in Aserbaidschan (Transkaukasus). Arch. Protistenkd. 77:187-191.

Yakimoff, W. L. & W. F. Gousseff. 1935. Eimeriose der Eichhornchen. Berl. Tieraerztl. Wochenschr. 51:740-741.

Yakimoff, W. L. & W. F. Gousseff. 1935a. Die Kokzidien der Hamster. Berl. Tieraerztl. Wochenschr. 51:485.

Yakimoff, W. L. & W. F. Gousseff. 1936. Nouvelles coccidies chez les gerboises. Ann. Parasitol. 14:447-448.

Yakimoff, W. L. & W. F. Gousseff. 1936a. Die Coccidien der schwarzen Ratten. Z. Parasitenkd. 8:504-508.

Yakimoff, W. L. & W. F. Gousseff. 1938. The coccidia of mice (*Mus musculus*). Parasitology. 30:1-3.

Yakimoff, W. L., W. F. Gousseff, & S. F. Suz'ko. 1945. Koktsidii ptits i mlekopitayushchikh Tadzhikistana. Akad. Nauk SSSR Tadzh. Filial Akad. Nauk SSSR. 14:172-183.

Yakimoff, W. L., P. S. Ivanova-Gobzem, & S. N. Machul'skii. 1936. Zur Frage der Infektion der Tiere mit heterogenen Kokzidien. IV and V Mitteilung. Zentralbl. Bakteriol. I. Orig. 137:299-302.

Yakimoff, W. L. & I. I. Sokoloff. 1935. *Eimeria beckeri*, n. sp. a new coccidium from the ground squirrel, *Citellus pygmaeus*. Vestn. Mikrobiol. Epidemiol. Parazitol. 13:331-334. (Also Iowa State Coll. J. Sci. 9:581-585.)

Yakimoff, W. L. & A. J. Sprinholtz-Schmidt. 1939. *Eimeria ussuriensis* n. sp., coccidie d'un rongeur d'extreme-Orient. Ann. Soc. Belge Med. Trop. 19:117-123.

Yakimoff, W. L. & S. K. Terwinsky. 1931. die Coccidiose der Zobeltieres. Arch. Protistenkd. 73:56-59.

Zajiček, M. D. 1955. Parasitarni invase u mladych nutrii. Sb. Vys. Šk. Zemd. Lesn. Fak. Brně, Řada B: Spisy Fak. Vet. Spis c. 232, Čislo 1:29-38.

Zaman, V. 1975. Revision of *Sarcocystis orientalis* Zaman and Colley, 1975. Southeast Asian J. Trop. Med. Public Health. 6:603.

Zaman, V. 1976. Host range of *Sarcocystis orientalis*. Southeast Asian J. Trop. Med. Public Health. 7:112.

Zaman, V. & F. C. Colley. 1975. Light and electron microscopic observations of the life cycle of *Sarcocystis orientalis* sp. n. in the rat (*Rattus norvegicus*). Z. Parasitenkd. 47:169-185.

Zaman, V. & F. C. Colley. 1976. Replacement of *Sarcocystis orientalis* Zaman and Colley, 1975 by *Sarcocystis singaporensis* n. sp. Z. Parasitenkd. 51:137.

Zasukhin, D. N., Ed. 1980. *Problema Toksoplazmoza*. "Meditsina," Moscow, USSR. 309 pp.

Zasukhin (Sassuchin), D. & T. Rauschenbach. 1932. Zum Studium der Darmprotozoenfauna der Näger in Süd-Osten RSFSR. II. Darmcoccidien des *Citellus pygmaeus* Pall. Arch. Protistenkd. 78:646-650.

Zasukhin, D. & V. Tiflov. 1932. Ento and ectoparasites of the steppe ground squirrel, *Citellus pygmaeus* Pall. Rev. Microbiol. 11:129-132.

Zasukhin, D. N. & V. E. Tiflov. 1933. Ento- und Ektoparasiten des Steppenziesels (*Citellus pygmaeus* Pall.) in Süd-Osten RSFSR. Z. Parasitenkd. 5:437-442.

Zolotarev, N. A. 1935. K voprosu o koktsidiyakh pushnykh zverei. Sb. Rabot Dagestan. Protozool. Nauch.-Issled. Oporn. Punkta Severok. VOS i Narkom. DASSR. 1: (cited by Musaev & Veisov, 1959).

Zolotarev, N. A. 1938. Zur Frage der Kokzidien in Zieseln. Wien. Tieraerztl. Monatsschr. 25:658-661.

Zwart, P. & W. J. Strik. 1961. Further observations on *Eimeria dolichotis*, a coccidium of the Patagonian cavy (*Dolichotis patagonica*). J. Protozool. 8:58-59.

INDEX

A

Acomys, 121
Adelina sp. Barnard, 54—55
Aeromys, 33
Agouti, 177
Akodon, 50
Allactaga, 170—172
Alticola, 87—88
Apodemus, 108—118
Arvicanthis, 123
Arvicola, 88—89
Asiatic chipmunk, see *Eutamias*

B

Baiomys, 50
Bandicota, 124
Bathyergidae, 183
Besnoitia besnoiti, 10, 19, 62, 81,
    160—161, 174
Besnoitia darlingi, 29, 62, 161
Besnoitia jellisoni, 19, 40, 49, 50, 62, 101,
    144, 162, 174
Besnoitia sp. Gunders, 69
Besnoitia sp. McKenna, 145, 162
Besnoitia spp. Matuschka, 63, 68, 82, 127,
    162
Besnoitia wallacei, 62, 82, 101, 143—146,
    162

C

Callosciurus, 31—32
Calomyscus, 58—59
Capromyidae, 178—182
Capromys, 178—179
Caryospora bigenetica, 146
Caryospora bubonis, 146—147
Caryospora microti, 99
Caryospora simplex, 147
Castor, 41
Castoridae, 41
Cavia, 173—175
Caviidae, 173—175
Chinchilla, 177—178
Chinchillidae, 177—178
Chipmunk, see *Eutamias; Tamias*
Chiropodomys, 118

Clethrionomys, 83—87
"Coccidium" arcay, 62
Cricetomys, 107—108
Cricetulus, 60—61
Cricetus, 59
Cryptosporidium muris, 87, 118, 162—163,
    174
Cryptosporidium parvum, 163—164
Ctenodactylidae, 183—185
Ctenodactylus, 183—185
Cynomys, 19—21

D

Dasymys, 123—124
Dasyprocta, 176—177
Dasyproctidae, 176—177
Dicrostonyx, 105—106
Dipodidae, 169—172
Dipodomys, 37—40
Dolichotis, 175—176
Dorisa arizonensis, 58
Dorisa bengalensis, 30—31
Dryomys, 166—168

E

Eastern chipmunk see *Tamias*
Echimyidae, 182—183
Eimeria abdildaevi, 166
Eimeria abidzhanovi, 82
Eimeria abusalimovi, 166—167
Eimeria abuschevi, 89—90
Eimeria achburunica, 69
Eimeria adiyamanensis, 63
Eimeria aeromysis, 33
Eimeria africana, 119—120
Eimeria agrarii, 108
Eimeria aguti, 176
Eimeria akeriana, 69
Eimeria akodoni, 50
Eimeria albigulae, 55
Eimeria alischerica, 127—128
Eimeria allactagae, 170
Eimeria amburdariana, 61
Eimeria andrewsi, 21—22
Eimeria antonellii, 55
Eimeria apionodes, 108
Eimeria apodemi, 108—109